Lecture Notes in Bioinformatics   11467

Subseries of Lecture Notes in Computer Science

More information about this series at http://www.springer.com/series/5381

Lenore J. Cowen (Ed.)

Research in Computational Molecular Biology

23rd Annual International Conference, RECOMB 2019
Washington, DC, USA, May 5–8, 2019
Proceedings

 Springer

Editor
Lenore J. Cowen
Tufts University
Cambridge, MA, USA

ISSN 0302-9743 ISSN 1611-3349 (electronic)
Lecture Notes in Bioinformatics
ISBN 978-3-030-17082-0 ISBN 978-3-030-17083-7 (eBook)
https://doi.org/10.1007/978-3-030-17083-7

Library of Congress Control Number: 2019936142

LNCS Sublibrary: SL8 – Bioinformatics

This Springer imprint is published by the registered company Springer Nature Switzerland AG
The registered company address is: Gewerbestrasse 11, 6330 Cham, Switzerland

Preface

This volume contains the 37 extended and short abstracts presented at the 23rd International Conference on Research in Computational Molecular Biology (RECOMB) 2019, which was hosted by the George Washington University in Washington, DC, during May 5–8.

These 37 contributions were selected from 175 submissions. There were 204 submissions made in total, but after removing submissions that were withdrawn or transferred to other conference tracks at the request of the authors, there were 175 submissions that were ultimately reviewed by the Program Committee (PC). In particular, each of the 175 submissions was assigned to three members of the PC for independent reviews, who in many cases solicited additional advice from external reviewers. Following the initial reviews, final decision were made after an extensive discussion of the submissions among the members of the PC. Reviews and discussions were conducted through the EasyChair Conference Management System.

While RECOMB 2019 did not allow parallel submissions, authors of accepted papers were given the option to publish short abstracts in these proceedings and submit their full papers to a journal. In addition, several accepted papers were invited to submit revised manuscripts for consideration for publication in *Cell Systems*. Papers accepted for oral presentation that were subsequently submitted to a journal are published as short abstracts and were deposited on the preprint server arxiv.org or biorxiv.org. All other papers that appear as long abstracts in the proceedings were invited for submission to the RECOMB 2019 special issue of the *Journal of Computational Biology*.

In addition to presentations of these contributed papers, RECOMB 2019 featured six invited keynote talks given by leading scientists. The keynote speakers were Carlos D. Bustamante (Stanford University), Rachel Kolodny (University of Haifa), Franziska Michor (Harvard University and Dana Farber Cancer Institute), Mihai Pop (University of Maryland), Eytan Ruppin (US National Cancer Institute), and Alfonso Valencia (Spanish National Bioinformatics Institute, and Barcelona Supercomputing Center).

RECOMB 2019 also featured highlight talks of computational biology papers that were published in journals during the previous 18 months. Of the 43 submissions to the Highlights track, 10 were selected for oral presentation at RECOMB. There was also a special invited panel session on Genomic Privacy.

Four RECOMB Satellite meetings took place in parallel directly preceding the main RECOMB meeting. The RECOMB Genetics Satellite was co-chaired by Itsik Pe'er (Columbia University), Simon Gravel (McGill University), and Seyoung Kim (Carnegie Mellon University). The RECOMB Satellite Workshop on Massively Parallel Sequencing (RECOMB-Seq) was co-chaired by Christina Boucher (University of Florida) and Vikas Bansal (University of California, San Diego). The RECOMB-Computational Cancer Biology Satellite meeting (RECOMB-CCB) was co-chaired by Max Leiserson (University of Maryland) and Rachel Karchin (Johns Hopkins University). The DREAM meeting with RECOMB 2019 was co-organized by

Laura Heiser (Oregon Health and Science University) and Gustavo Stolovitzky (IBM Research and Icahn School of Medicine at Mount Sinai).

The organization of this conference was a dedicated community effort with many colleagues contributing their time and expertise. I thank the Steering Committee for their input, especially the chair, Bonnie Berger (MIT), for their wisdom and guidance, as well as Ben Raphael (Princeton University), the program chair of RECOMB 2018 for his help, advice, and support throughout the process. I thank Mona Singh (Princeton University) for chairing the Highlights track, and Rob Patro (Stony Brook University) for chairing the Posters track. I thank co-chairs Max Alekseyev (The George Washington University) and Teresa Przytycka (US National Institutes of Health), and all the members of their Organizing Committee including Pavel Avdeyev (The George Washington University), Rebecca Sarto Basso (University of California, Berkeley), Chanson Benjamin (The George Washington University), Jan Hoinka (US National Institutes of Health), and Damian Wojtowicz (US National Institutes of Health). Damian Wojtowicz also served as chair of the Student Travel Fellowship Award Committee.

A very special thanks to Yann Ponty (CNRS/LIX, Ecole Polytechnique) who as chair of the Publications Committee served as proceedings chair for this volume, including chasing down final versions and checking copyright forms and final camera-ready copy. Thanks to all PC members and external reviewers who completed their reviews in a very tight timeframe despite their busy schedules, the authors of the papers, highlights, and posters for their scientific contributions, and all the attendees for their enthusiastic participation in the conference.

We thank all our conference sponsors for their support, who at press time for this volume included: Akamai Technologies, *Computation* (MDPI journal), Journal of Computational Biology (Mary Ann Liebert, Inc.), Natera, Springer, and The George Washington University. A very special thanks to our student travel fellowship sponsors, the US National Science Foundation (NSF) and the International Society for Computational Biology (ISCB).

May 2019   Lenore J. Cowen

Organization

General Chairs

Max Alekseyev   The George Washington University, USA
Teresa Przytycka   National Institutes of Health, USA

Program Committee Chair

Lenore J. Cowen   Tufts University, USA

Steering Committee

Vineet Bafna   University of California, San Diego, USA
Bonnie Berger (Chair)   Massachusetts Institute of Technology, USA
Eleazar Eskin   University of California, Los Angeles, USA
Teresa Przytycka   National Institutes of Health, USA
Cenk Sahinalp   Indiana University, USA
Roded Sharan   Tel Aviv University, Israel

Program Committee

Derek Aguiar   University of Connecticut, USA
Tatsuya Akutsu   Kyoto University, Japan
Can Alkan   Bilkent University, Turkey
Rolf Backofen   Albert Ludwigs University, Freiburg, Germany
Vineet Bafna   University of California, San Diego, USA
Nuno Bandeira   University of California, San Diego, USA
Ziv Bar-Joseph   Carnegie Mellon University, USA
Anastasia Baryshnikova   Calico Life Sciences, USA
Niko Beerenwinkel   ETH Zurich, Switzerland
Bonnie Berger   Massachusetts Institute of Technology, USA
Mathieu Blanchette   McGill University, Canada
Michael Brudno   University of Toronto, Canada
Sebastian Böcker   Friedrich Schiller University Jena, Germany
Tony Capra   Vanderbilt University, USA
Cedric Chauve   Simon Fraser University, Canada
Lenore J. Cowen (Chair)   Tufts University, USA
Nadia El-Mabrouk   University of Montreal, Canada
Irit Gat-Viks   Tel Aviv University, Israel
Dario Ghersi   University of Nebraska, Omaha, USA
Anna Goldenberg   Hospital for Sick Kids and University of Toronto,
                                 Canada

Fereydoun Hormozdiari	University of California, Davis, USA
Sorin Istrail	Brown University, USA
Tao Jiang	University of California, Riverside, USA
John Kececioglu	University of Arizona, USA
Manolis Kellis	Massachusetts Institute of Technology, USA
Carl Kingsford	Carnegie Mellon University, USA
Gunnar W. Klau	Heinrich Heine University, Düsseldorf, Germany
Mehmet Koyuturk	Case Western Reserve University, USA
Smita Krishnaswamy	Yale University, USA
Jens Lagergren	KTH Royal Institute of Technology, Sweden
Mark Leiserson	University of Maryland, College Park, USA
Ming Li	University of Waterloo, Canada
Po-Ru Loh	Harvard Medical School, USA
Paul Medvedev	The Pennsylvania State University, USA
Bernard Moret	EPFL, Switzerland
Sara Mostafavi	University of British Columbia, Canada
Veli Mäkinen	University of Helsinki, Finland
William Stafford Noble	University of Washington, USA
Lior Pachter	California Institute of Technology, USA
Laxmi Parida	IBM, USA
Robert Patro	Stony Brook University, USA
Yann Ponty	CNRS and École Polytechnique, France
Nataša Pržulj	University College, London, UK
Mireille Régnier	CNRS and École Polytechnique, France
Knut Reinert	Freie Universität Berlin, Germany
Cenk Sahinalp	Indiana University, USA
Michael Schatz	Johns Hopkins University, USA
Alexander Schönhuth	Centrum Wiskunde and Informatica, The Netherlands
Russell Schwartz	Carnegie Mellon University, USA
Roded Sharan	Tel Aviv University, Israel
Mona Singh	Princeton University, USA
Donna Slonim	Tufts University, USA
Sagi Snir	University of Haifa, Israel
Jens Stoye	Bielefeld University, Germany
Fengzhu Sun	University of Southern California, USA
Wing-Kin Sung	National University of Singapore, Singapore
Ewa Szczurek	University of Warsaw, Poland
Haixu Tang	Indiana University, Bloomington, USA
Glen Tesler	University of California, San Diego, USA
Fabio Vandin	University of Padua, Italy
Martin Vingron	Max Planck Institute for Molecular Genetics, Germany
Jérôme Waldispühl	McGill University, Canada
Tandy Warnow	University of Illinois at Urbana-Champaign, USA
Sebastian Will	University of Vienna, Austria
Jinbo Xu	Toyota Technological Institute at Chicago, USA
Yuzhen Ye	Indiana University, Bloomington, USA

Alex Zelikovsky   Georgia State University, USA
Jianyang Zeng   Tsinghua University, China
Louxin Zhang   National University of Singapore, Singapore

Additional Reviewers

Aganezov, Sergey
Alanko, Jarno
Alkhnbashi, Omer
Almodaresi, Fatemeh
Altieri, Federico
Amodio, Matthew
Arvestad, Lars
Ayati, Marzieh
Bai, Xin
Balvert, Marleen
Ben-Bassat, Ilan
Benner, Philipp
Bittremieux, Wout
Bruner, Ariel
Budach, Stefan
Burkhardt, Daniel
Burzykowski, Tomasz
Cameron, Christopher
Castro, Egbert
Chaisson, Mark
Chen, Yong
Cho, Hyunghoon
Conrad, Tim
Costa, Ivan
Cowman, Tyler
Cristea, Simona
Crovella, Mark
Csuros, Miklos
Cunial, Fabio
da Silva, Israel
Davila Velderrain, Jose
DeBlasio, Dan
Ding, Jun
Dirmeier, Simon
Doerr, Daniel
Drysdale, Erik
Durham, Timothy
Dührkop, Kai

El-Kebir, Mohammed
Ernst, Jason
Fischer, Mareike
Fleischauer, Markus
Frishberg, Amit
Fröhlich, Holger
Gambette, Philippe
Gaudelet, Thomas
Gervits, Asia
Gigante, Scott
Gross, Barak
Guo, Jun-Tao
Gursoy, Gamze
Haiminen, Niina
Han, Wontack
Harel, Tom
Harrison, Robert
He, Liang
He, Yuan
Heider, Dominik
Heller, David
Hescott, Benjamin
Hoffmann, Steve
Hormozdiari, Farhad
Hou, Lei
Hu, Xiaozhe
Huson, Daniel
Huynh, Linh
Itzhacky, Nitay
Jahn, Katharina
Jain, Siddhartha
Jun, Seong-Hwan
Karpov, Nikolai
Kelleher, Jerome
Kim, Jongkyu
Knyazev, Sergey
Kobren, Shilpa
Nadimpalli

Kockan, Can
Kovaka, Sam
Krug, Joachim
Kuipers, Jack
Lafond, Manuel
Lasker, Keren
Lee, Heewook
Lei, Haoyun
Levitan, Anton
Levovitz, Chaya
Li, Yue
Lin, Andy
Lin, Dejun
Liu, Jie
Liu, Liang
Lopez-Rincon, Alejandro
Lu, Yang
Ludwig, Marcus
Lugo-Martinez, Jose
Ly, Lam-Ha
Ma, Cong
Maaskola, Jonas
Mahmoud, Medhat
Malcha, Ortal
Malik, Laraib
Malikic, Salem
Malod-Dognin, Noel
Mandric, Igor
Marass, Francesco
Marçais, Guillaume
Mautner, Stefan
Maxwell, Sean
Mcpherson, Andrew
Mehringer, Svenja
Mezlini, Aziz
Miladi, Milad
Minkin, Ilia
Mohammadi, Shahin

Mohammed, Noman
Molloy, Erin
Naumenko, Sergey
Navlakha, Saket
Nazeen, Sumaiya
Ng, Bernard
Nguyen, Duong
Norri, Tuukka
Noutahi, Emmanuel
Nyquist, Sarah
Nzabarushimana, Etienne
Oesper, Layla
Oliver, Carlos
Park, Jisoo
Park, Yongjin
Pellegrina, Leonardo
Peng, Jian
Pirkl, Martin
Platt, Daniel
Posada-Céspedes, Susana
Przytycka, Teresa
Pullman, Benjamin
Qu, Fangfang
Raden, Martin
Rahmann, Sven
Rajaraman, Ashok
Rajkumar, Utkrisht
Ramani, Arun
Raphael, Ben
Rashid, Sabrina
Rashidi Mehrabadi, Farid
Reinharz, Vladimir
Renard, Bernhard
Richard, Hugues
Robinson, Welles

Rogovskyy, Artem
Rojas, Carlos
Roth, Andrew
Ruffalo, Matthew
Röhl, Annika
Sahlin, Kristoffer
Salmela, Leena
Sarazzin-Gendron, Roman
Sarkar, Hirak
Sarmashghi, Shahab
Sason, Itay
Sauerwald, Natalie
Scholz, Markus
Schreiber, Jacob
Schulz, Marcel
Schulz, Tizian
Scornavacca, Celine
Sedlazeck, Fritz
Shajii, Ariya
Shi, Alvin
Shooshtari, Parisa
Singer, Jochen
Singh, Ritambhara
Skums, Pavel
Smith, Kevin
Srivastava, Avi
Stamboulian, Moses
Standage, Daniel
Stanfield, Zachary
Stanislas, Virginie
Steiger, Edgar
Steuerman, Yael
Sun, Chen
Tao, Yifeng
Thomas, Marcus

Tomescu, Alexandru I.
Tong, Alex
Tonon, Andrea
Tozzo, Veronica
Tran, Dinh
Tran, Hieu
Tremblay-Savard, Olivier
Tung, Laura
Turinski, Andrei
van Dijk, David
Videm, Pavankumar
von Haeseler, Arndt
Wan, Lin
Wang, Bo
Wang, Weili
Wang, Wenhao
Wang, Ying
West, Sean
Windels, Sam
Wittler, Roland
Wu, Yu-Wei
Yadav, Vinod
Yankovitz, Gal
Yardimci, Galip
Yilmaz, Serhan
Zakeri, Mohsen
Zeng, Haoyang
Zhang, Sai
Zhang, Yang
Zheng, Hongyu
Zhu, Kaiyuan
Zhu, Shanfeng
Zhu, Zifan

Contents

Short Papers

An Efficient, Scalable and Exact Representation of High-Dimensional Color Information Enabled via de Bruijn Graph Search

Fatemeh Almodaresi[1], Prashant Pandey[1(✉)], Michael Ferdman[1], Rob Johnson[1,2], and Rob Patro[1]

[1] Computer Science Department, Stony Brook University, Stony Brook, USA
{falmodaresit,ppandey,mferdman,rob.patro}@cs.stonybrook.edu
[2] VMware Research, Palo Alto, USA
robj@vmware.com

Abstract. The colored de Bruijn graph (cdbg) and its variants have become an important combinatorial structure used in numerous areas in genomics, such as population-level variation detection in metagenomic samples, large scale sequence search, and cdbg-based reference sequence indices. As samples or genomes are added to the cdbg, the color information comes to dominate the space required to represent this data structure.

In this paper, we show how to represent the color information efficiently by adopting a hierarchical encoding that exploits correlations among color classes—patterns of color occurrence—present in the de Bruijn graph (dbg). A major challenge in deriving an efficient encoding of the color information that takes advantage of such correlations is determining which color classes are close to each other in the high-dimensional space of possible color patterns. We demonstrate that the dbg itself can be used as an efficient mechanism to search for approximate nearest neighbors in this space. While our approach reduces the encoding size of the color information even for relatively small cdbgs (hundreds of experiments), the gains are particularly consequential as the number of potential colors (i.e. samples or references) grows into the thousands.

We apply this encoding in the context of two different applications; the implicit cdbg used for a large-scale sequence search index, Mantis, as well as the encoding of color information used in population-level variation detection by tools such as Vari and Rainbowfish. Our results show significant improvements in the overall size and scalability of representation of the color information. In our experiment on 10,000 samples, we achieved more than 11× better compression compared to RRR.

1 Introduction

The colored de Bruijn graph (cdbg) [1], an extension of the classical de Bruijn graph [2–4], is a key component of a growing number of genomics tools.

© Springer Nature Switzerland AG 2019
L. J. Cowen (Ed.): RECOMB 2019, LNBI 11467, pp. 1–18, 2019.
https://doi.org/10.1007/978-3-030-17083-7_1

Augmenting the traditional de Bruijn graph with "color" information provides a mechanism to associate meta-data, such as the raw sample or reference of origin, with each k-mer. Coloring the de Bruijn graph enables it to be used in a wide range of applications, such as large-scale sequence search [5–9] (though some [6–8] do not explicitly couch their representations in the language of the cdbg), population-level variation detection [10–12], traversal and search in a pan-genome [11], and sequence alignment [13]. The popularity and applicability of the cdbg has spurred research into developing space-efficient and high-performance data-structure implementations.

An efficient and fast representation of cdbg requires optimizing both the de Bruijn graph and the color information. While there exist efficient and scalable methods for representing the topology of the de Bruijn graph [4,14–18] with fast query time, a scalable and exact representation of the color information has remained a challenge. Recently, Mustafa et al. [19] has tackled this challenge by relaxing the exactness constraints—allowing the returned color set for a k-mer to contain extra samples with some controlled probability—but it is not immediately clear how this method can be made exact.

Specifically, existing exact color representations suffer from large sizes and a fast growth rate that leads them to dominate the total representation size of the cdbg with even a moderate number of input samples (see Fig. 3b). As a result, the color information grows to dominate the space used by all these indexes and limits their ability to scale to large input data sets.

Iqbal et al. introduced cdbgs [1] and proposed a hash-based representation of the de Bruijn graph in which each k-mer is additionally tagged with the list of reference genomes in which it is contained. Muggli et al. reduced the size of the cdbg in VARI [10] by replacing the hash map with BOSS [16] (a BWT-based [20] encoding of the de Bruijn graph that assigns a unique ID to each k-mer) and using a boolean matrix indexed by the unique k-mer ID and genome reference ID to indicate occurrence. They reduced the size of the occurrence matrix by applying off-the-shelf compression techniques RRR [21] and Elias-Fano [22] encoding. Rainbowfish [12] shrank the color table further by ensuring that rows of the color matrix are unique, mapping all k-mers with the same color information to a single row, and assigning row indices based on the frequency of each occurrence pattern. However, despite these improvements, the scalability of the resulting structure remains limited because even after eliminating redundant colors, the space for the color table grows quickly to dominate the total space used by these data structures.

We observe that, in real biological data, even when the number of distinct color classes is large, many of them will be near each other in terms of the set of samples or references they encode. That is, the color classes tend to be highly correlated rather than uniformly spread across the space of possible colors. There are intuitive reasons for such characteristics. For example, we observe that adjacent k-mers in the de Bruijn graph are extremely likely to have either identical or similar color classes, enabling storage of small deltas instead of the complete color classes. This is because k-mers adjacent in the de Bruijn graph are likely to be adjacent (and hence present) in a similar set of input samples. In the context of sequence-search, because genomes and transcriptomes are largely preserved across organs, individuals, and even across related species, we expect two k-mers that occur together in one sample to be highly correlated in their occurrence across many samples. Thus, we can take advantage of this correlation when devising an efficient encoding scheme for the cdbg's associated color information.

In this paper, we develop a general scheme for efficient and scalable encoding of the color information in the cdbg by encoding color classes (i.e. the patterns of occurrence of a k-mer in samples) in terms of their differences (which are small) with respect to some "neighboring" color class. The key technical challenge, solved by our work, is efficiently searching for the neighbors of color classes in the high-dimensional space of colors by leveraging the observation that similar color classes tend to be topologically close in the underlying de Bruijn graph. We construct a weighted graph on the color classes in the cdbg, where the weight of each edge corresponds to the space required to store the delta between its endpoints. Finding the minimum spanning tree (MST) of this graph gives a minimal delta-based representation. Although reconstructing a color class on this representation requires a walk to the MST root node, abundant temporal locality on the lookups allows us to use a small cache to mitigate the performance impact, yielding query throughput that is essentially the same as when all color classes are represented explicitly.

An alternative would have been to try to limit the depth (or diameter) of the MST. This problem is heavily studied in two forms: the unrooted bounded-diameter MST problem [23] and the rooted hop-constrained MST problem [24]. Neither is in APX, i.e. it is not possible to approximate them to within any constant factor [25]. Althaus et al. gave an $O(\log n)$ approximation assuming the edge weights form a metric [24]. Khuller et al. show that, if the edge *lengths* are the same as the edge *weights*, then there is an efficient algorithm for finding a spanning tree that is within a constant of optimal in terms of both diameter and weight [26]. Marathe et al. show that in general we can find trees within $O(\log n)$ of the minimum diameter and weight [27]. We can't use Khuller's approach (because our edge lengths are not equal to our edge weights), and even a $O(\log n)$ approximation would give up a potentially substantial amount of space.

We showcase the generality and applicability of our color class table compression technique by demonstrating it in two computational biology applications: sequence search and variation detection. We compare our novel color class table representation with the representation used in Mantis [5], a state-of-the-art large-scale sequence-search tool that uses a cdbg to index a set of sequencing samples, and the representation used in Rainbowfish [12], a state-of-the-art index to facilitate variation detection over a set of genomes. We show that our approach maintains the same query performance while achieving over 11× and 2.5× storage savings relative to the representation previously used by these tools.

2   Methods

This section describes our compact cdbg representation. We first define cdbgs and briefly describe existing compact cdbg representations. We then outline the high-level idea behind our compact representation and explain how we use the de Bruijn graph to efficiently build our compact representation. Finally, we describe implementation details and optimizations to our query algorithm.

2.1   Colored de Bruijn graphs

De Bruijn graphs are widely used to represent the topological structure of a set of k-mers [2, 18, 28–33]. The de Bruijn graph induced by a set of k-mers is defined below.

Definition 1. *Given a set E of k-mers, the de Bruijn graph induced by E has edge set E, where each k-mer (or edge) connects its two $(k-1)$-length substrings (or vertices).*

Cdbgs extend the de Bruijn graph by assigning a *color class* $C(x)$ to each edge (or node) x of the de Bruijn graph. The color class $C(x)$ is a set drawn from some universe U. Examples of U and $C(x)$ are

- Sometimes, U is a set of reference genomes, and $C(x)$ is the subset of reference genomes containing k-mer x [10, 12, 13, 34].
- Sometimes, U is a set of *reads*, and $C(x)$ is the subset of reads containing x [35–37].
- Sometimes, U is a set of sequencing experiments, and $C(x)$ is the subset of sequencing experiments containing x [5–8].

The goal of a cdbg representation is to store E and C as compactly as possible[1], while supporting the following operations efficiently:

- *Point query.* Given a k-mer x, determine whether x is in E.
- *Color query.* Given a k-mer $x \in E$, return $C(x)$.

Given that we can perform point queries, we can traverse the de Bruijn graph by simply querying for the 8 possible predecessor/successor edges of an edge. This enables us to implement more advanced algorithms, such as bubble calling [1].

Many cdbg representations typically decompose, at least logically, into two structures: one structure storing a de Bruijn graph and associating an ID with each k-mer, and one structure mapping these IDs to the actual color class [10, 12, 38]. The individual color classes can be represented as bit-vectors, lists, or via a hybrid scheme [39]. This information is typically compressed [21, 40, 41].

Our paper follows this standard approach, and focuses exclusively on reducing the space required for the structure storing the color information. We propose

---

[1] The nodes of the de Bruijn graph are typical stored implicitly, because the node set is simply a function of E.

a compact representation that, given a color ID, can return the corresponding color efficiently. Although we pair our color table representation with the de Bruijn graph structure representation of the counting quotient filter [38] as used in Mantis [5], our proposed color table representation can be paired with other de Bruijn graph representations.

2.2 A Similarity-Based cdbg Representation

The key observation behind our compressed color-class representation is that the color classes of k-mers that are adjacent in the de Bruijn graph are likely to be very similar. Thus, rather than storing each color class explicitly, we can store only a few color classes explicitly and, for all the remaining color classes, we store only their differences from other color classes. Because the differences are small, the total space used by the representation will be small.

Motivated by the observation above, we propose to find an encoding of the color classes that takes advantage of the fact that most color classes can be represented in terms of only a small number of edits (i.e., flipping the parity of only a few bits) with respect to some neighbor in the high-dimensional space of the color classes. This idea was first explored by Bookstein and Klein [42] in the context of information retrieval. Bookstein and Klein showed how to exploit the implicit clustering among bitmaps in IR to achieve excellent reduction in storage space to represent those bitmaps using an MST as the underlying representation. Unfortunately, the approach taken by Bookstein and Klein cannot be directly used in our problem, since it requires computing and optimizing upon the full Hamming distance graph of the bitvectors being represented, which is not tractable for the scale of data we are analyzing. Hence, what we need is a method to efficiently discover an incomplete and highly-sparse Hamming distance graph that, nonetheless, supports a low-weight spanning tree. We describe below how we apply and modify this approach in the context of the set of correlated bit vectors (i.e. color classes) that we wish to encode.

We construct our compressed color class representation as follows (see Fig. 1). For each edge x of the de Bruijn graph, let $C(x)$ be the color class of x. Let $\mathcal{C}$ be the set of all color classes that occur in the de Bruijn graph. We first construct an undirected graph with vertex set $\mathcal{C}$ and edge set reflecting the adjacency relationship implied by the de Bruijn graph. In other words, there is an edge between color classes c_1 and c_2 if there exist adjacent edges (i.e. incident on the same node) x and y in the de Bruijn graph such that $c_1 = C(x)$ and $c_2 = C(y)$. These edges indicate color classes that are likely to be similar, based on the structure of the de Bruijn graph. We then add a special node $\emptyset$ to the color class graph, which is connected to every node. We set the weight of every edge in the color class graph to be the Hamming distance between its two endpoints (where we view color classes as bit vectors and $\emptyset$ is the all-zeros bit vector).

We then compute a minimum spanning tree of the color class graph and root the tree at the special $\emptyset$ node. Note that, because the $\emptyset$ node is connected to every other node in the graph, the graph is connected and hence an MST is

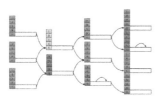
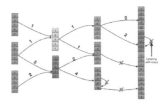

(a) A colored de Bruijn graph. Each rectangle node represents a kmer. Each vector represents a color class (equal color classes have the same color).

(b) The color class graph from the cdbg. There is an edge between each pair of color classes that which correspond to adjacent k-mers in cdbg. Weights on the edges represent the Hamming distances of the color class vectors.

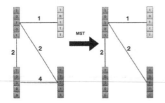

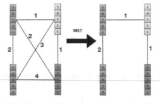

(c) The color class graph we achieve from 1b by removing duplicate edges and its corresponding MST.

(d) The complete color class graph and its derived MST which has the minimum achievable total weight.

Fig. 1. Encoding color classes by finding the MST of the color class graph, an undirected graph derived from cdbg. The order of the process is a, b and c. The arrows in a and b show the direction of edges in the de Bruijn graph which is a directed graph. The optimal achievable MST is shown in d for comparison. Since we never observe the edge between any k-mers from color classes green and yellow in cdbg, we won't have the edge between color classes green and yellow and therefore, our final MST is not equal to the best MST we can get from a complete color class graph. (Color figure online)

guaranteed to exist. By using a minimum spanning tree, we minimize the total size of the differences that we need to store in our compressed representation.

We then store the MST as a table mapping each color class ID to the ID of its parent in the MST, along with a list of the differences between the color class and its parent. For convenience we can view the list of differences between color class c_1 and color class c_2 as a bit vector $c_1 \oplus c_2$, where $\oplus$ is the bit-wise exclusive-or operation. To reconstruct a color class given its ID i, we simply xor all the difference vectors we encounter while walking from i to the root of the MST.

2.3 Implementation of the MST Data Structure

Assuming we have $|\mathcal{C}|$ color classes, $|U|$ colors, and an MST with total weight of w over the color class graph, we store all the information required to retrieve the original color bit-vector for each color class ID based on the MST structure into three data structures:

– **Parent vector**: This vector contains $|\mathcal{C}|$ slots, each of size $\lceil \log_2 \mathcal{C} \rceil$. The value stored in index i represents the parent color class ID of the color class with index i in the MST.
– **Delta vector**: This vector contains w slots, each of size $\lceil \log_2 |U| \rceil$. For each pair of parent and child in the parent vector, we compute a vector of the indices at which they differ. The delta vector is the concatenation of these per-edge delta vectors, ordered by the ID of the source of the edge. Note that the per-edge delta vectors will not all be of the same length, because some edges have larger weight than others. Thus, we need an auxiliary data structure to record the boundaries between the per-edge deltas within the overall delta vector.
– **Boundary bit-vector**: This vector contains w bits, where a set bit indicates the boundary between two delta sets within the delta vector. To find the starting position, within the delta vector, of the per-edge delta list for the MST edge with source ID i, we perform $select(i)$ on the boundary vector. Select returns the position of the ith one in the boundary vector.

Query of the MST-Based Representation. Figure 2 shows how queries proceed using this encoding. We start with an empty accumulator bit vector and a color class ID i for which we want to compute the corresponding color class. We perform a select query for i and $i + 1$ in the boundary bit-vector to get the boundaries of i's difference list in the delta vector. We then iterate over its difference list and flip the indicated bits in our accumulator. We then set $i \leftarrow$ PARENT$[i]$ and repeat until i becomes 0, which indicates that we have reached the root. At this point, the accumulator will be equal to the bit vector for color class i.

2.4 Integration in Mantis

Once constructed, our MST-based color class representation is a drop-in replacement for the current color class representations used in several existing tools, including Mantis [5] and Rainbowfish [12]. Their existing color class tables support a single operation—querying for a color class by its ID—and our MST-based representation supports exactly the same operation.

For this paper, we integrated our MST-based representation into Mantis. The same space savings can be achieved in other tools, particularly Rainbowfish, which has a similar color-class encoding as Mantis.

Construction. We construct our MST-based color-class representation as follows. First, we run Mantis to build its default representation of the cdbg. We then build the color-class graph by walking the de Bruijn graph and adding all the corresponding edges to the color-class graph. The edge set is typically much smaller than the de Bruijn graph (because many de Bruijn graph edges may map to the same edge in the color-class graph), so this can be done in RAM. Note that we do not compute the weights of the edges during this pass, because that would require having all of the large color-class bit vectors in memory in order to compute their Hamming distance.

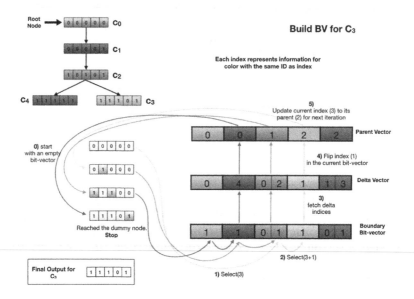

Fig. 2. The conceptual MST (top-left), the data structure to store the color information in the format of an MST (right). This figure also illustrates the steps required to build one of the color vectors (C_3) at the leaf of the tree. Note that the query process shown here does not depict the caching policy we apply in practice.

In the second pass, we traverse the edge set and compute the weight of each edge. To minimize RAM usage during this phase, we sort the edges and iterate over them in a "blocked" fashion. Specifically, Mantis stores the color class bit vectors on-disk sequentially by ID, grouped into blocks of roughly 6 GBs each. We sort the edges lexicographically by their source and destination block. We then load all pairs of blocks and compute the weights of all the edges between the two blocks currently in memory. At all times, we need only two blocks of color class vectors in memory. Given the weighted graph, we compute the MST and make one final pass to determine the relevant delta lists and encode our final MST structure.

Parallelization. We note that, after having constructed the Mantis representation, most phases of the MST construction algorithm are trivially parallelized. MST construction decomposes into three phases: (1) color-class graph construction, (2) MST computation, and (3) color-class representation generation. We parallelize graph construction and color-class representation generation. The MST computation itself is not parallelized.

We parallelized the determination of edges in the color-class graph by assigning each thread a range of the $k-mer$-to-color-class-ID map. Each thread explores the neighbors of the k-mers that appear in its assigned range, and any redundant edges are deduplicated when all threads are finished. Similarly, we parallelized the computation of edge weights and the extraction of the delta vectors that correspond to each edge in the MST. Given the list of edges sorted lexicographically by their endpoints (determined during the first phase), it is straightforward to partition the work for processing batches of edges across many threads. It is possible, of course, that the batches will display different workloads and that some threads will complete their assigned work before others. We have not yet made any attempt to optimize the parallel construction of the MST in this regard, though many such optimizations are likely possible.

Accelerating Queries with Caching. The encoded MST is not a balanced tree, so decoding a color bit-vector might require walking a long path to the root, which negatively impacts the query time. Attempting to explicitly minimize the depth or diameter of the MST is, as discussed in Sect. 1, not generally approximable within a constant factor. However, considering the fact that the frequency distribution of the color classes is very skewed, some of the color classes are more popular or have more children and, therefore, are in the path of many more nodes. We take advantage of these data characteristics by caching the most recent queried color bit-vectors. Every time we walk up the tree, if the color bit-vector for a node is already in the cache, our query algorithm stops at that point and applies all the deltas to this bit-vector instead of the zero bit-vector of the root. This caching approach significantly improves the query time, resulting in the final query time required to decode a color class being marginally faster than direct RRR access.

The cache policy is designed with the tree structure of our color-class representation in mind. Specifically, we want to cache nodes near the leaves, but not so close to the leaves that we end up caching essentially the entire tree. Also, we don't want to cache infrequently queried nodes. Thus we use the following caching policy: all queried nodes are cached. Furthermore, we cache interior nodes visited during a query as follows. If a query visits a node that has been visited by more than 10 other queries and is more than 10 hops away from the currently queried item, then we add that node to the cache. If a query visits more than one such node, we add the first one encountered.

In our experiments, we used a cache of 10,000 nodes and managed the cache using a FIFO policy.

2.5 Comparison with Brute-Force and Approximate-Nearest-Neighbor-Based Approaches

Our MST-based color-class representation uses the de Bruijn graph as a hint as to which color classes are likely to be similar. This leads to the natural question: how good are the hints provided by the de Bruijn graph?

One could imagine alternatively constructing the MST on the complete color-class graph. This would yield the absolutely lowest-weight spanning tree on the color classes. Unforunately, no MST algorithm runs in less than $\Omega(|E|)$ time, so this would make our construction time quadratic in the number of color classes. The number of color classes in our experiments range from 10^6 to 10^9, so the number of edges in the complete color-class graph would be on the order of 10^{12} to 10^{18}, or possibly even more, making this algorithm impractical for the largest data sets considered in this paper.

Alternatively, we could try to use an approximate nearest-neighbor algorithm to find pairs of color classes with small Hamming distance. As an experiment, we implemented an approximate nearest neighbor algorithm that bucketed color classes by their projection into a smaller-dimensional subspace. Nearest-neighbor queries were computed by searching within the queried item's bucket. Results were disappointing. Even on small data sets, the average distance between the queried item and the returned neighbor was several times larger than the average distance found using the neighbors suggested by the de Bruijn graph. Thus, we did not pursue this direction further.

3   Evaluation

In this section we evaluate our MST-based representation of the color information in the cdbg. All our experiments use Mantis with our integrated MST-based color-class representation.

Evaluation Metrics. We evaluate our MST-based representation on the following parameters:

- **Scalability.** How does our MST-based color-class representation scale in terms of space with increasing number of input samples, and how does it compare to the existing representations of Mantis?
- **Construction time.** How long does it take – in addition to the original construction time for building cdbg – to build our MST-based color-class representation?
- **Query performance.** How long does it takes to query the cdbg using our MST-based color-class representation?

3.1   Experimental Procedure

System Specifications. Mantis takes as input a collection of *squeakr* files [43]. Squeakr is a k-mer counter that takes as input a collection of fastq files and produces as output, a single file with a compact hash table mapping each k-mer to the number of times it occurs in the input files. As is standard in evaluations of large-scale sequence search indexes, we do not benchmark the time required to construct these filters.

The data input to the construction process was stored on 4-disk mirrors (8 disks total). Each is a Seagate 7200rpm 8 TB disk (ST8000VN0022). They were formatted using ZFS and exported via NFS over a 10 Gb link. We used different systems to run and evaluate time, memory, and disk requirements for the two steps of preprocessing and index building as was done by Prashant et al. [5].

For index building and query benchmarks, we ran all the experiments on the same system used in Mantis [5], an Intel(R) Xeon(R) CPU (E5-2699 v4 @2.20 GHz with 44 cores and 56 MB L3 cache) with 512 GB RAM and a 4TB TOSHIBA MG03ACA4 ATA HDD running Ubuntu 16.10 (Linux kernel 4.8.0-59-generic). Constructing the main index was done using a single thread, and the MST construction was performed using 16 threads. Query benchmarks were also performed using a single thread.

Data to Evaluate Scalability and Comparison to Mantis. We integrated and evaluated our MST-based color-class representation within Mantis, so we briefly review Mantis here. Mantis builds an index on a collection of unassembled raw sequencing data sets. Each data set is called a *sample*. The Mantis index enables fast queries of the form, "Which samples contain this k-mer," and "Which samples are likely to contain this string of bases?" Mantis takes as input one *squeakr* file per sample [43]. A squeakr file is a compact hash table mapping each k-mer to the number of times it occurs within that sample. Squeakr also has the ability to serialize a hash that simply represents the set of k-mers present at or above some user-provided threshold; we refer to these as filtered Squeakr files. Using the filtered Squeakr files vastly reduces the required intermediate storage space, and also decreases the construction time required for Mantis considerably. For example, for the breast, blood, and brain dataset (2586 samples), the unfiltered Squeakr files required ~ 2.5 TB of space while the filtered files require only ~ 108 GB. To save intermediate storage space and speed index construction, we built our Mantis representation from these filtered Squeakr files.

Given the input files, Mantis constructs an index consisting of two files: a map from k-mer to color-class ID, and a map from color-class ID to the bit vector encoding that color class. The first map is stored as a *counting quotient filter* (CQF), which is the same compact hash table used by Squeakr. The color-class map is an RRR-compressed bit vector.

Recall that our construction process is implemented as a post-processing step on the standard Mantis color-class representation. For construction times, we report only this post-processing step. This is because our MST-based color-class representation is a generic tool that can be applied to many cdbg representations other than Mantis, so we want to isolate the time spent on MST construction.

To test the scalability of our new color class representation, we used a randomly-selected set of 10,000 paired-end, human, bulk RNA-seq short-read experiments downloaded from European Nucleotide Archive (ENA) [44] in gzipped FASTQ format. Additionally, we have built the proposed index for 2,586 sequencing samples from human blood, brain, and breast tissues (BBB) originally used by [6] and also used in the subsequent work [7, 8, 39], including Mantis [5], as a point of comparison with these representations. The set of 10,000 experiments does not overlap with the BBB samples. The full list of 10,000 experimental identifiers can be obtained from https://github.com/COMBINE-lab/color-mst/blob/master/input_lists/nobbb10k_shuffled.lst. The total size of all these experiments (gzipped) is 25.23 TB.

In order to eliminate spurious k-mers that occur with insignificant abundance within a sample, the squeakr files are filtered to remove low-abundance k-mers. We adopted the same cutoff policy originally proposed by Solomon and Kingsford [6], by discarding k-mers that occur less than some threshold number of time. The thresholds are determined according to the size (in bytes) of the gzipped sample, and the thresholds are given in Table 1. We adopt a value of $k = 23$ for all experiments.

Table 1. Minimum number of times a k-mer must appear in an experiment in order to be counted as abundantly represented in that experiment (taken from the SBT paper). Note, the k-mers with count of "cutoff" *are* included at each threshold.

Min size	Max size	Cutoff	# of experiments with specified threshold
0	≤ 300MB	1	2,784
>300MB	≤500MB	3	798
>500MB	≤1GB	10	1,258
>1GB	≤3GB	20	2,296
>3GB	∞	50	2,864

3.2   Evaluation Results

Scalability of the New Color Class Representation. Figure 3a and Table 2 show how the size of our MST-based color-class representation scales as we increase the number of samples indexed by Mantis. For comparison, we also give the size of Mantis' RRR-compression-based color-class representation. Figure 3a also plots the size of the CQF that Mantis uses to map k-mers to color class IDs. We can draw several conclusions from this data:

- The MST-based representation is an order-of-magnitude smaller than the RRR-based representation.
- The gap between the RRR-based representation and the MST-based representation grows as we increase the number of input samples. This suggests that the MST-based representation grows asymptotically slower than the RRR-based representation.
- The MST-based color-class representation is, for large numbers of samples, about 5× smaller than the CQF. This means that representing the color classes is no longer the scaling bottleneck.

Table 2 also shows the scaling rate of all elements of the MST representation, in addition to the ratio of MST over the color bit-vector. As expected, the list of deltas dominate the MST representation both in terms of total size and in terms of growth. Table 2 also shows the average edge weight of the edges in the MST. The edge weight grows approximately proportional to $\Theta(\log(\# \text{ of samples}))$ (i.e. every time we double the number of samples, the average edge weight increases by almost exactly 1). This suggests that our de Bruijn graph-based algorithm is able to find pairs of similar color classes.

To better understand the scaling of the different components of a cdbg representation, we plot the sizes of the RRR-based color-class representations and MST-based representations on a log-log scale in Fig. 3b. Based on the data, the RRR-based representation appears to grow in size at a rate of roughly $\Theta(n^{1.5})$, whereas the new MST-based representation grows roughly at a rate of $\Theta(n^{1.2})$. This explains why the RRR-based representation grows to dwarf the CQF (which grows roughly linearly) and become the bottleneck to scaling to larger data sets,

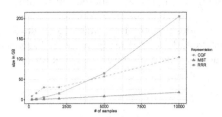

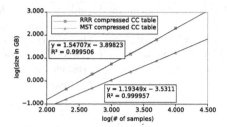

(a) Sizes of the RRR and MST-based color class representations with respect to the number of samples indexed from the human bulk RNA-seq data set. The counting quotient filter component is the Mantis representation of the de Bruijn graph.

(b) Empirical asymptotic analysis of the growth rates of the sizes of RRR-based color class representation and the MST-based color class representation. The RRR-based representation grows at a rate of $\approx \Theta(n^{1.5})$, where n is the number of samples. The MST-based representation grows at a rate of $\approx \Theta(n^{1.2})$.

Fig. 3. Size of the MST-based color-class representation vs. the RRR-based color-class representation.

Table 2. Space required for RRR and MST-based color class encodings over different numbers of samples (sizes in GB) and time and memory required to build MST. Central columns break down the size of individual MST components.

Dataset	# samples	RRR matrix	MST						Expected edge weight	$\frac{size(MST)}{size(RRR)}$
			Total space	Parent vector	Delta vector	Boundary bit-vector	Build memory (GB)	Build time (hh:mm:ss)		
H. sapiens RNA-seq samples	200	0.42	0.15	0.08	0.06	0.01	8	0:05:42	2.42	0.37
	500	1.89	0.46	0.2	0.24	0.03	16	0:12:15	3.42	0.24
	1,000	5.14	1.03	0.37	0.6	0.06	29	0:25:03	4.39	0.2
	2,000	14.2	2.35	0.71	1.5	0.14	29	0:51:58	5.38	0.17
	5,000	59.89	7.21	1.72	5.1	0.39	59	3:52:34	6.61	0.12
	10,000	190.89	16.28	3.37	12.06	0.86	111	10:17:42	7.68	0.085
Blood, Brain, Breast (BBB)	2586	15.8	2.66	0.63	1.88	0.16	29	00:57:43	6.98	0.17
E. coli strain reference genomes	5,598	2.06	0.83	0.02	0.76	0.06	6	00:03:15	7.8	0.4

Table 3. The MST construction time for 1000 experiments using different number of threads. Memory stays the same across all the runs.

# of threads	1	2	4	8	16	32
Run time (hh:mm:ss)	02:47:08	01:38:26	01:02:42	00:31:57	00:22:00	00:14:17

whereas the MST-based representation does not. With the MST-based representation, the CQF itself is now the bottleneck.

Table 4. Query time and resident memory for mantis using the MST-based representation for color information and the original mantis (using RRR-compressed color classes) over 10, 000 experiments. The "query" column provides just the time taken to execute all queries (as would be required if the index was already loaded in e.g. a server-based search tool). Note that, in resident memory usage for the MST-based representation, the counting quotient filter always dominates the total required memory.

	Mantis with MST			Mantis		
	Index load + query	Query	Space	Index load + query	Query	Space
10 Transcripts	1 min 10 s	0.3 s	118 GB	32 min 59 s	0.5 s	290 GB
100 Transcripts	1 min 17 s	8 s	119 GB	34 min 33 s	11 s	290 GB
1000 Transcripts	2 min 29 s	79 s	120 GB	46 min 4 s	80 s	290 GB

Finally, the last two rows in Table 2 show the size of the RRR- and MST-based color-class representations for the human blood, brain, breast (BBB) and *E. coli* data sets respectively. BBB is the data set used in SBT and its subsequent tools [7,8,39], as well as in Mantis [5] and *E. coli* is the data set analyzed in the Rainbowfish paper. This dataset, which has been obtained from GenBank [45], consists of 5,598 distinct *E. coli* strains. We specifically chose this dataset since Rainbowfish has already demonstrated a large improvement in size for it compared to Vari [10].

As the table shows, our MST-based color-class representation is able to effectively compress genomic color data in addition to RNA-seq color data.

Index Building Evaluation. The "Memory" and "Build time" columns in Table 2 show the memory and time required to build our MST-based color-class representation from Mantis' RRR-based representation respectively. All builds used 16 threads. Table 3 shows how the MST construction time for a 1000 sample dataset scales as a function of the number of build threads. The memory consumption is not affected by number of threads and remains fixed for all trials.

Overall, the MST construction time is only a tiny fraction of the overall time required to build the Mantis index from raw fastq files. The vast bulk of the time is spent processing the fastq files to produce filtered squeakrs. This step was performed on a cluster of 150 machines over roughly one week. Thus MST construction represents less than 1% of the overall index build time. The memory required to build the MST is dependent on the size of the CQF and grows proportional to that. In fact, due to the multi-pass construction procedure, the peak MST construction memory is essentially the size of the CQF plus a relatively small (and adjustable) amount of working memory. For the run over $10k$ experiments, where the CQF size was the largest ($98G$), the peak memory required to build MST is $111G$.

Query Evaluation. We evaluate query speed in the following manner. We select random subsets, of increasing size, of transcripts from the human transcriptome, and query the Mantis index to determine the set of experiments containing each of these transcripts. Mantis answers transcript queries as follows. For each k-mer in the transcript, it computes the set of samples containing that k-mer. It then reports a sample as containing a transcript if the sample contains more than Θ fraction of the k-mers in the transcript, where Θ is a user-adjustable parameter. Note that, for Mantis, the Θ threshold is applied at the very end. Mantis first computes, for each sample, the fraction of k-mers that occur in that sample, and then filters as a last step. Thus the query times reported here are valid for any Θ.

Table 4 reports the query performance of both the RRR and MST-based Mantis indexes. Despite the vastly-reduced space occupied by the MST-based index, and the fact that the color class decoding procedure is more involved, query in the MST-based index is slightly faster than querying in the RRR-based index. The average query time in both RRR-based and MST-based index is 0.08 s/query.

Once the index has been loaded into RAM, Mantis queries are much faster than the three SBT-based large-scale sequence search data structures, and our MST-based color-class representation doesn't change that.

4   Discussion and Conclusion

We have introduced a novel exact representation of the color information associated with the cdbg. Our representation yields large improvements in terms of representation size when compared to previous state-of-the-art approaches. While our MST-based representation is much smaller, it still provides rapid query and can, for example, return the query results for a transcript across an index of 10,000 RNA-seq experiments in ~0.08 s/query. Further, the size benefit of our proposed representation over that of previous approaches appears to grow with the number of color classes being encoded, meaning it is not only much smaller, but also much more scalable. Finally, the representation we propose is, essentially, a stand-alone encoding of the cdbg's associated color information, making this representation conceptually easy to integrate with any tool or method that needs to store color information over a large de Bruijn graph.

Though it is not clear how much further the color information can be compressed while maintaining a lossless representation, this is an interesting theoretical question. It may be fruitful to approach this question from the perspective suggested by Yu et al. [46], of evaluating the metric entropy, fractal dimension, and information-theoretic entropy of the space of color classes. Practically, however, we have observed that, at least in our current system, Mantis, for large-scale sequence search, the counting quotient filter, which is used to store the topology of the de Bruijn graph and to associate color class labels with each k-mer, has become the new scalability bottleneck. Here, it may be possible to reduce the space required by this component by making use of some of the same observations we relied upon to allow efficient color class neighbor search. For example, because many adjacent k-mers in the de Bruijn graph share the same color class ID, it is likely possible to encode this label information sparsely across the de Bruijn graph, taking advantage of the coherence between topologically nearby k-mers. Further, to allow scalability to truly-massive datasets, it will likely be necessary to make the system hierarchical, or even to adopt a more space-efficient (and domain-specific) representation of the underlying de Bruijn graph. Nonetheless, because we have designed our color class representation as essentially orthogonal to the de Bruijn graph representation, we anticipate that we can easily integrate this approach with improved representations of the de Bruijn graph.

Mantis with the new MST-based color class encoding is written in C++17 and is available at https://github.com/splatlab/mantis.

Acknowledgments and Declarations. This work was supported by the US National Science Foundation grants BIO-1564917, CCF-1439084, CCF-1716252, CNS-1408695, National Institutes of Health grant R01HG009937. The experiments were conducted with equipment purchased through NSF CISE Research Infrastructure Grant Number 1405641. RP is a co-founder of Ocean Genomics.

References

1. Iqbal, Z., Caccamo, M., Turner, I., Flicek, P., McVean, G.: De novo assembly and genotyping of variants using colored de Bruijn graphs. Nat. Genet. **44**, 226–232 (2012). https://doi.org/10.1038/ng.102810.1038/ng.1028
2. Pevzner, P.A., Tang, H., Waterman, M.S.: An Eulerian path approach to DNA fragment assembly. Proc. National Acad. Sci. **98**(17), 9748–9753 (2001)
3. Pevzner, P.A., Tang, H.: Fragment assembly with double-barreled data. Bioinformatics **17**(Suppl. 1), s225–s233 (2001)
4. Chikhi, R., Limasset, A., Jackman, S., Simpson, J.T., Medvedev, P.: On the representation of de bruijn graphs. In: Sharan, R. (ed.) RECOMB 2014. LNCS, vol. 8394, pp. 35–55. Springer, Cham (2014). https://doi.org/10.1007/978-3-319-05269-4_4
5. Prashant, P., Fatemeh, A., Bender, M.A., Ferdman, M., Johnson, R., Patro, R.: Mantis: a fast, small, and exact large-scale sequence-search index. Cell Syst. **7**(2), 201–207.e4 (2018). https://doi.org/10.1016/j.cels.2018.05.021
6. Solomon, B., Kingsford, C.: Fast search of thousands of short-read sequencing experiments. Nat. Biotechnol. **34**(3), 300–302 (2016)
7. Solomon, B., Kingsford, C.: Improved search of large transcriptomic sequencing databases using split sequence bloom trees. In: Sahinalp, S.C. (ed.) RECOMB 2017. LNCS, vol. 10229, pp. 257–271. Springer, Cham (2017). https://doi.org/10.1007/978-3-319-56970-3_16
8. Sun, C., Harris, R.S., Chikhi, R., Medvedev, P.: AllSome sequence bloom trees. In: Sahinalp, S.C. (ed.) RECOMB 2017. LNCS, vol. 10229, pp. 272–286. Springer, Cham (2017). https://doi.org/10.1007/978-3-319-56970-3_17
9. Bradley, P., den Bakker, H., Rocha, E., McVean, G., Iqbal, Z.: Real-time search of all bacterial and viral genomic data. BioRxiv, p. 234955 (2017)
10. Muggli, M.D., et al.: Succinct colored de bruijn graphs. Bioinformatics **33**, 3181–3187 (2017)
11. Holley, G., Wittler, R., Stoye, J.: Bloom Filter Trie: an alignment-free and reference-free data structure for pan-genome storage. Algorithms Mol. Biol. **11**(1), 3 (2016)
12. Almodaresi, F., Pandey, P., Patro, R.: Rainbowfish: a succinct colored de Bruijn graph representation. In: LIPIcs-Leibniz International Proceedings in Informatics, vol. 88. Schloss Dagstuhl-Leibniz-Zentrum fuer Informatik (2017)
13. Liu, B., Guo, H., Brudno, M., Wang, Y.: deBGA: read alignment with de Bruijn graph-based seed and extension. Bioinformatics **32**(21), 3224–3232 (2016a)

14. Chikhi, R., Rizk, G.: Space-efficient and exact de bruijn graph representation based on a bloom filter. In: Raphael, B., Tang, J. (eds.) WABI 2012. LNCS, vol. 7534, pp. 236–248. Springer, Heidelberg (2012). https://doi.org/10.1007/978-3-642-33122-0_19

15. Salikhov, K., Sacomoto, G., Kucherov, G.: Using cascading bloom filters to improve the memory usage for de brujin graphs. Algorithms Mol. Biol. **9**(1), 2 (2014)

16. Bowe, A., Onodera, T., Sadakane, K., Shibuya, T.: Succinct de bruijn graphs. In: Raphael, B., Tang, J. (eds.) WABI 2012. LNCS, vol. 7534, pp. 225–235. Springer, Heidelberg (2012). https://doi.org/10.1007/978-3-642-33122-0_18

17. Crawford, V., Kuhnle, A., Boucher, C., Chikhi, R., Gagie, T., Hancock, J.: Practical dynamic de bruijn graphs. Bioinformatics **34**, 4189–4195 (2018)

18. Pandey, P., Bender, M.A., Johnson, R., Patro, R.: deBGR: an efficient and near-exact representation of the weighted de bruijn graph. Bioinformatics **33**(14), i133–i141 (2017)

19. Mustafa, H., Schilken, I., Karasikov, M., Eickhoff, C., Rätsch, G., Kahles, A.: Dynamic compression schemes for graph coloring. Bioinformatics, p. bty632 (2018). https://doi.org/10.1093/bioinformatics/bty632

20. Burrows, M., Wheeler, D.J.: A block-sorting lossless data compression algorithm (1994)

21. Raman, R., Raman, V., Srinivasa Rao, S.: Succinct indexable dictionaries with applications to encoding k-ary trees and multisets. In: Proceedings of the Thirteenth Annual ACM-SIAM Symposium on Discrete Algorithms, pp. 233–242. Society for Industrial and Applied Mathematics (2002)

22. Elias, P.: Efficient storage and retrieval by content and address of static files. J. ACM (JACM) **21**(2), 246–260 (1974)

23. Raidl, G.R.: Exact and heuristic approaches for solving the bounded diameter minimum spanning tree problem. Ph.D. thesis (2008)

24. Althaus, E., Funke, S., Har-Peled, S., Könemann, J., Ramos, E.A., Skutella, M.: Approximating k-hop minimum-spanning trees. Oper. Res. Lett. **33**(2):115–120 (2005). https://doi.org/10.1016/j.orl.2004.05.005. http://www.sciencedirect.com/science/article/pii/S0167637704000719. ISSN 0167-6377

25. Manyem, P., Stallmann, M.F.M.: Some approximation results in multicasting. Technical report, Raleigh, NC, USA (1996)

26. Khuller, S., Raghavachari, B., Young, N.E.: Balancing minimum spanning and shortest path trees. CoRR, cs.DS/0205045 (2002). http://arxiv.org/abs/cs.DS/0205045

27. Marathe, M.V., Ravi, R., Sundaram, R., Ravi, S.S., Rosenkrantz, D.J., Hunt III, H.B.: Bicriteria network design problems. CoRR, cs.CC/9809103 (1998). http://arxiv.org/abs/cs.CC/9809103

28. Simpson, J.T., Wong, K., Jackman, S.D., Schein, J.E., Jones, S.J.M., Birol, I.: ABySS: a parallel assembler for short read sequence data. Genome Res. **19**(6), 1117–1123 (2009)

29. Schulz, M.H., Zerbino, D.R., Vingron, M., Birney, E.: Oases: robust de novo RNA-seq assembly across the dynamic range of expression levels. Bioinformatics **28**(8), 1086–1092 (2012)

30. Zerbino, D.R., Birney, E.: Velvet: algorithms for de novo short read assembly using de Bruijn graphs. Genome Res. **18**(5), 821–829 (2008)

31. Grabherr, M.G., et al.: Full-length transcriptome assembly from RNA-seq data without a reference genome. Nature Biotechnol. **29**(7), 644–652 (2011)

32. Chang, Z., et al.: Bridger: a new framework for de novo transcriptome assembly using RNA-seq data. Genome Biol. **16**(1), 30 (2015)

33. Liu, J., et al.: Binpacker: packing-based de novo transcriptome assembly from RNA-seq data. PLOS Comput. Biol. **12**(2), e1004772 (2016b)
34. Almodaresi, F., Sarkar, H., Srivastava, A., Patro, R.: A space and time-efficient index for the compacted colored de Bruijn graph. Bioinformatics **34**(13), i169–i177 (2018)
35. Turner, I., Garimella, K.V., Iqbal, Z., McVean, G.: Integrating long-range connectivity information into de Bruijn graphs. Bioinformatics **34**(15), 2556–2565 (2018). https://doi.org/10.1093/bioinformatics/bty157
36. Alipanahi, B., Muggli, M.D., Jundi, M., Noyes, N., Boucher, C.: Resistome SNP calling via read colored de Bruijn graphs. bioRxiv, p. 156174 (2018)
37. Alipanahi, B., Kuhnle, A., Boucher, C.: Recoloring the colored de Bruijn graph. In: Gagie, T., Moffat, A., Navarro, G., Cuadros-Vargas, E. (eds.) SPIRE 2018. LNCS, vol. 11147, pp. 1–11. Springer, Cham (2018b). https://doi.org/10.1007/978-3-030-00479-8_1
38. Pandey, P., Bender, M.A., Johnson, R., Patro, R.: A general-purpose counting filter: making every bit count. In: Proceedings of the 2017 ACM International Conference on Management of Data, pp. 775–787. ACM (2017)
39. Yu, Y., et al.: SeqOthello: querying RNA-seq experiments at scale. Genome Biol. **19**(1), 167 (2018). https://doi.org/10.1186/s13059-018-1535-9. ISSN 1474-760X
40. Ottaviano, G., Venturini, R.: Partitioned Elias-Fano Indexes. In: Proceedings of the 37th International ACM SIGIR Conference on Research and Development in Information Retrieval, pp. 273–282. ACM (2014)
41. Ziv, J., Lempel, A.: A universal algorithm for sequential data compression. IEEE Trans. Inf. Theory **23**(3), 337–343 (1977)
42. Bookstein, A., Klein, S.T.: Compression of correlated bit-vectors. Inf. Syst. **16**(4), 387–400 (1991)
43. Pandey, P., Bender, M.A., Johnson, R., Patro, R.: Squeakr: an exact and approximate k-mer counting system. Bioinformatics, btx636 (2017). https://doi.org/10.1093/bioinformatics/btx636
44. NIH. SRA (2017). https://www.ebi.ac.uk/ena/browse. Accessed 06 Nov 2017
45. O'Leary, N.A., et al.: Reference sequence (RefSeq) database at NCBI: current status, taxonomic expansion, and functional annotation. Nucleic Acids Res. gkv1189 (2015)
46. Yu, Y.W., Daniels, N.M., Danko, D.C., Berger, B.: Entropy-scaling search of massive biological data. Cell systems **1**(2), 130–140 (2015)

Identifying Clinical Terms in Free-Text Notes Using Ontology-Guided Machine Learning

Aryan Arbabi[1,2,3], David R. Adams[4], Sanja Fidler[1,3],
and Michael Brudno[1,2,3(✉)]

[1] Department of Computer Science, University of Toronto,
Toronto, ON, Canada
brudno@cs.toronto.edu
[2] Center for Computational Medicine, Hospital for Sick Children,
Toronto, ON, Canada
[3] Vector Institute, Toronto, ON, Canada
[4] Section on Human Biochemical Genetics,
National Human Genome Research Institute,
National Institutes of Health, Bethesda, MD, USA

Abstract. Objective: Automatic recognition of medical concepts in unstructured text is an important component of many clinical and research applications and its accuracy has a large impact on electronic health record analysis. The mining of such terms is complicated by the broad use of synonyms and non-standard terms in medical documents. Here we presented a machine learning model for concept recognition in large unstructured text which optimizes the use of ontological structures and can identify previously unobserved synonyms for concepts in the ontology.

Materials and Methods: We present a neural dictionary model which can be used to predict if a phrase is synonymous to a concept in a reference ontology. Our model, called Neural Concept Recognizer (NCR), uses a convolutional neural network and utilizes the taxonomy structure to encode input phrases, then rank medical concepts based on the similarity in that space. It also utilizes the biomedical ontology structure to optimize the embedding of various terms and has fewer training constraints than previous methods. We train our model on two biomedical ontologies, the Human Phenotype Ontology (HPO) and SNOMED-CT.

Results: We tested our model trained on HPO on two different data sets: 288 annotated PubMed abstracts and 39 clinical reports. We also tested our model trained on the SNOMED-CT on 2000 MIMIC-III ICU discharge summaries. The results of our experiments show the high accuracy of our model, as well as the value of utilizing the taxonomy structure of the ontology in concept recognition.

Conclusion: Most popular medical concept recognizers rely on rule-based models, which cannot generalize well to unseen synonyms. Also, most machine learning methods typically require large corpora of annotated text that cover all classes of concepts, which can be extremely difficult to get for biomedical ontologies. Without relying on a large-scale labeled training data or requiring any custom training, our model can efficiently generalize to new synonyms and performs as well or better than state-of-the-art methods custom built for specific ontologies.

© Springer Nature Switzerland AG 2019
L. J. Cowen (Ed.): RECOMB 2019, LNBI 11467, pp. 19–34, 2019.
https://doi.org/10.1007/978-3-030-17083-7_2

Keywords: Concept recognition · Ontologies · Named entity recognition · Phenotyping · Human Phenotype Ontology

1 Introduction

Automatic recognition of medical concepts in unstructured text is a key component of biomedical information retrieval systems, having applications such as analysis of the unstructured text in electronic health records (EHR) [1–3] or knowledge discovery from biomedical literature [4, 5].

Many medical terminologies are structured as ontologies, adding the relations between terms, and often including several synonyms for each concept. One of the most widely used ontologies in the medical space is SNOMED-CT [6], which provides structured relationships for over 300,000 medical concepts. SNOMED-CT is commonly used in Electronic Health Record Systems to help summarize patient encounters and is fully integrated with ICD-9 billing codes used in the US and many other jurisdictions. The Human Phenotype Ontology (HPO) [7] is an arrangement of the terms used to describe the visible manifestations (phenotypes) of human genetic diseases. The HPO has ~ 12,000 terms, and has become the standard ontology used in Rare Disease research and clinical genetics, having been adopted by IRDiRC [8], ClinGen [9] and many other projects. Both SNOMED-CT and HPO, like most other ontologies, provide a number of synonyms for each term, they usually miss many valid synonymous terms, as manually curating every term that refers to a concept is extremely difficult, if possible.

There have many concept recognition and text annotation tools developed for biomedical text. Examples of popular general purpose ones are NCBO annotator [10], OBO annotator [11], MetaMap [12] and Apache cTAKES [13]. Other tools have been developed focusing on more specific domains, such as BioLark [14] which is developed for automatic recognition of terms from the HPO, and Lobo et al. [15] which combines a machine learning approach with manual validation rules. These systems usually consist of a pipeline of natural language processing components including tokenizer, part-of-speech tagger, sentence boundary detector and named entity recognizer (NER)/annotator. Generally, the NER/annotator component of these tools are based on text matching, dictionary look-ups and rule-based methods, which usually require significant engineering effort, and are often unable to handle novel synonyms absent in the ontology.

On the other hand, in the more general domain of natural language processing, many machine learning based text classification and NER tools have been recently introduced [16–18]. Typically, these methods do not need the manual rule-based engineering effort, however they are dependent on large annotated text data for training. Popular among them is a model known as LSTM-CRF, in which Long-short-term-memory (LSTM) [19], a variation of recurrent neural networks widely used for processing sequences such as text, is used to extract rich representations of the tokens in a sentence, followed by a conditional random field (CRF) [20] on top of these representations to recognize named entities.

While these methods address a similar problem, they cannot be used directly for concept recognition, as the number of named entity classes is typically much fewer than the concepts in medical ontologies. For instance, CoNLL-2003 [21], one of the data sets widely used for evaluations of such methods, contains only 4 classes: locations, persons, organizations, and miscellaneous. As a result, these methods typically have a large number of training and test examples for each class, while in our setting we are trying to recognize tens or hundreds of thousands of terms and may have only a few or even no examples of a specific term. In this setting, where the training data does not fully cover all of the classes, methods based on dictionary look-up might have some advantage, as they can identify a concept in a given text by simply matching it to a synonym available in their dictionary, without requiring training data annotated with that concept.

In this paper we develop a hybrid approach called Neural Concept Recognizer (NCR), by introducing a neural dictionary model that learns to generalize to novel synonyms for concepts. Our model is trained on the information provided by the ontology, including the concept names, synonyms, and the taxonomic relations between the concepts, and can be used to rank the concepts a given phrase can match as a synonym. Our model consists of two main components: an encoder which maps an input phrase to a vector representation, and an embedding table consisting of the vector representations learned for the ontology concepts. The classification is done based on the similarity between the phrase vector and the concept vectors. To allow for the use of our model to also detect concepts from longer texts, we scan the input text with fixed-size windows and report a phrase as matching a concept if it is above a threshold that is chosen from an appropriate validation dataset.

We trained our neural dictionary model on the Human Phenotype Ontology (HPO) and used it to recognize concepts from two data sets including 228 PubMed abstracts and 39 clinical reports of patients with rare genetic diseases. Additionally, we also used a subset of SNOMED-CT containing concepts that have matching terms in ICD-9, and experiment on 2000 ICU discharge summaries from MIMIC-III data-set [22]. In both settings we trained our model solely on the ontology data and did not use the text corpora except for setting the recognition sensitivity threshold from a small validation set. Our experiments show the high accuracy of our model, on par with or better than hand-trained methods. Our tool has already been used in two applications. It has been integrated with the PhenoTips tool to suggest concepts for a clinical report [23], and to automatically recognize occurrences of phenotypes in a clinical report for subsequent data visualization [24]. Although the main focus of this work is recognizing HPO and SNOMED-CT concepts, our method can be easily trained on other biomedical ontologies.

1.1 Related Works

Recently, several machine learning methods have been used in biomedical NER or concept recognition. Habibi et al. [25] trained the LSTM-CRF NER model introduced by Lample et al. [16] for recognizing five entity classes of genes/proteins, chemicals, species, cell lines and diseases. They tested their model on several biomedical corpora and achieved better results compared to previous rule-based methods. In another recent work, Vani et al. [26] introduced a novel RNN-based model, and showed its efficiency on predicting ICD-9 codes in clinical notes. Both of these methods require a training corpus annotated with the concepts (loosely annotated in the case of Vani et al. [26]).

Curating such annotated corpus is more difficult for typical biomedical ontologies, as the corpus has to cover thousands of classes. For example, HPO contains 11442 concepts (classes), while the only (to the best of our knowledge) publicly available corpus annotated with HPO concepts [14] contains 228 PubMed abstracts with only 607 unique annotations that are not an exact match of a concept name or a synonym. Thus, training a method to recognize the presence of concepts in biomedical text requires a different approach when there is a large number of concepts.

The concepts in an ontology often have a hierarchical structure (i.e. a taxonomy), which can be utilized in representation learning. Hierarchies have been utilized in several recent machine learning approaches. Deng et al. [27] proposed a CRF based method for image classification that takes into account inheritance and exclusion relations between the labels. Their CRF model transfers knowledge between classes by summing the weights along the hierarchy, leading to improved performance. Vendrov et al. [28] introduced the order-embedding penalty to learn representations of hierarchical entities, and used it for image-caption retrieval task. Gaussian embeddings were introduced by Neelakantan et al. [29] which instead of single point vectors learns a high-dimensional Gaussian distribution which can model entailment. Most recently, Nickel et al. [30] showed learning representations in hyperbolic space can improve performance for hierarchical representations.

2   Methods

The methods section is organized as follows: in Sect. 2.1 we describe the neural dictionary model that computes the likelihood that a given phrase matches each concept from an ontology. Following, in Sect. 3.2 we show how to apply the model to larger text fragments, such as a full sentence, which may have multiple (or no) terms.

2.1   Overview of the Neural Dictionary Model

The neural dictionary model receives as input a word or a phrase and finds the probability of the concepts in the ontology matching it. The model consists of a text encoder, which is a neural network that maps the query phrase into vector representation, and an embedding matrix with rows corresponding the ontology concepts. We use the dot product of the query vector and a concept vector as the measure of similarity. Figure 1 shows an overview of our model and the following subsections describe our model in more details.

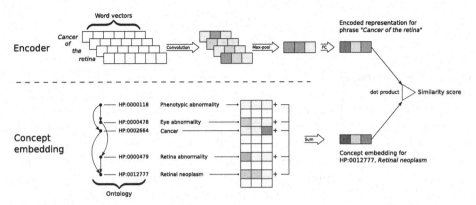

Fig. 1. Architecture of the neural dictionary model. The encoder is shown at the top and the procedure for computing the embedding for a concept is illustrated at the bottom. Encoder: a query phrase is first represented by its word vectors, which are then projected by a convolution layer into a new space. Then, a max-over-time pooling layer is used to aggregate the set of vectors into a single one. Afterwards, a fully connected layer (referred by FC) maps this vector into the final representation of the phrase. Concept embedding: a matrix of raw embeddings is learned, where each row represents one concept. The final embedding of a concept is retrieved by summing the raw embeddings for that concept and all of its ancestors in the ontology.

Encoder. We use word embeddings to represent the input words, learned in a pre-processing step by running fastText [31] on publicly available MEDLINE/PubMed abstracts. The goal of this unsupervised step is to map semantically similar words (e.g. synonyms) to close vectors. We selected fastText for this task mainly because it takes into account the subword information, which is important in medical domain where there are many semantically close words with slight morphologic variations.

Inspired by Kim et al. [32], our encoder projects these word vectors into another space using a convolution neural network. We have used a much simpler network, consisting of a single convolution layer, with a filter size of one word. Although this choice of filter size has the disadvantage of losing the word order information, in our settings this was outweighed by the benefit of having fewer network parameters to learn. In the next step, these projected vectors are aggregated into a single vector v, using a max-over-time pooling operation, as shown in equation below:

$$v = \max_t\left\{\text{ELU}\left(Wx^{(t)} + b\right)\right\},$$

where $x^{(t)}$ is the vector for the tth word in the phrase, W and b are the weight matrix and the bias vector of the convolution filter, and ELU [33] is the activation function we used in the convolution layer. It should also be noted that the max operation used in the equation above is an element-wise operation that takes the maximum value of each feature across projected word vectors. Finally, a fully connected layer with the weights U is applied on v followed by a ReLU activation and $l2$ normalization. The result e is used as the encoded vector representation of the phrase:

$$e = \frac{\text{ReLU}(Uv)}{\text{Relu}\|(Uv)\|_2}.$$

Concept Representations. Our model also learns representations for the concepts and measures the similarity between an input phrase and the concepts by computing the dot product between these representations and the encoded phrase e.

We denote these representations by the matrix H, where each row corresponds to one concept. Our model does not learn H directly, instead it learns a matrix $\tilde{H}$, where each row $\tilde{H}_c$ represents the features of concept c that are "novel" compared to its ancestors. Then H can be derived by multiplying $\tilde{H}_c$ by the taxonomy's ancestry matrix A, where $A_{i,j}$ is 1 if and only if the concept j is an ancestor i (including $i = j$):

$$H = A\tilde{H}.$$

Finally, the classification is done by computing the dot product followed by a softmax layer as follows:

$$p(c|e) \propto \exp(H_c e).$$

The taxonomy information can be ignored by setting A to the identity matrix I. In this scenario the model would behave like an ordinary softmax classifier with the weights H.

Training Procedure. Training is done on the names and synonyms provided by the ontology. If a concept has multiple synonyms, each synonym-concept pair is considered as a separate training example. We train our model by minimizing the cross-entropy loss between the softmax output and the class labels using Adam optimizer [34]. The parameters learned during the training are the encoder parameters W and U, and the concept representations through $\tilde{H}$.

2.2   Concept Recognition in a Sentence

To use our neural dictionary model to recognize concepts in a sentence or larger text, we extract all n-grams of one to seven words in the text and use the neural dictionary model to match each n-gram to a concept. We filter irrelevant n-grams by removing the candidates whose matching score (the softmax probability provided by the neural dictionary model), is lower than a threshold. This threshold is chosen based on the performance of the method (f-measure) on a validation set. We also use random n-grams from an unrelated corpus (in our case Wikipedia) as negative examples labeled with a dummy *none* concept, when training the neural dictionary model. The lengths of these n-grams were uniformly selected to be between 1 and 10.

After all the n-grams satisfying the conditions are captured, a post-processing step is performed to make the results consistent. For every pair of overlapping captured n-grams, if both n-grams matched the same concept we retain the smaller n-gram. Otherwise if they were matched to different concepts, we favor choosing the longer n-gram, as this reduces the bias of choosing shorter more general concepts in the presence of a more specific concept. For example, when annotating the sentence "The patient was diagnosed with conotruncal heart defect.", our method will favor choosing the longer more specific concept "conotruncal heart defect", rather than the more general concept "heart defect".

3 Results

To evaluate our model, we applied it to a number of medical texts, and trained on two ontologies (HPO and SNOMED-CT). We also evaluated the model on two different tasks. In the first task, the model ranks concepts matching an input isolated phrase (synonym classification), and in the second task concepts are recognized and classified from a document (concept recognition).

To assess the effectiveness of the techniques used in our model, we trained four variations of our model as follows:

- **NCR:** This is the full model, with the same architecture as described in Sect. 2.1. The training data for this model includes negative examples.
- **NCR-H:** In this version the model ignores the taxonomic relations by setting the ancestry matrix A to the identity matrix I.
- **NCR-N:** Similar to the original NCR, this version utilizes the taxonomic relations. However, this model has not been trained on negative samples.
- **NCR-HN:** This refers to a variation which both ignores the taxonomy and has not been trained on negative examples.

3.1 Data Sets

In most of our experiments we used the Human Phenotype Ontology (HPO) to train the neural dictionary model. The version of HPO (2016 release) we used (to maintain consistency with previous work) contains a total of 11,442 clinical phenotypic abnormalities seen in human disease and provides a total of 19,202 names and synonyms for them, yielding an average of 1.67 names per concept.

We evaluated the accuracy of our model trained on HPO on two different data sets:

- **PubMed:** This data set contains 228 PubMed article abstracts, gathered and manually annotated with HPO concepts by Groza et al. [14].
- **UDP:** The second set includes 39 clinical reports provided by NIH Undiagnosed Diseases Program (UDP) [35]. Each case contains the medical history of a patient in unstructured text format and a list of phenotypic findings, recorded as a set of HPO concepts, gathered by the examining clinician from the patient encounter.

In order to examine the effectiveness of the model on different ontologies, we also trained our model on a subset of SNOMED-CT, which is a comprehensive collection of medical concepts and includes their synonyms and taxonomy. We evaluated the trained model for concept recognition on a subset of MIMIC-III [22] described in the following:

- **MIMIC:** We used a subset of 2000 Intensive Care Unit discharge summaries from MIMIC-III. The discharge summaries are in unstructured text format and are accompanied with a list of disease diagnosis terms in the format of ICD-9 (International Classification of Diseases) codes.

Since SNOMED-CT provides a more sophisticated hierarchy than ICD-9 and there exists a mapping between the two, we used a subset of SNOMED-CT concepts that either map to, or are an ancestor of an ICD-9 concept. We only considered the 1,292 most frequent ICD-9 concepts that have a minimum of 50 occurrences in MIMIC-III, resulting in a total of 11,551 SNOMED-CT concepts.

3.2 Synonym Classification Results

In this experiment we evaluated our method's performance in matching isolated phrases with ontology concepts. For this purpose, we extracted 607 unique phenotypic phrases, which did not have an exact-match among the names and synonyms in HPO, from the 228 annotated PubMed abstracts. We used our model to classify HPO concepts for these phrases and ranked them by their score.

In addition to the four variations of our model, we compared with another method based on Apache Solr, customized to suggest HPO terms for phenotypic queries. This tool is currently being used as a component of the phenotyping software PhenoTips [23]. The results of this experiment are provided in Table 1. We measured the fraction of the predictions where the correct label was among the top-1 and top-5 recalled concepts. NCR significantly outperforms PhenoTips in this experiment. While NCR-N slightly outperforms regular NCR in this test, the experiments here contained no queries without phenotypic terms, which is the task that NCR-N was built to model.

Table 1. Synonym classification experiments on 607 phenotypic phrases extracted from 228 PubMed abstracts. R@1 and R@5 accuracies represent recall using top-1 and top-5 results from each method.

Method	Accuracy (%)	
	R@1	R@5
PhenoTips	28.9	49.3
NCR	49.1	73.6
NCR-H	45.8	68.4
NCR-N	52.2	73.5
NCR-HN	50.1	70.5

3.3 Concept Recognition Results

We evaluated the four versions of NCR for concept recognition and also compared with four rule-based methods, cTAKES [13] and BioLarK [14], NCBO annotator [10], OBO annotator [11]. The NCBO annotator is a general concept recognition tool with access to hundreds of biomedical ontologies, including HPO, cTAKES is a more general medical knowledge extraction system primarily designed for SNOMED-CT, while BioLarK and the OBO annotator are concept recognizers primarily tailored for HPO. Another method called IHP [15] was recently introduced for identifying HPO terms in unstructured text, using machine learning for named entity recognition and a rule-based approach for further extending them. However, their method is not comparable to us as to the best of our knowledge their provided tool only reported the text-spans that are a phenotype and did not classify or rank matching HPO terms.

In order to choose a score threshold for filtering irrelevant concepts, we used 40 random PubMed abstracts as a validation set and compared the Micro F1-score given different threshold values; the selected thresholds were 0.85, 0.8, 0.8 and 0.75 for NCR, NCR-H, NCR-N and NCR-HN respectively. Since the UDP dataset contained fewer reports (39 in total), we did not choose a separate UDP validation set and used the same threshold determined for the PubMed abstracts. We tested our methods on the remaining 188 PubMed abstracts and the 39 UDP reports and calculated micro and macro versions of precision, recall and F1-Score, as shown in following equations:

$$\text{Micro Recall} = \frac{\sum_d |R_d \cap L_d|}{\sum_d |L_d|},$$

$$\text{Micro Precision} = \frac{\sum_d |R_d \cap L_d|}{\sum_d |R_d|},$$

$$\text{Macro Recall} = \frac{1}{|D|} \sum_d \frac{|R_d \cap L_d|}{|L_d|},$$

$$\text{Macro Precision} = \frac{1}{|D|} \sum_d \frac{|R_d \cap L_d|}{|R_d|}.$$

In these equations, D is the set of all documents, and R_d and L_d notate the set of reported concepts and label concepts for the document d, respectively. We also calculated a less strict version of accuracy measurements, that takes the taxonomic relations of the concepts into consideration. To do so, we extended the reported set and the label set for each document to include all of their ancestor concepts, which we notate by $E(R_d)$ and $E(L_d)$ respectively and calculated an extended version of the precision and recall, as well as the Jaccard Index of the extended sets. The following equations show how these accuracies are derived.

$$\text{Extended Recall} = \frac{1}{|D|} \sum_d \frac{|E(R_d) \cap L_d|}{|L_d|},$$

$$\text{Extended Precision} = \frac{1}{|D|} \sum_d \frac{|R_d \cap E(L_d)|}{|R_d|},$$

$$\text{Jaccard Index} = \frac{1}{|D|} \sum_d \frac{|E(R_d) \cap E(L_d)|}{|E(R_d) \cup E(L_d)|}.$$

The results of these experiments are available in Table 2. In both experiments, based on the measurements of the Jaccard index and all three versions of micro, macro and extended F1-scores, NCR has considerably higher accuracy compared to all other baselines. Furthermore, by comparing the NCR and NCR-H, it can be observed that using the hierarchy information has improved the accuracy of the model. Finally, with the narrow exception of the micro F1-score, comparison of NCR and NCR-N shows that using negative examples during the training has slightly improved the overall accuracy.

Table 2. Concept recognition results for four variations of NCR, BioLark and cTAKES. NCR models were trained on HPO. Two set of experiments were performed, including 188 PubMed abstracts and 39 UDP clinical notes. In the heading, PR refers to the precision.

Method		Micro (%)			Macro (%)			Extended (%)			Jaccard (%)
		PR	Recall	F1	PR	Recall	F1	PR	Recall	F1	
PubMed	BioLarK	78.5	60.5	68.3	71.3	60.7	65.6	79.0	68.2	73.2	74.2
	cTAKES	72.2	55.6	62.8	64.4	56.1	60.0	**82.5**	61.9	70.7	71.0
	OBO	78.3	53.7	63.7	72.1	53.3	61.3	80.4	64.1	71.3	72.8
	NCBO	81.6	44.0	57.2	67.7	43.3	52.9	80.7	50.0	61.7	62.7
	NCR	**79.5**	**62.1**	**69.7**	**74.2**	62.2	**67.7**	81.6	69.0	**74.8**	**75.9**
	NCR-H	69.9	60.5	64.9	64.4	62.2	63.3	71.3	**70.5**	70.9	69.6
	NCR-N	78.7	61.6	69.1	73.0	61.6	66.8	79.8	68.4	73.7	74.5
	NCR-HN	66.0	60.0	62.9	62.3	61.5	61.9	68.8	69.3	69.0	66.5
UDP	BioLarK	27.6	21.0	23.9	28.7	21.6	24.6	43.5	26.1	32.6	29.5
	cTAKES	31.5	18.9	23.6	34.9	20.2	25.6	49.5	22.9	31.3	27.3
	OBO	26.8	20.5	23.2	28.8	20.1	23.7	38.8	28.0	32.5	31.3
	NCBO	**33.4**	16.9	22.5	**37.1**	19.9	25.9	**52.8**	23.2	32.2	27.2
	NCR	24.8	25.5	25.1	27.8	26.4	**27.1**	43.0	30.8	**35.9**	**31.6**
	NCR-H	24.3	25.6	24.9	26.0	26.0	26.0	39.6	31.5	35.1	30.4
	NCR-N	24.1	26.6	**25.3**	25.8	27.0	26.4	39.2	31.3	34.8	31.2
	NCR-HN	22.2	**28.3**	24.9	24.7	**29.0**	26.7	38.1	**33.9**	35.8	29.8

To evaluate the effectiveness of the techniques employed in NCR on a different ontology, we trained the four variations of our model on the SNOMED-CT subset, using 200 MIMIC reports as the validation set and the remaining 1800 ones as test set. We mapped each reported SNOMED-CT concept to the corresponding ICD-9 code and calculated the accuracy measurements, available in Table 3. The results show that utilizing the hierarchy information and including negative examples in the training has improved both micro and macro F1-scores. Since the labels are only available as ICD-9 codes, which do not hold a sufficiently rich hierarchical structure similar to HPO and

SNOMED-CT, the Jaccard index and the extended accuracy measurements were less meaningful and were not calculated. We also ran the original cTAKES, which is optimized for SNOMED-CT concepts, on the 1800 test documents and filtered its reported SNOMED-CT results to ones that have a corresponding ICD-9. While cTAKES had a high recall, the overall F1 scores were lower than NCR.

Table 3. Results for concept recognition experiments on 1800 MIMIC documents. The NCR models were trained on a subset of SNOMED-CT ontology.

Method	Micro (%)			Macro (%)		
	Precision	Recall	F1	Precision	Recall	F1
cTAKES	9.1	**37.0**	14.6	8.7	**36.3**	14.0
NCR	**11.2**	22.6	**15.0**	**10.9**	22.9	**14.8**
NCR-H	9.3	32.3	14.5	9.0	32.1	14.0
NCR-N	9.6	29.3	14.4	9.3	29.5	14.1
NCR-HN	8.8	31.4	13.7	8.4	31.2	13.2

3.4   Qualitative Results

To better understand how utilizing the hierarchy information affects the model, we used t-SNE to embed and visualize the learned concept representations for the rows of matrix H for NCR-N (using hierarchy) and NCR-NH (not using the hierarchy), trained on HPO. These representations are illustrated in Fig. 2 where colors are assigned to concepts based on their high-level ancestor (the 23 children of the root). If a concept had multiple high-level ancestors we chose one randomly. As it is evident in the plots the representations learned for NCR-N are better clustered compared to NCR-NH.

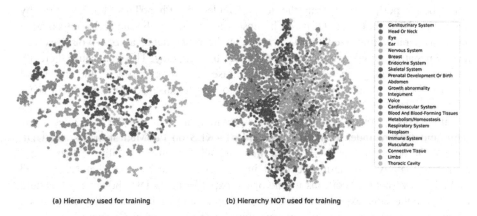

(a) Hierarchy used for training   (b) Hierarchy NOT used for training

Fig. 2. Visualization of the representations learned (embedded into 2 dimensions by t-SNE) for HPO concepts. The colors denote the high-level ancestors of the concepts. The plot on the left shows the representations learned in NCR-N where the taxonomy information was used in training and the plot on the right shows representations learned for NCR-HN, where the taxonomy was ignored. (Color figure online)

An interesting observation in the representations learned for NCR-N is that concepts in categories such as "Neoplasm" (colored in dark grey), which share children with many other categories, are located in the center of the plot, close to many other categories, while a category like "Abnormality of ear" (colored in orange) has formed its own cluster far from center and separated from other categories.

To further investigate the false positives reported by NCR, we manually investigated the false positives reported by our method in three clinical reports randomly chosen from the UDP dataset. We looked at false positives from the extended version of evaluations, which includes concepts reported by our method where neither it nor any of its descendants are in the label set. This brought a total number of 61 unique false positives for the three documents. Based on a manual analysis of these terms conducted by a medical expert on rare genetic diseases (co-author DRA), 44.2% of the reported false positives were actually correctly adding more information to the closest phenotype reported in the label set. One example is "Congenital hypothyroidism on newborn screening", which while our method correctly recognized "Congenital hypothyroidism", the closest concept in the extended label set was "Abnormality of the endocrine system". In an additional 13.1% of cases our model had correctly reported a more specific concept than what was present in the patient record, but the concept was sufficiently close to a specified phenotype not to be considered a novel finding. Furthermore, 19.6% of the reported false positives were in fact mentioned in the text, though as negations, such as "Group fiber atrophy was not seen" and in 6.5% of them the reported phenotype was mentioned but not confidently diagnosed, such as "possible esophagitis and gastric outlet delay".

4   Discussion

In the synonym classification task, as evident in Table 1, all variations of NCR have a much better performance than the tool provided by PhenoTips. Furthermore, by comparing NCR and NCR-H, it can be observed that using the hierarchy information has considerably improved the accuracy. On the other hand, as one would expect, having negative samples in the training reduces the accuracy. This is because there are no negatives among the 607 phenotypic phrases in this experiment and the actual purpose for including them in training was to reduce false positives in the concept recognition task.

In the concept recognition experiments provided in Table 2, NCR had a better F1-score and Jaccard index than BioLarK and cTAKES on PubMed abstracts and UDP reports. On both datasets, NCR had a higher recall, showing its ability to better generalize to synonymous terms occurred in the text. It can be seen that NCBO and cTAKES have better precisions in one of the experiments (UDP), however we should note that in the same experiments NCR has achieved a much better recall rate and when taking both precision and recall into account, NCR has a higher F1-score.

Among different variations of NCR, using the hierarchy information has always led to higher F1-scores and Jaccard index. Having negative samples during training has also generally improved accuracy, however in some cases the difference has been small and even on UDP data set NCR-N had slightly higher Micro F1-score.

While the PubMed abstracts were manually annotated with HPO concepts by Groza et al. [14] the text provided for UDP is not annotated and there is no explicit association between the provided HPO terms and phenotypic phrases in the text. However, since both the text and the terms refer to the same patient, there exists a weaker implicit correspondence. This can explain the overall higher accuracy of all methods on PubMed data compared to UDP. As a result, these performance measurements would be more meaningful when observed in a relative manner, which show the better performance of NCR compared to the baselines.

The experiments on MIMIC data, where the model was trained on SNOMED-CT, resulted in much lower accuracy compared to the two experiments performed for HPO. In addition to the problem of implicit correspondence between labels and actual occurrences in the text, in this experiment we used a mapping between ICD-9 and SNOMED-CT terms, which can introduce other inconsistencies. On the other hand, for the sake of evaluating the techniques employed in our model on another ontology, it can be observed in Table 3 that utilizing the SNOMED-CT hierarchy indeed improves the F1-scores.

In addition to the quantitative results showing the advantage of using the hierarchy information, our visualization of the concept representations in Fig. 2 shows that the representations learned for NCR-N better cluster compared to NCR-HN. Although in theory NCR-N has the flexibility to learn representations identical to NCR-HN, the way our model utilizes the taxonomy entangles the embedding of related concepts during the training, which in practice has led to better separated clusters. Because of this entanglement, if the optimizer updates the raw embedding $\tilde{H}_c$ of a concept c, systematically all descendents of c would "inherit" this update in their final representation in H.

NCR has already been used in several applications in practice. Currently a version of NCR trained on HPO is deployed as a component of PhenoTips software [23], being used in both annotation of clinical notes and term suggestion for manually entered phenotypes. Another example is PhenoLines [24], a software for visualizing disease subtypes, relying on a mapping between HPO and UMLS [36] terms. NCR was effectively used to help improving the coverage of their mapping. The code for NCR is available at https://github.com/ccmbioinfo/NeuralCR under the MIT license.

5   Conclusion

In this paper we presented neural dictionary model, which ranks matching concepts for a query phrase and can be used for concept recognition in larger text. Unlike other machine learning based concept recognition tools our training is solely done on the ontology data (except the unsupervised learning of the word vectors) and it does not require any annotated corpus. This is important as it is usually difficult to manually curate a corpus that scales to the large number of concepts in a medical ontology. Another novelty of our model is our approach in utilizing the taxonomic relations between concepts, that based on our experiments improves synonym classification.

NCR uses convolutional neural networks to encode query phrases into vector representations and computes their similarity to embeddings learned for ontology concepts. The model benefits from knowledge transfer between child and parent concepts by summing the raw embeddings of a concept's ancestors to compute its final embedding.

We tested our neural dictionary model by classifying 607 phenotypic phrases and our model achieved a considerably higher accuracy compared to another method designed for this task and the baseline versions of our model that do not use the taxonomy information. We also tested our method for concept recognition on full text by experimenting on four data sets. In one setting we trained our model on the HPO ontology and tested on two data sets, including 188 PubMed paper abstracts and 39 UDP clinical, while in another setting we trained the model on a subset of SNOMED-CT medical concepts and tested on 1800 MIMIC ICU discharge notes. Our results showed the efficiency of our methods in both settings. One major challenge for the concept recognition task is how to filter candidates that do not match any class in the ontology. In our experiments we approached this challenge by adding negative samples from Wikipedia in the training. Although this improved the results, it did not fully solve the problem, as for instance in the HPO setting, there can be many relevant medical terms in a clinical text which are not phenotypic, while are not available in the negative examples.

Although our experiments have shown the high accuracy of our model in classifying synonyms, we believe there is much room for improvement in the overall concept recognition method, especially the way the n-grams are selected and filtered. Another interesting direction for future work is to investigate the possibility of using unsupervised methods for encoding phrases, such as skip-thought vectors [37], to utilize the massive amount of available unannotated biomedical corpora for better generalization of classifying synonymous phrases and concept recognition.

Acknowledgements. We thank Michael Glueck for his valuable comments and discussions. We also thank Tudor Groza for his helpful comments and for providing us the BioLarK API used for the experiments.

References

1. Simmons, M., Singhal, A., Lu, Z.: Text mining for precision medicine: bringing structure to EHRs and biomedical literature to understand genes and health. In: Shen, B., Tang, H., Jiang, X. (eds.) Translational Biomedical Informatics. AEMB, vol. 939, pp. 139–166. Springer, Singapore (2016). https://doi.org/10.1007/978-981-10-1503-8_7
2. Jonnagaddala, J., Dai, H.-J., Ray, P., Liaw, S.-T.: Mining electronic health records to guide and support clinical decision support systems. In: Healthcare Ethics and Training: Concepts, Methodologies, Tools, and Applications, pp. 184–201. IGI Global (2017)
3. Luo, Y., et al.: Natural language processing for EHR-based pharmacovigilance: a structured review. Drug Saf. **40**(11), 1075–1089 (2017)
4. Gonzalez, G.H., Tahsin, T., Goodale, B.C., Greene, A.C., Greene, C.S.: Recent advances and emerging applications in text and data mining for biomedical discovery. Brief. Bioinform. **17**(1), 33–42 (2015)
5. Piñero, J., et al.: DisGeNET: a discovery platform for the dynamical exploration of human diseases and their genes. Database **2015** (2015)

6. SNOMED-CT. https://www.nlm.nih.gov/healthit/snomedct/
7. Köhler, S., et al.: The human phenotype ontology in 2017. Nucleic Acids Res. **45**(D1), D865–D876 (2017)
8. Lochmüller, H., et al.: 'IRDiRC Recognized Resources': a new mechanism to support scientists to conduct efficient, high-quality research for rare diseases. Eur. J. Hum. Genet. **25** (2), 162–165 (2017)
9. Rehm, H.L., et al.: ClinGen—the clinical genome resource. N. Engl. J. Med. **372**(23), 2235–2242 (2015)
10. Jonquet, C., Shah, N.H., Musen, M.A.: The open biomedical annotator. Summit Transl. Bioinform. **2009**, 56 (2009)
11. Taboada, M., Rodríguez, H., Martínez, D., Pardo, M., Sobrido, M.J.: Automated semantic annotation of rare disease cases: a case study. Database (Oxford) **2014** (2014)
12. Aronson, A.R.: Effective mapping of biomedical text to the UMLS Metathesaurus: the MetaMap program. In: Proceedings of the AMIA Symposium, p. 17 (2001)
13. Savova, G.K., et al.: Mayo clinical Text Analysis and Knowledge Extraction System (cTAKES): architecture, component evaluation and applications. J. Am. Med. Inform. Assoc. **17**(5), 507–513 (2010)
14. Groza, T., et al.: Automatic concept recognition using the Human Phenotype Ontology reference and test suite corpora. Database **2015**, bav005 (2015)
15. Lobo, M., Lamurias, A., Couto, F.M.: Identifying human phenotype terms by combining machine learning and validation rules. Biomed. Res. Int. **2017**, Article no. 8565739 (2017)
16. Lample, G., Ballesteros, M., Subramanian, S., Kawakami, K., Dyer, C.: Neural architectures for named entity recognition. arXiv Preprint arXiv:1603.01360 (2016)
17. Huang, Z., Xu, W., Yu, K.: Bidirectional LSTM-CRF models for sequence tagging. arXiv Preprint arXiv:1508.01991 (2015)
18. Ma, X., Hovy, E.: End-to-end sequence labeling via bi-directional LSTM-CNNs-CRF. arXiv Preprint arXiv:1603.01354 (2016)
19. Hochreiter, S., Schmidhuber, J.: Long short-term memory. Neural Comput. **9**(8), 1735–1780 (1997)
20. Lafferty, J., McCallum, A., Pereira, F.C.N.: Conditional random fields: probabilistic models for segmenting and labeling sequence data (2001)
21. Tjong Kim Sang, E.F., De Meulder, F.: Introduction to the CoNLL-2003 shared task: language-independent named entity recognition. In: Proceedings of the Seventh Conference on Natural Language Learning at HLT-NAACL 2003, vol. 4, pp. 142–147 (2003)
22. Johnson, A.E.W., et al.: MIMIC-III, a freely accessible critical care database. Sci. Data **3** (2016)
23. Girdea, M., et al.: PhenoTips: patient phenotyping software for clinical and research use. Hum. Mutat. **34**(8), 1057–1065 (2013)
24. Glueck, M., et al.: PhenoLines: phenotype comparison visualizations for disease subtyping via topic models. IEEE Trans. Vis. Comput. Graph. **24**(1), 371–381 (2018)
25. Habibi, M., Weber, L., Neves, M., Wiegandt, D.L., Leser, U.: Deep learning with word embeddings improves biomedical named entity recognition. Bioinformatics **33**(14), i37–i48 (2017)
26. Vani, A., Jernite, Y., Sontag, D.: Grounded recurrent neural networks. arXiv Preprint arXiv: 1705.08557 (2017)
27. Deng, J., et al.: Large-scale object classification using label relation graphs. In: Fleet, D., Pajdla, T., Schiele, B., Tuytelaars, T. (eds.) ECCV 2014. LNCS, vol. 8689, pp. 48–64. Springer, Cham (2014). https://doi.org/10.1007/978-3-319-10590-1_4
28. Vendrov, I., Kiros, R., Fidler, S., Urtasun, R.: Order-embeddings of images and language. arXiv Preprint arXiv:1511.06361 (2015)

29. Neelakantan, A., Roth, B., McCallum, A.: Compositional vector space models for knowledge base inference. In: 2015 AAAI Spring Symposium Series (2015)
30. Nickel, M., Kiela, D.: Poincaré embeddings for learning hierarchical representations. arXiv Preprint arXiv:1705.08039 (2017)
31. Bojanowski, P., Grave, E., Joulin, A., Mikolov, T.: Enriching word vectors with subword information. arXiv Preprint arXiv:1607.04606 (2016)
32. Kim, Y.: Convolutional neural networks for sentence classification. arXiv Preprint arXiv: 1408.5882 (2014)
33. Clevert, D.-A., Unterthiner, T., Hochreiter, S.: Fast and accurate deep network learning by exponential linear units (ELUs). arXiv Preprint arXiv:1511.07289 (2015)
34. Kingma, D., Ba, J.: Adam: a method for stochastic optimization. arXiv Preprint arXiv:1412. 6980 (2014)
35. Tifft, C.J., Adams, D.R.: The National Institutes of Health undiagnosed diseases program. Curr. Opin. Pediatr. **26**(6), 626 (2014)
36. Bodenreider, O.: The Unified Medical Language System (UMLS): integrating biomedical terminology. Nucleic Acids Res. **32**(90001), 267D–270D (2004)
37. Kiros, R., et al.: Skip-thought vectors. In: Advances in Neural Information Processing Systems, pp. 3294–3302 (2015)

ModHMM: A Modular Supra-Bayesian Genome Segmentation Method

Philipp Benner[✉] and Martin Vingron

Department of Computational Molecular Biology,
Max Planck Institute for Molecular Genetics, Ihnestraße 73, 14195 Berlin, Germany
{benner,vingron}@molgen.mpg.de

Abstract. Genome segmentation methods are powerful tools to obtain cell type or tissue specific genome-wide annotations and are frequently used to discover regulatory elements. However, traditional segmentation methods show low predictive accuracy and their data-driven annotations have some undesirable properties. As an alternative, we developed ModHMM, a highly modular genome segmentation method. Inspired by the supra-Bayesian approach, it incorporates predictions from a set of classifiers. This allows to compute genome segmentations by utilizing state-of-the-art methodology. We demonstrate the method on ENCODE data and show that it outperforms traditional segmentation methods not only in terms of predictive performance, but also in qualitative aspects. Therefore, ModHMM is a valuable alternative to study the epigenetic and regulatory landscape across and within cell types or tissues. The software is freely available at https://github.com/pbenner/modhmm.

1 Introduction

A single organism may consist of a remarkable diversity of cell types all sharing the same genotype. To understand how this diversity arises, current research in molecular biology has focused much attention on the functioning of transcriptional regulation. Genome-wide measurements of epigenetic marks and RNA expression have recently become available for many cell types and tissues [8]. These data provide a first glimpse at the regulatory program on a genome-wide scale. It is used to annotate regulatory elements and to locate important switches that control cell identity [22].

Genome segmentations are frequently used as a starting point for the identification and analysis of regulatory elements within cell types or tissues. By combining data from multiple experiments a genome segmentation method assigns a chromatin state to each genomic position. This may include regulatory elements such as active or repressed promoters and enhancers, active transcription or regions without an apparent function. The set of chromatin states a segmentation method is able to detect heavily depends on the choice of features. Typically a variety of histone modification ChIP-seq experiments is used possibly in combination with measurements of chromatin accessibility. However, other

© Springer Nature Switzerland AG 2019
L. J. Cowen (Ed.): RECOMB 2019, LNBI 11467, pp. 35–50, 2019.
https://doi.org/10.1007/978-3-030-17083-7_3

sources of information may be used as well, including DNA methylation, CpG content or evolutionary conservation. Chromatin states are characterized by specific combinations of such features. For instance, promoters are often conserved elements and accessible for transcription factors where the flanking nucleosomes are marked by H3K4me3. On the other hand, most enhancers are less conserved and marked by H3K4me1. Enrichment of H3K27ac is found at active promoters and enhancers, whereas H3K27me3 is known to be a repressive mark observed at bivalent promoters and poised enhancers [2,6,19,20].

Most common segmentation methods are instances of Hidden Markov Models (HMMs) where the observed data is assumed to be caused by an unobserved sequence of hidden states with Markov dependency structure. A frequently used implementation of this type is ChromHMM [12], which however relies on binarized data. A more advanced HMM based method is EpiCSeg [31] that addresses this shortcoming by using negative multinomial distributions to model observations. The handling of both methods is seemingly easy. Parameters are estimated unsupervisedly without the need for a training set using a maximum likelihood approach. Afterwards, by inspection of estimated parameters each hidden state is identified with one or more chromatin states. While this approach is very easy to apply, it also bears several risks. The specific combination of features known to mark a chromatin state and their spatial distribution is often not well reflected by the model. Hence, supervised methods specialized in the detection of regulatory elements typically perform much better. To obtain good classification performances of unsupervised HMMs the number of hidden states often exceeds the actual observed number of chromatin states. This leads to highly fragmented genome segmentations where single chromatin states are represented by multiple states of the HMM. Figure 1 illustrates this problem on a promoter of a transcribed gene. An optimal segmentation would detect the region as a single active promoter with a transcribed region to the left. However, typical segmentations obtained with ChromHMM or EpiCSeg instead show a highly fragmented promoter region. Another drawback of the unsupervised HMM approach is the low flexibility of the model offering no glaring way to improve a poor segmentation. ChromHMM and EpiCSeg have only two parameters, namely the number of hidden states and the genomic bin size, whose effect on the resulting segmentation is highly unpredictable. Furthermore, to determine the optimal set of parameters it is necessary to learn and evaluate a large number of different models, effectively negating the presumed simplicity.

ChroModule [45] is a supervised alternative that models the spatial distribution of features at chromatin states with left-right structured HMMs that are commonly used in speech recognition [36]. However, the construction of a model requires a training set for each chromatin state. As an alternative to the HMM based methods, Segway [21,22] allows to compute segmentations based on arbitrary hidden processes, as long as the model can be represented as a dynamic Bayesian network. It operates on a single base-pair resolution and with its default model computes segmentations that are even more fine-grained than those of ChromHMM and EpiCSeg [21]. Segway models are typically trained on

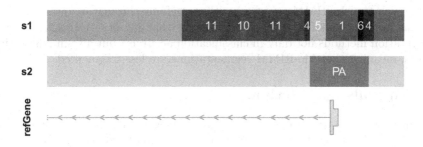

Fig. 1. Genome segmentations. Typical genome segmentation (s1) where the promoter is fragmented into many different segments. In this example, the optimal segmentation (s2) shows a single active promoter segment (PA) with a transcribed region to the left and no signal to the right.

a small fraction of the available data, due to the computational complexity of the inference algorithm and the high data resolution.

We propose here a new modular segmentation method based on HMMs called ModHMM that addresses some of these shortcomings in the following way. First of all, we recognize that jointly learning all parameters of an HMM in an unsupervised way is overly ambitious and leads to poor results. Instead, we propose to assemble the segmentation method piece by piece allowing us to guide the learning process as much as possible. Second, our objective is to construct a method that may benefit from the ample variety of well-performing classifiers that have been developed for most regulatory elements. Inspired by the supra-Bayesian approach [15,16,23,28–30], we propose here to construct an HMM that acts as a decision maker who integrates assessments from several experts. Each expert or classifier is specialized in the detection of a single chromatin state, possibly by considering only a subset of the available features. The classifiers may also model the spatial distribution of features near functional elements to improve prediction accuracy.

Hence, our segmentation method consists of an HMM combined with a set of chromatin state classifiers. As opposed to traditional segmentation methods, our HMM does not take feature tracks (i.e. ChIP-seq/ATAC-seq tracks) as input, but instead regards the genome-wide probability assessments of the chromatin state classifiers as observations. We constructed the method in a highly modular way, allowing to easily improve segmentations by replacing single classifiers. To facilitate the usage of ModHMM, we constructed a default set of chromatin state classifiers. The parameters of supervised classifiers are commonly estimated on some training set. However, constructing a training set for each classifier would not only be a tedious task, it would also shift control over the resulting segmentation to the composition of such training sets. Instead, we do not rely on a training set but engineer each classifier by translating contemporary knowledge of chromatin states into a probabilistic model.

We evaluate ModHMM equipped with its default chromatin state classifiers on promoter and enhancer test sets and show that it outperforms traditional segmentation methods not only in classification accuracy, but also in qualitative aspects, meaning that ModHMM segmentations are less fragmented.

2   Materials and Methods

We consider chromatin states that are most relevant for the analysis of gene regulation and that are typically found in genome annotation studies [13,17,22]. This includes active promoters (PA) and enhancers (EA), primed (PR) and bivalent (BI) regions as well as regions of active (TR) and low transcription (TL). In addition, we model heterochromatic regions marked by H3K27me3 (R1) or H3K9me3 (R2) and regions where either no signal (NS) or a control signal (CL) is observed. Enhancers and promoters are detected based on ATAC-seq [3,4] data measuring chromatin accessibility in combination with histone marks H3K4me1 and H3K4me3. To measure the activity of promoters and enhancers we use histone mark H3K27ac [9]. While active promoters and enhancers can be accurately discriminated by the ratio of histone marks H3K4me1 and H3K4me3, we observed that the prediction accuracy is much lower for bivalent promoters and poised enhancers (Sect. 3.3). Therefore, we decided to merge both chromatin states into a single bivalent state (BI), which is marked by H3K27me3 and H3K4me1 or H3K4me3. Similarly, we define primed states (PR) as accessible regions marked by H3K4me1 or H3K4me3 but showing no H3K27ac and H3K27me3 signal. We also model regions solely marked by either H3K27me3 (R1) or H3K9me3 (R2). Histone mark H3K27me3, catalyzed by the polycomb repressive complex 2, is involved in gene silencing and associated with constitutive heterochromatin [26,32]. On the other hand, histone mark H3K9me3 is associated with constitutive heterochromatin, predominantly formed in gene-empty regions [39]. Transcribed regions are typically marked by H3K36me3 [43], however, we decided to use polyA RNA-seq data instead since it is a more direct and less noisy measurement. Our model also accounts for low levels of transcription (TL) that frequently occur at repressed genes or in intergenic regions, for instance near certain types of enhancers that generate unidirectional polyA+ eRNAs [24].

ModHMM is a highly modular genome segmentation method that incorporates predictions from a set of classifiers. In contrast to unsupervised methods, we manually construct most parts of the model. It consists of two components, the HMM and the set of chromatin state classifiers, both will be outlined in the following. For the chromatin state classifiers we consider a default set of engineered classifiers.

2.1   Hidden Markov Model

ModHMM implements a hidden Markov model, which consists of an unobserved Markov process generating a series of hidden states, each emitting a single observation [7]. In order to define the HMM, we first must assign each chromatin state to one or more hidden states of the HMM and decide on a set of feasible transitions of the unobserved Markov process. Finally, we present the emission model for incorporating genome-wide predictions of the classifiers and show that transition rates must be further constrained in order to construct a well-functioning model.

Hidden States and Feasible Transitions. Unsupervised HMM based segmentations methods, including ChromHMM and EpiCSeg, learn transition rates using a maximum likelihood approach and initially allow transitions between any two states. To guide the learning process, it is often helpful to enforce a predefined structure on the transition matrix [14]. In genetics such structured HMMs have been utilized before, for instance, for the prediction of gene structures from DNA sequences [5]. In our case, the structure of the HMM should encode any prior knowledge about the context in which chromatin states appear in the genome and may for instance be used to implement a model for actively transcribed genes. However, one has to be cautious not to enforce an overly simplistic model. For instance, an HMM that requires each gene to have exactly one promoter and a single transcribed region would not be realistic and result in wrong predictions. The converse, a model that is excessively complex would have equally poor predictive performance. Therefore, we decided on an HMM with minimal structure, as depicted in Fig. 2. Some chromatin states are represented by multiple hidden states of the HMM. For instance, active enhancers (EA) are represented by hidden states EA, EA_1, and EA_2, in order to model different contexts in which enhancers may appear. The HMM structure enforces that each transcribed region must be flanked by at least one active promoter and each active promoter must be flanked by a transcribed region. It also forbids that transcribed regions are flanked by active enhancers and primed or bivalent regions. In addition, transitions between active promoters, active enhancers, primed and bivalent states are forbidden.

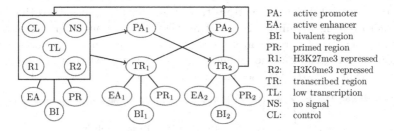

Fig. 2. ModHMM state diagram. Some chromatin states are represented by multiple hidden states. For instance, active enhancers (EA) are represented by hidden states EA, EA_1, and EA_2. If two states are connected by an undirected edge, transitions in both directions are allowed. Self-transitions are in general allowed but omitted in the figure. A box is used to group states that are fully connected. If an arrow points to the box, transitions to any of the states in that box are admissible. Crossing edges are connected if marked with a circle.

Emissions. The ModHMM segmentation method is inspired by the supra-Bayesian approach that integrates predictions of an expert committee from which a decision maker reaches a final decision. The expert committee consists of a set of classifiers, each specialized in the detection of a single chromatin state $s \in \mathcal{S} = \{\text{PA}, \text{EA}, \text{BI}, \dots\}$. The output of the classifiers are the genome-wide predictions of chromatin states, i.e. prediction $c_t(s)$ yields the assessment of the expert for chromatin state s that genomic position t is in this state. In the supra-Bayesian approach, expert predictions are treated as observations and a separate model, the decision maker, is constructed to reach a final decision. Here, we decided to implement the decision maker as an HMM in which each chromatin state s is associated with one or more hidden states $s' \in \mathcal{S}' = \{\text{PA}, \text{EA}, \text{EA}_1, \text{EA}_2, \dots\}$. For the emission model several choices would be conceivable. For instance, the family of beta distributions is frequently used to model probabilities. However, to reduce the number of parameters that must be estimated from data we decided to use a likelihood model that contains no free parameters and incorporates the classifier predictions as they are. We define the emission distribution of state s' in terms of the density function

$$f_{s'}(x) \propto x,$$

where $x = c_t(s)$ and s' is associated with chromatin state s.

Transition Rates. In typical HMMs, transition rates are estimated from data and reflect context-dependent state prevalences. The situation is different in our case, where classifiers are used to optimally discriminate between chromatin states and to account for prevalances. By naïvely integrating classifiers into an HMM, results of classification may get overruled by transition rates, making it more difficult to combine classifiers into a well-functioning model. Still, transition rates are valuable parameters and we use them to model the expected length of chromatin states.

Two types of constraints are imposed on the transition matrix Σ. First, the structure of the HMM forbids certain transitions resulting in entries that must remain zero during learning. Second, transition rates should only account for the average length of chromatin states. To accomplish this, it is necessary to constrain all non-zero off-diagonal entries within a row to share the same value. More specifically, ModHMM uses the transition matrix $\Sigma = (\sigma_{ij})$ with

$$\sigma_{ij} = \begin{cases} \delta_i & \text{if } i = j, \\ \nu_i & \text{if } i \neq j \text{ and transition is feasible}, \\ 0 & \text{otherwise}. \end{cases}$$

The parameter δ_i models the expected length of the ith chromatin state while ν_i represents the transition rate into another state. In addition to the above constraints, we also assume that the expected length of chromatin states is context-independent, i.e. $\delta_i = \delta_j$ for $i, j \in \{\text{EA}, \text{EA}_1, \text{EA}_2\}$, $i, j \in \{\text{BI}, \text{BI}_1, \text{BI}_2\}$, $i, j \in \{\text{PR}, \text{PR}_1, \text{PR}_2\}$, and $i, j \in \{\text{TR}_1, \text{TR}_2\}$. For each row i, diagonal entries δ_i

and off-diagonal entries ν_i must be chosen such that the row sum is equal to one. The constraints that are imposed on the transition matrix $\Sigma = (\sigma_{ij})$ complicate the estimation step, which requires a modified Baum-Welch algorithm.

2.2   Default Chromatin State Classifiers

ModHMM takes as input the genome-wide predictions of a set of classifiers. In principle, any type of classifier can be used, however, ModHMM implements a default classifier set in order to simplify usage. These engineered classifiers require no training data and consist of two layers. First, a single-feature classifier is constructed for each feature that determines the probability of enrichment at each genomic position. These single-feature classifiers are then combined into a set of naïve Bayesian multi-feature classifiers [11,33,34,38], each specialized in the detection of a single chromatin state.

Single-feature Classifiers. The purpose of a single-feature classifier is to assess the enrichment of a feature genome-wide. Since it assigns a probability to each genomic position, we may also interpret this step as a normalization step that decouples the definition of the engineered multi-feature classifiers from the actual observations. Such classifiers are the basic ingredient of many peak calling methods, implementing a large variety of different statistical models [44], ranging from Poisson [35,42] or local Poisson [46] to hidden Markov models [41]. Our approach differs in that we do not assume the same model for all features but rather account for the high heterogeneity. More specifically, we first compute the coverage along the genome in 200 bp bins. The coverage values are then modeled by a feature-specific mixture distribution. Consider an event $\{X_t^\varphi = x\}$ with coverage value x from a feature $\varphi \in \{\text{ATAC}, \text{H3K4me1}, \dots\}$ at bin t. We assume that

$$X_t^\varphi \sim \sum_{k \in F \cup B} \pi_k p_k$$

where p_k is the kth component of the feature-specific mixture distribution with weight π_k. F and B partition the set $\{p_k\}$ into foreground $\{p_k \mid k \in F\}$ and background components $\{p_k \mid k \in B\}$. Whether a component belongs to the foreground or background is a subjective choice and must be determined by visual inspection of the data. The parameters of the mixture distribution are estimated by maximum likelihood using the expectation maximization algorithm [10].

Once a mixture distribution for a feature φ is determined, the probability that a given bin t of the genome with coverage value x is enriched is given by the posterior probability

$$q_t(\varphi) = \frac{\sum_{k \in F} \pi_k p_k(x)}{\sum_{k \in F \cup B} \pi_k p_k(x)}.$$

In this way, a single-feature classifier is constructed for all features. Every such classifier consists of a mixture of Poisson, geometric, and delta distributions.

Multi-feature Classifiers. Multi-feature classifiers are defined as simple combinations of single-feature classifiers. Consider the case of active promoters that are known to be accessible and marked by H3K27ac and H3K4me3 as well as a high H3K4me3-to-me1 ratio (H3K4me3/1). A classifier for active promoters should assign high probabilities to regions where those three features co-occur. Therefore, a naïve Bayesian promoter model is given by

$$c_t(\text{PA}) =$$
$$q_t(\text{ATAC}) \cdot q_t(\text{H3K27ac}) \cdot q_t(\text{H3K4me3}) \cdot q_t(\text{H3K4me3/1}) \cdot \bar{q}_t(\text{Control})$$

where

$$\bar{q}_t(\text{Control}) = 1 - q_t(\text{Control})$$

enforces that no peak is observed in the control data set. The classifier can be improved by also considering the spatial structure of features. For instance, histone modifications are typically more broadly distributed than ATAC-seq peaks. For such features it is necessary to also consider surrounding bins and ask for the probability that any one of the bins is enriched. For a feature φ the probability that any one out of three adjacent bins is enriched is given by

$$q_{t-1:t+1} = q_{t-1} + \bar{q}_{t-1} \cdot q_t + \bar{q}_{t-1} \cdot \bar{q}_t \cdot q_{t+1} ,$$

where function arguments have been omitted for better readability. Some features may require a more detailed modeling of the spatial structure. For instance, H3K4me1 is symmetrically distributed around regulatory elements as opposed to H3K4me3 that shows a higher enrichment at promoters towards transcribed regions due to its role in preinitiation complex formation [27]. The symmetric structure of a feature is captured by

$$s_{t-1:t+1} = q_{t-1} \cdot q_{t+1} + \overline{q_{t-1} \cdot q_{t+1}} \cdot q_t$$

where

$$\overline{q_{t-1} \cdot q_{t+1}} = q_{t-1} \cdot \bar{q}_{t+1} + \bar{q}_{t-1} \cdot q_{t+1} + \bar{q}_{t-1} \cdot \bar{q}_{t+1} .$$

The classifier requires an enrichment at bins $t-1$ and $t+1$ or an enrichment at the center bin t.

In this fashion, a multi-feature classifier is constructed for every chromatin state. The classifiers are then assigned to states of the HMM, where some classifiers may also be shared among several states. This is for instance the case for the active enhancer states EA, EA_1, and EA_2, as well as the bivalent states BI, BI_1, and BI_2. A full specification of the classifiers is given in Table 1.

3   Results

We compared our method, equipped with its default set of engineered chromatin state classifiers, with two other segmentation methods, namely ChromHMM

Table 1. Multi-feature classifier definitions.

	PA	EA	PR		BI		TL	TR	CL	NS	R1	R2
ATAC	$\checkmark_c$	$\checkmark_c$	$\checkmark_c$	$\checkmark_c$			$\times_c$	$\times_c$		$\times_c$		
H3K27ac	$\checkmark_s$	$\checkmark_s$	$\times_a$	$\times_a$						$\times_c$		
H3K4me1		$\checkmark_{s,3}$		$\checkmark_{s,3}$		$\checkmark_{s,3}$	$\times_c$	$\times_c$		$\times_c$	$\times_c$	$\times_c$
H3K4me3	$\checkmark_a$			$\checkmark_a$		$\checkmark_a$	$\times_c$	$\times_c$		$\times_c$	$\times_c$	$\times_c$
H3K4me3/1	$\checkmark_a$	$\times_a$										
H3K27me3			$\times_a$	$\times_a$	$\checkmark_s$	$\checkmark_s$				$\times_c$	$\checkmark_c$	
H3K9me3										$\times_c$		$\checkmark_c$
RNA									$\checkmark_c$	$\times_c$		
RNA (low)					$\checkmark_c$							
Control	$\times_a$	$\times_a$	$\times_a$	$\times_a$	$\times_a$	$\times_a$			$\checkmark_c$	$\times_c$	$\times_c$	$\times_c$

$\checkmark_c$: bin t is enriched $[q_t]$; $\checkmark_a$: at least one bin out of $\{t-2, \ldots, t+2\}$ is enriched $[q_{t-2:t+2}]$; $\checkmark_s$: symmetric enrichments at bins $\{t-2, \ldots, t+2\}$ $[s_{t-2:t+2}]$; $\checkmark_{s,3}$: symmetric enrichments at bins $\{t-3, \ldots, t+3\}$ $[s_{t-3:t+3}]$; $\times_a$: no enrichment in all bins $i \in \{t-2, \ldots, t+2\}$ $[\bar{q}_{t-2:t+2}]$; $\times_c$: bin t shows no enrichment $[\bar{q}_t]$.

[12] and EpiCSeg [31]. ChromHMM is the most popular segmentation method, although it uses Bernoulli emission distributions for which the data must first be binarized into enriched and nonenriched regions. To also incorporate how strongly genomic regions are enriched, EpiCSeg uses negative multinomial emissions. Compared to the multinomial distribution, the negative multinomial better models the variability observed in most ChIP-seq data. We decided to omit a comparison to Segway, since its segmentations are even more fine-grained than those of ChromHMM and EpiCSeg [21]. We also omit a comparison to Chro-Module [45] since no software package was published by the authors. For all three methods a bin size of 200 bps was used. The ModHMM segmentation is computed as the most likely sequence of hidden states, i.e. the Viterbi path. ChromHMM and EpiCSeg use the posterior decoding algorithm to compute segmentations.

We evaluated all three methods on ENCODE [8] data from mouse embryonic liver at day 12.5. For a first qualitative comparison, Fig. 3 shows a small region within chromosome 1 that contains three actively transcribed genes, as indicated by the data. The ModHMM segmentation correctly detects the promoters and transcribed regions. In addition, there is a primed region located in one of the gene bodies. For ChromHMM and EpiCSeg the number of states were determined by maximizing classification performance (Sect. 3.1). The segmentations of both methods are highly fragmented and much more difficult to interpret. States that appear close to the promoter are also found at the primed region. This is due to the lack of an appropriate model for the spatial distribution of features around regulatory elements. To study this observation further, we looked at state equivalences between the three methods. In general, there is a low overlap between states. Since we used more states for ChromHMM and

EpiCSeg, it is clear that a single ModHMM state must be represented by multiple states from ChromHMM and EpiCSeg. However, there are also multiple recombinations, for instance, one ChromHMM state corresponds to ModHMM active promoter (PA) and enhancer (EA) states.

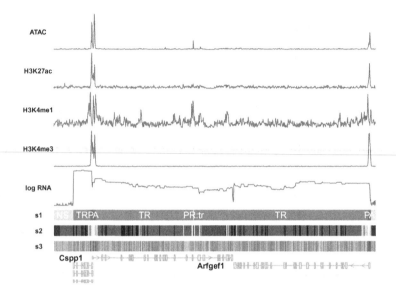

Fig. 3. Qualitative comparison of segmentation methods. s1: ModHMM, s2: ChromHMM, s3: EpiCSeg. The primed states of ModHMM within transcribed regions (PR$_1$, PR$_2$) are both abbreviated as PR:tr.

3.1   Enhancer Predictions

To compare the predictive performance of segmentation methods we constructed a test set with active enhancers identified by the FANTOM consortium [1]. Enhancers are experimentally detected as origins of bidirectional capped transcripts using CAGE [40]. We took all enhancers that showed at least 5 CAGE reads in liver at embryonic day 12 resulting in only 537 regions. With such a low detection threshold on the number of counts many false positives ought to be expected. Indeed, about half of the regions did not show the desired histone modification patterns. To filter false positives, we clustered the enhancer regions using deepTools2 [37]. We dropped all clusters that either showed no histone marks or high levels of H3K4me3, resulting in 265 positive regions. The detection of active enhancers based on chromatin marks is difficult, mostly because promoters show a very similar pattern. Therefore, we constructed a test set consisting of 2650 regions, 1/10th are the filtered FANTOM enhancers, 8/10th promoters and 1/10th random genomic regions.

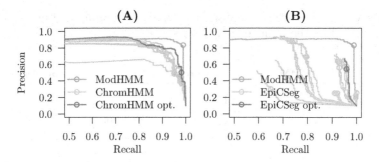

Fig. 4. Classification performances of active enhancer regions in mouse embryonic liver at day 12.5. (A) Performance of ModHMM and ChromHMM. Lines show performances evaluated using posterior marginals, while dots mark the performances of segmentations. For ChromHMM models with an even number of states between 10 and 30 were tested. The precision-recall curve is evaluated for several states and all possible combinations. For each model only the best curve is shown in gray. The optimal (opt.) curve is highlighted in blue. (B) Performance summary of ModHMM and EpiCSeg similar to (A). (Color figure online)

While ModHMM has a well defined enhancer state, ChromHMM and EpiC-Seg are unsupervised methods where states must be assigned a function after training. This assignment is often difficult especially for EpiCSeg where also the intensity of enrichment is modeled. To avoid wrong assignments, we consider for each model 2–3 putative enhancer states that are most abundant at the positive enhancer regions. Performance is then evaluated based on these states including all possible combinations. The best performance is then reported, potentially giving ChromHMM and EpiCSeg a strong advantage over ModHMM.

Results are summarized in Fig. 4. For all three methods we used posterior marginals of one or several states to compute precision-recall curves. In addition, we computed the classification performances of Viterbi paths. ModHMM shows the highest area under the precision-recall curve. The best ChromHMM model consists of 22 states and surprisingly outperforms the best EpiCSeg model with 20 states. The Viterbi path of ModHMM optimally balances precision and recall yielding the highest F-score. In contrast, especially the segmentations of ChromHMM show a poor balance of precision and recall with a maximum precision of around 60%.

3.2   Promoter State Frequencies

To understand why ModHMM performs better than ChromHMM and EpiCSeg, we looked at segmentations around active promoters. We used UCSC refGenes to obtain an initial set of transcription start sites (TSSs). Promoters are defined as 2 kbp windows around the TSS. From this set we took regions that have a clear ATAC-seq, H3K27ac, and RNA-seq signal. Regions enriched with H3K27me3 were filtered out. For each segmentation method, we computed at every position relative to the TSS the frequency of every state.

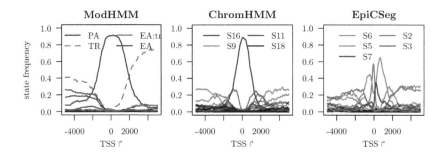

Fig. 5. State frequencies at promoters. Promoters that belong to genes on the reverse strand are inverted so that the gene body is right of the TSS. For ChromHMM and EpiCSeg, states that are frequently observed at enhancers are colored in red. (Color figure online)

For ModHMM we observe a clear enrichment of the active promoter state (PA) around transcription start sites (Fig. 5). The active transcription state (TR) is flanking this region in most cases, while other states are rarely observed. For ChromHMM active promoters are modeled by state 16, which represents enrichment in ATAC-seq, H3K27ac, H3K4me3, and RNA-seq. However, the region represented by this state is much narrower and it is frequently flanked by a diverse set of states. One of them is state 9, which models enrichment in ATAC-seq, H3K27ac, H3K4me1 and H3K4me3. It is also frequently found at enhancers that show enrichment in H3K4me3 above the binarization threshold set by ChromHMM. The situation is similar for EpiCSeg, however, the promoter is fragmented into several states modeling different levels of ATAC-seq and H3K4me3 enrichment. State 2 is frequently flanking promoters, which also occurs at enhancers since it models enrichment in ATAC-seq, H3K27ac, and H3K4me1 but low enrichment in H3K4me3. Peaks of H3K4me3 tend to be more localized than H3K4me1 peaks, so that regions close to promoters typically show characteristics of enhancers. Both ChromHMM and EpiCSeg do not model the spatial distribution of features around promoters and enhancers and therefore often fail to correctly discriminate between them.

3.3   Bivalent State

During the development of ModHMM, we observed that the H3K4me1-to-me3 ratio has low predictive power for discriminating between bivalent promoters and poised enhancers. This led us to represent both chromatin states by a single bivalent state. To quantify this observation, we consider all bivalent regions in the ModHMM segmentation of mouse embryonic liver at day 12.5. The H3K4me1-to-me3 ratio is then used to separate promoters from the remaining regions (i.e.

putative poised enhancers). All bivalent regions overlapping annotated UCSC refGene promoters (500 bp regions around transcription start sites) are defined as true positives. This leads to an area under the precision recall curve (PR-AUC) of around 0.84. The minimal PR-AUC achieved by a random classifier is around 0.63. As a comparison, we took all regions of the ModHMM segmentation that are either labeled as active promoter or enhancer. Here, the same procedure leads to a PR-AUC of about 0.96, whereas a random classifier achieves a performance of 0.46.

4   Discussion

Traditional genome segmentation methods, such as ChromHMM, EpiCSeg, or Segway, are unsupervised methods and can be used to detect known and unknown patterns in genomics data. They have been extensively used in the past to analyze the epigenetic landscape of a large variety of cell types and tissues [13, 22, 25]. However, nowadays much is known about the epigenetic landscape and the features that mark regulatory elements. This extensive knowledge questions, at least to some extent, the traditional approach to genome segmentation. Instead, we used this knowledge to construct a segmentation method that outperforms the traditional methods in several aspects. ModHMM has a higher prediction accuracy and the segmentations show a better balance of precision and recall. With each hidden state of ModHMM a classifier is associated that detects a well-defined chromatin state. This leads to segmentations that are superior in qualitative aspects. Functional elements, such as active promoters or enhancers, are typically contained in a single segment, which is not the case for ChromHMM and EpiCSeg.

Inspired by the supra-Bayesian approach, ModHMM integrates predictions of a set of experts or classifiers. Using the output of classifiers as input to the HMM has certain advantages over classical HMMs that model observations directly. For instance, a classifier may cherry-pick only a subset of the available data. The coverage of RNA-seq reads in a single genomic bin already provides enough information to decide whether the region is transcribed. On the other hand, classification of active promoters and enhancers requires data from multiple features and several surrounding bins.

ModHMM uses a default set of engineered classifiers to detect chromatin states. The basis of which is a single-feature enrichment analysis with a mixture model tailored to each feature (Sect. 2.2). This is unique to ModHMM as most peak calling methods implement a single model. Applying ModHMM to a new data set requires to perform the enrichment analysis de novo. Alternatively, ModHMM may quantile normalize a new data set to a known reference for which single-feature models already exist. Unlike ChromHMM and EpiCSeg, ModHMM has well-defined hidden states that do not change when applied across different cell types or tissues. This makes ModHMM ideal for differential analysis.

Compared to traditional segmentation methods, ModHMM is much more flexible and provides many leverage points to construct high-quality segmentations. For instance, any of the chromatin state classifiers from the default set can be replaced by more accurate alternatives, allowing to incorporate predictions from state-of-the-art methods such as REPTILE [18]. To improve a given segmentation, ModHMM allows visual inspection of all classifier predictions, which may serve as a powerful tool to decide which classifiers must be replaced.

Acknowledgements. We thank Anna Ramisch, Tobias Zehnder, and Verena Heinrich for their comments on the manuscript and many inspiring discussions.

PB was supported by the German Ministry of Education and Research (BMBF, grant no. 01IS18037G).

References

1. Andersson, R., et al.: An atlas of active enhancers across human cell types and tissues. Nature **507**(7493), 455 (2014)
2. Barski, A., et al.: High-resolution profiling of histone methylations in the human genome. Cell **129**(4), 823–837 (2007)
3. Buenrostro, J.D., Giresi, P.G., Zaba, L.C., Chang, H.Y., Greenleaf, W.J.: Transposition of native chromatin for fast and sensitive epigenomic profiling of open chromatin, DNA-binding proteins and nucleosome position. Nat. Methods **10**(12), 1213 (2013)
4. Buenrostro, J.D., Wu, B., Chang, H.Y., Greenleaf, W.J.: ATAC-seq: a method for assaying chromatin accessibility genome-wide. Curr. Protoco. Mol. Biol. **109**(1), 21–29 (2015)
5. Burge, C., Karlin, S.: Prediction of complete gene structures in human genomic DNA. J. Mol. Biol. **268**(1), 78–94 (1997)
6. Calo, E., Wysocka, J.: Modification of enhancer chromatin: what, how, and why? Mol. cell **49**(5), 825–837 (2013)
7. Cappé, O., Moulines, E., Rydén, T.: Inference in Hidden Markov Models, vol. 6. Springer, Heidelberg (2005). https://doi.org/10.1007/0-387-28982-8
8. Consortium, E.P., et al.: An integrated encyclopedia of DNA elements in the human genome. Nature **489**(7414), 57 (2012)
9. Creyghton, M.P., et al.: Histone H3K27ac separates active from poised enhancers and predicts developmental state. Proc. Nat. Acad. Sci. **107**(50), 21931–21936 (2010)
10. Dempster, A.P., Laird, N.M., Rubin, D.B.: Maximum likelihood from incomplete data via the EM algorithm. J. Roy. Stat. Soc.: Ser. B (Methodol.) **39**(1), 1–22 (1977)
11. Duda, R.O., Hart, P.E.: Pattern Classification and Scene Analysis. A Wiley-Interscience Publication, New York (1973)
12. Ernst, J., Kellis, M.: ChromHMM: automating chromatin-state discovery and characterization. Nat. Methods **9**(3), 215 (2012)
13. Ernst, J., Kellis, M.: Chromatin-state discovery and genome annotation with ChromHMM. Nat. Protoc. **12**(12), 2478 (2017)
14. Galassi, U., Giordana, A., Saitta, L.: Structured hidden markov model: a general framework for modeling complex sequences. In: Basili, R., Pazienza, M.T. (eds.) AI*IA 2007. LNCS (LNAI), vol. 4733, pp. 290–301. Springer, Heidelberg (2007). https://doi.org/10.1007/978-3-540-74782-6_26
15. Gelfand, A.E., Mallick, B.K., Dey, D.K.: Modeling expert opinion arising as a partial probabilistic specification. J. Am. Stat. Assoc. **90**(430), 598–604 (1995)

16. Genest, C., Zidek, J.V., et al.: Combining probability distributions: a critique and an annotated bibliography. Stat. Sci. **1**(1), 114–135 (1986)
17. Gorkin, D., et al.: Systematic mapping of chromatin state landscapes during mouse development. bioRxiv p. 166652 (2017)
18. He, Y., et al.: Improved regulatory element prediction based on tissue-specific local epigenomic signatures. Proc. Nat. Acad. Sci. **114**(9), E1633–E1640 (2017)
19. Heintzman, N.D., et al.: Distinct and predictive chromatin signatures of transcriptional promoters and enhancers in the human genome. Nat. Genet. **39**(3), 311 (2007)
20. Heinz, S., Romanoski, C.E., Benner, C., Glass, C.K.: The selection and function of cell type-specific enhancers. Nat. Rev. Mol. Cell Biol. **16**(3), 144 (2015)
21. Hoffman, M.M., Buske, O.J., Wang, J., Weng, Z., Bilmes, J.A., Noble, W.S.: Unsupervised pattern discovery in human chromatin structure through genomic segmentation. Nat. Methods **9**(5), 473 (2012)
22. Hoffman, M.M., et al.: Integrative annotation of chromatin elements from encode data. Nucleic Acids Res. **41**(2), 827–841 (2012)
23. Jacobs, R.A.: Methods for combining experts' probability assessments. Neural Comput. **7**(5), 867–888 (1995)
24. Koch, F., et al.: Transcription initiation platforms and GTF recruitment at tissue-specific enhancers and promoters. Nat. Struct. Mol. Biol. **18**(8), 956 (2011)
25. Kundaje, A., et al.: Integrative analysis of 111 reference human epigenomes. Nature **518**(7539), 317 (2015)
26. Kuzmichev, A., Nishioka, K., Erdjument-Bromage, H., Tempst, P., Reinberg, D.: Histone methyltransferase activity associated with a human multiprotein complex containing the Enhancer of Zeste protein. Genes Dev. **16**(22), 2893–2905 (2002)
27. Lauberth, S.M., et al.: H3K4me3 interactions with TAF3 regulate preinitiation complex assembly and selective gene activation. Cell **152**(5), 1021–1036 (2013)
28. Lindley, D.: The improvement of probability judgements. J. Roy. Stat. Soc. Ser. A (Gen.) **145**, 117–126 (1982)
29. Lindley, D.: Reconciliation of discrete probability distributions. In: J. Bernardo, M. DeGroot, D. Lindley, A. Smith (eds.) Bayesian statistics 2: Proceedings of the Second Valencia International Meeting, pp. 375–390. Valencia University Press (1985)
30. Lindley, D.V., Tversky, A., Brown, R.V.: On the reconciliation of probability assessments. J. Roy. Stat. Soc. Ser. A (Gen.) **142**, 146–180 (1979)
31. Mammana, A., Chung, H.R.: Chromatin segmentation based on a probabilistic model for read counts explains a large portion of the epigenome. Genome Biol. **16**(1), 151 (2015)
32. Margueron, R., Reinberg, D.: The polycomb complex PRC2 and its mark in life. Nature **469**(7330), 343 (2011)
33. Maron, M.E.: Automatic indexing: an experimental inquiry. J. ACM (JACM) **8**(3), 404–417 (1961)
34. Mitchell, T.M.: Machine Learning. McGraw-Hill Boston, MA (1997)
35. Mortazavi, A., Williams, B.A., McCue, K., Schaeffer, L., Wold, B.: Mapping and quantifying mammalian transcriptomes by RNA-seq. Nature Methods **5**(7), 621 (2008)
36. Rabiner, L.R.: A tutorial on hidden markov models and selected applications in speech recognition. Proc. IEEE **77**(2), 257–286 (1989)
37. Ramírez, F., et al.: deepTools2: a next generation web server for deep-sequencing data analysis. Nucleic Acids Res. **44**(W1), W160–W165 (2016)

38. Russell, S.J., Norvig, P.: Artificial Intelligence: A Modern Approach. Pearson Education Limited, Malaysia (2016)
39. Saksouk, N., Simboeck, E., Déjardin, J.: Constitutive heterochromatin formation and transcription in mammals. Epigenet. Chromatin **8**(1), 3 (2015)
40. Shiraki, T., et al.: Cap analysis gene expression for high-throughput analysis of transcriptional starting point and identification of promoter usage. Proc. Nat. Acad. Sci. **100**(26), 15776–15781 (2003)
41. Spyrou, C., Stark, R., Lynch, A.G., Tavaré, S.: BayesPeak: Bayesian analysis of ChIP-seq data. BMC Bioinf. **10**(1), 299 (2009)
42. Valouev, A., et al.: Genome-wide analysis of transcription factor binding sites based on ChIP-seq data. Nat. Methods **5**(9), 829 (2008)
43. Wagner, E.J., Carpenter, P.B.: Understanding the language of Lys36 methylation at histone H3. Nature Rev. Mol. Cell Biol. **13**(2), 115 (2012)
44. Wilbanks, E.G., Facciotti, M.T.: Evaluation of algorithm performance in ChIP-seq peak detection. PloS One **5**(7), e11471 (2010)
45. Won, K.J., et al.: Comparative annotation of functional regions in the human genome using epigenomic data. Nucleic Acids Res. **41**(8), 4423–4432 (2013)
46. Zhang, Y., et al.: Model-based analysis of ChIP-seq (MACS). Genome Biol. **9**(9), R137 (2008)

Learning Robust Multi-label Sample Specific Distances for Identifying HIV-1 Drug Resistance

Lodewijk Brand⬤, Xue Yang, Kai Liu⬤, Saad Elbeleidy, Hua Wang^(✉)⬤,
and Hao Zhang

Department of Computer Science, Colorado School of Mines, Golden, CO 80401, USA
{lbrand,selbeleidy,hzhang}@mines.edu, edyxueyx@gmail.com,
liukaizhijia@gmail.com, huawangcs@gmail.com

Abstract. Acquired immunodeficiency syndrome (AIDS) is a syndrome caused by the human immunodeficiency virus (HIV). During the progression of AIDS, a patient's the immune system is weakened, which increases the patient's susceptibility to infections and diseases. Although antiretroviral drugs can effectively suppress HIV, the virus mutates very quickly and can become resistant to treatment. In addition, the virus can also become resistant to other treatments not currently being used through mutations, which is known in the clinical research community as cross-resistance. Since a single HIV strain can be resistant to multiple drugs, this problem is naturally represented as a multi-label classification problem. Given this multi-class relationship, traditional single-label classification methods usually fail to effectively identify the drug resistances that may develop after a particular virus mutation. In this paper, we propose a novel multi-label Robust Sample Specific Distance (RSSD) method to identify multi-class HIV drug resistance. Our method is novel in that it can illustrate the relative strength of the drug resistance of a reverse transcriptase sequence against a given drug nucleoside analogue and learn the distance metrics for all the drug resistances. To learn the proposed RSSDs, we formulate a learning objective that maximizes the ratio of the summations of a number of ℓ_1-norm distances, which is difficult to solve in general. To solve this optimization problem, we derive an efficient, non-greedy, iterative algorithm with rigorously proved convergence. Our new method has been verified on a public HIV-1 drug resistance data set with over 600 RT sequences and five nucleoside analogues. We compared our method against other state-of-the-art multi-label classification methods and the experimental results have demonstrated the effectiveness of our proposed method.

Keywords: Human immunodeficiency virus · Drug resistance · Multi-label classification

© Springer Nature Switzerland AG 2019
L. J. Cowen (Ed.): RECOMB 2019, LNBI 11467, pp. 51–67, 2019.
https://doi.org/10.1007/978-3-030-17083-7_4

1 Introduction

According to estimations by the World Health Organization, around 35 million people suffer from the Human immunodeficiency virus (HIV). HIV is a serious virus that attacks cells in the human immune system. During the later stages of the virus it can critically weaken the immune system and increase the patient's susceptibility to serious infection and disease. Fortunately, with the advent of antiretroviral therapies, we have been able to stem the progression of HIV and extend the lifespan of individuals affected by the virus. Unfortunately, the high mutation rates of HIV Type 1 (HIV-1) can produce viral strains that adapt very quickly to new drugs [24]. The mutation of HIV-1 during antiretroviral treatments can lead to a phenomenon called "cross-resistance" [7,23]. Cross-resistance of HIV-1 occurs when the virus develops resistance against the drugs which are currently being used in addition to other drugs that have not yet been used in the treatment of a particular patient. This can make the treatment of HIV-1 significantly more difficult, because a collection of drugs may not be effective after the initial treatment regimen due to the cross-resistance phenomenon observed in HIV-1. In order to address this problem, it is important that we develop automatic methods that can associate genetic strains of HIV to their corresponding drug resistances.

Recently, experimental testing of viral resistance in patients has been widely used in research as well as in clinical settings to gain insight into the ways in which the drug resistance evolves. For example, large-scale pharmacogenomic screens have been conducted to explore the relationships between drug resistances and genomic sequences [21]. Furthermore, many clinical trials have been performed to discover mutation rates of the genetic subtypes of HIV-1 and how they develop resistances against various drug treatments [19]. In addition to these experimental phenotypic studies, computational approaches that use various machine learning methods offer the possibility to predict drug resistance in HIV-1 by using short sequence information of the viral genotype, such as the genetic sequence of the viral reverse transcriptase (RT). For example, Rhee et al. [22] used five different machine learning methods, including decision trees, artificial neural networks, support-vector machines, least-square regression and least-angle regression, to investigate drug resistance in HIV-1 based on the RT sequences. Besides, genotype and phenotype features of HIV-1 extracted from RT sequences have been studied to predict drug resistance [9]. Additionally, a Bayesian algorithm that combines kernel-based nonlinear dimensionality reduction and binary classification has been proposed to predict drug susceptibility of HIV within a multi-task learning framework [5]. A critical drawback of these existing studies lies in the fact that they routinely consider HIV-1 drug resistance prediction as a *single-label* classification problem. This approach has been recognized to be inappropriate since HIV strains can develop resistances against multiple drugs at once due to their high mutation rate [7,23]. To tackle this difficulty, in this paper we propose to solve the problem of HIV-1 drug resistance prediction as a *multi-label classification* problem.

Multi-label classification is an emerging research topic in machine learning driven by the advances of modern technologies in recent years [27–32,39]. As a generalization of traditional single-label classification that requires every data sample to belong to one and only one class, multi-label classification relaxes this restriction and allows a data sample to belong to multiple different classes at the same time. As a result, the classes in single-label classification problems are mutually exclusive, while those in multi-label classification problems are interdependent on one another. Although the labeling relaxation in multi-label classification problems have brought a number of successes in a variety of real-world applications [29,30,32], it also causes labeling ambiguity that inevitably complicates the problem [27,28]. In the context of predicting drug resistance developed by HIV-1, some HIV strains can develop the capability to resist multiple drugs, including those currently being used and those that have not yet been applied in a clinical setting. As a result, it is often unclear how to utilize a data sample that belongs to multiples classes to train a classifier for a given class [27,28]. A simple strategy to solve this problem is to use such data samples as the training data for all the classes to which they belong [27,29], which is equivalent to assume that every data sample contributes equally to a trained classification model [28]. However, this is not true in most real-world multi-label classification problems. For example, some RT sequences natively resist against a certain drug. On the other hand, the same RT sequences can develop resistances against other drugs through mutations, which is assumed to be not as strong as native resistances. Simply put, in order to create an effective multi-label classifier to predict HIV-1 resistances, it is critical to clarify the labeling ambiguity on data samples that belong to multiple classes and learn an appropriate scaling factor when we train the classifiers for different classes [28].

In this paper we propose a novel Robust Sample Specific Distance (RSSD) for multi-label data to predict HIV-1 drug resistance, which, as illustrated in Fig. 1, is able to explicitly rank the relevance of a training sample with respect to a specific class and characterize the second-order data-dependent statistics of all the classes. To learn the sample relevances and the class-specific distance metrics, we formulate a learning objective that simultaneously maximizes and minimizes the summations of the ℓ_1-norm distances. To solve the optimization problem of our objective, using the same method in our previous works [6,15], we derive an efficient iterative algorithm with theoretically guaranteed convergence, which, different from our previous works [35,37], is a *non-greedy* algorithm such that it has a better chance to find the optima of the proposed objective. In addition, as an important theoretical contribution of this paper, our new algorithm solves the general optimization problem that maximizes the ratio of the summations of the ℓ_1-norm distances in a non-greedy way, which can find many applications to improve a number of machine learning models. We applied our new method to predict the HIV-1 drug resistance on a public benchmark data set. The experimental results have shown that our new RSSD method outperforms other state-of-the-art competing methods.

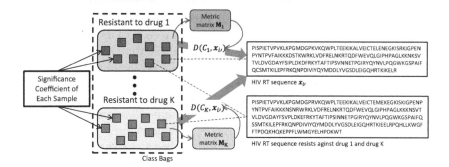

Fig. 1. The illustration of the proposed RSSD method. The small squares in the same color represent the data samples (RT sequences) that belong to one same class (*e.g.*, resistance to a specific nucleoside analogue). Two HIV RT sequences are listed in the right panel, which correspond to the data samples shown by the small squares (connected by the dash lines). The top sequence in the right column only resists against drug 1, while the bottom sequence resists against both drug 1 and drug K, *i.e.*, it is a multi-label data sample. Ideally, the learned Significance Coefficients for each data sample should be different with respect to different classes. For example, the bottom RT sequence is associated with s_{i1} for class 1 and s_{iK} for class K, which could be different depending on how the resistances evolved. (Color figure online)

2   Learning Robust Sample Specific Distances (RSSDs) for Multi-label Classification

In this section, we first formalize the problem of predicting the drug resistance of HIV-1. Then we derive a novel RSSD to solve the problem following previous works [26, 33–35, 38] that solve multi-instance problems.

Throughout this paper, we write matrices as bold uppercase letters and vectors as bold lowercase letters. The ℓ_1-norm of a vector $\mathbf{v}$ is defined as $\|\mathbf{v}\|_1 = \sum_i |v_i|$ and the ℓ_2-norm of $\mathbf{v}$ is defined as $\|\mathbf{v}\|_2 = \sqrt{\sum_i v_i^2}$. Given a matrix $\mathbf{M} = [m_{ij}]$, we denote its Frobenius norm as $\|\mathbf{M}\|_F$ and we define its ℓ_1-norm as $\|\mathbf{M}\|_1 = \sum_i \sum_j |m_{ij}|$. The trace of $\mathbf{M}$ is defined as $\text{tr}(\mathbf{M}) = \sum_i m_{ii}$.

In a multi-label classification problem, we are given a data set with n samples (n RT sequences) $\{\mathbf{x}_i, \mathbf{y}_i\}_{i=1}^n$ and K classes (resistances to K target nucleoside analogues). Here $\mathbf{x}_i \in \Re^d$, and $\mathbf{y}_i \in \{0, 1\}^K$ such that $\mathbf{y}_i(k) = 1$ if $\mathbf{x}_i$ belongs to the k-th class, and $\mathbf{y}_i(k) = 0$ otherwise. Our goal is to learn from the training data $\{\mathbf{x}_i, \mathbf{y}_i\}_{i=1}^n$ a classifier that is able to predict which nucleoside analogues (drug variants) a HIV-1 RT sequence is resistant to.

2.1   The Class-to-Sample (C2S) Distance

To learn the distance from a class to a data sample, we first represent each class as a bag that consists of all samples that belong to this class, *i.e.*, $C_k = \{\mathbf{x}_i | i \in \pi_k\}$, where $\pi_k = \{i | Y_{ik} = 1\}$ is the set of indices of all training samples that belong to the k-th class. The number of samples in C_k is denoted as m_k, *i.e.*, $|C_k| = m_k$. Note that, in a single-label classification problem, a data sample precisely belongs to one and only one class at a time. It follows that $\sum_{i=1}^{K} Y_{ik} = 1$ and $C_k \cap C_l = \emptyset$ $(\forall k \neq l)$. In contrast, in a multi-label classification problem, a data sample may belong to more than one class at the same time. It can happen that $C_k \cap C_l \neq \emptyset$ $(\exists k \neq l)$, *i.e.*, different class bags may overlap and an individual data sample $\mathbf{x}_i$ may appear in multiple class bags.

We first define the elementary distance from a sample $\mathbf{x}_i$ in the k-th class bag C_k to a data sample $\mathbf{x}_{i'}$ as the squared Euclidean distance between the two involved vectors in the d-dimensional Euclidean space:

$$d_k(\mathbf{x}_i, \mathbf{x}_{i'}) = \|\mathbf{x}_i - \mathbf{x}_{i'}\|_2^2, \quad \forall\, i \in \pi_k, \forall\, k \;\; 1 \leq k \leq K. \tag{1}$$

We then compute the C2S distance from C_k to $\mathbf{x}_{i'}$ by summing all the elementary distances from the samples that belong to the k-th class to the data sample $\mathbf{x}_{i'}$ as following:

$$D(C_k, \mathbf{x}_{i'}) = \sum_{\mathbf{x}_i \in C_k} d_k(\mathbf{x}_i, \mathbf{x}_{i'}) = \sum_{\mathbf{x}_i \in C_k} \|\mathbf{x}_i - \mathbf{x}_{i'}\|_2^2. \tag{2}$$

2.2   Parameterized C2S Distance

Because the C2S distance in Eq. (2) does not take into account the resistance strength against a certain nucleoside analogue, we further develop it by weighting the samples in a class bag by their relevance to this class.

Due to the ambiguous associations between the samples and the labels in a multi-label classification problem [27,28], some samples in a class may characterize that particular class more strongly than the others from the statistical point of view. For example [23], some viral RT sequences may develop a stronger drug resistance, while other viral RT sequences may be less resistant to a drug but may still be considered to be resistant. We must capture both of these in order for our method to be effective. As a result, we should assign less weight to less resistant RT sequences when determining whether to apply the "resistant" label to a query viral RT sequence.

Because we assume that counter-resistance against a target nucleoside analogue does not exist, we define $s_{ik} \geq 0$ as a nonnegative constant that assess relative importance of $\mathbf{x}_i$ with respect to the k-th class, by which we can further develop the C2S distance as following:

$$D(C_k, \mathbf{x}_{i'}) = \sum_{\mathbf{x}_i \in C_k} s_{ik}^2 \|\mathbf{x}_i - \mathbf{x}_{i'}\|^2. \tag{3}$$

Because s_{ik} reflects the relative importance of a sample $\mathbf{x}_i$ when we train a classifier for the k-th class, we call it the Significance Coefficient (SC) of $\mathbf{x}_i$ with respect to the k-th class. Obviously, the SCs quantitatively assess the resistances developed by the training viral RT sequences against the target nucleoside analogues during the learning process.

2.3 Parameterized C2S Distance Refined by Class Specific Distance Metrics

The RSSD defined in Eq. (3) is simply a weighted Euclidean distance that does not take into account the information conveyed by the input data other than the first-order statistics. Similar to many other statistical models in machine learning, using the Mahalanobis distances with appropriate distance metrics is recommended in order to capture the second-order statistics of the input data. Instead of learning one single global distance metric for all the classes as in many existing statistical studies, we propose to learn K different class-specific distance metrics $\{\mathbf{M}_k \succ 0\}_{k=1}^K \in \Re^{d \times d}$, one for each class. Thus we further develop the parameterized C2S distance as:

$$D(C_k, \mathbf{x}_{i'}) = \sum_{\mathbf{x}_i \in C_k} s_{ik}^2 (\mathbf{x}_i - \mathbf{x}_{i'})^T \mathbf{M}_k (\mathbf{x}_i - \mathbf{x}_i'). \tag{4}$$

Because the class-specific distance metric $\mathbf{M}_k$ is a positive definite matrix, we can reasonably write it as $\mathbf{M}_k = \mathbf{W}_k \mathbf{W}_k^T$, where $\mathbf{W}_k \in \Re^{d \times r}$ is an orthonormal matrix such that $\mathbf{W}_k^T \mathbf{W}_k = \mathbf{I}$. Here we can also reasonably assume that $d > r$, because $\mathbf{M}_k = \mathbf{W}_k \mathbf{W}_k^T$ is a $d \times d$ matrix and its maximum rank is d. Thus we can rewrite Eq. (4) as follows:

$$\begin{aligned} D(C_k, \mathbf{x}_{i'}) &= \sum_{\mathbf{x}_i \in C_k} s_{ik}^2 (\mathbf{x}_i - \mathbf{x}_{i'})^T \mathbf{W}_k \mathbf{W}_k^T (\mathbf{x}_i - \mathbf{x}_{i'}) \\ &= \sum_{\mathbf{x}_i \in C_k} \left\| \mathbf{W}_k^T (\mathbf{x}_i - \mathbf{x}_{i'}) s_{ik} \right\|_2^2. \end{aligned} \tag{5}$$

A critical problem of $D(C_k, \mathbf{x}_{i'})$ defined in Eq. (5) lies in that it computes the summation of a number of squared ℓ_2-norm distances. These squared terms are notoriously known to be sensitive to both outlying samples and features [2,37]. Due to the cross-resistance phenomenon [7], this problem is particularly critical for identifying HIV-1 drug resistance. To promote the robustness of $D(C_k, \mathbf{x}_{i'})$ against outliers, following many previous works [11,12,17,18,36,37,40], we define it using the ℓ_1-norm distance as follows:

$$D(C_k, \mathbf{x}_{i'}) = \sum_{\mathbf{x}_i \in C_k} \left\| \mathbf{W}_k^T (\mathbf{x}_i - \mathbf{x}_{i'}) s_{ik} \right\|_1, \tag{6}$$

which we call the proposed *Robust Sample Specific Distance (RSSD)*.

To use RSSD defined in Eq. (6), we need to learn two sets of parameters s_{ik} and $\mathbf{W}_k$ for every class. Following the most broadly used machine learning strategy to maximize data discriminativity for classification, such as Fisher's linear discriminant [4], for a given class C_k we simultaneously maximize the overall RSSDs from every class bag to all its non-belonging samples and minimize the overall RSSDs from every class bag to all the samples belonging to that class:

$$\max \frac{\sum_{\mathbf{x}_{i'} \notin C_k} \sum_{\mathbf{x}_i \in C_k} \left\| \mathbf{W}_k^T (\mathbf{x}_i - \mathbf{x}_{i'}) s_{ik} \right\|_1}{\sum_{\mathbf{x}_{i'} \in C_k} \sum_{\mathbf{x}_i \in C_k} \left\| \mathbf{W}_k^T (\mathbf{x}_i - \mathbf{x}_{i'}) s_{ik} \right\|_1}, \quad s.t. \ \mathbf{W}_k^T \mathbf{W}_k = \mathbf{I}, s_{ik} \geq 0. \tag{7}$$

Algorithm 1. Algorithm to solve Eq. (8).

---
1. Randomly initialize $v^0 \in \Omega$ and set $t = 1$.
while *not converge* **do**
   2. Calculate $\lambda^t = \frac{h(v^{t-1})}{m(v^{t-1})}$.
   3. Find a $v^t \in \Omega$ satisfying $h(v^t) - \lambda^t m(v^t) > h(v^{t-1}) - \lambda^t m(v^{t-1}) = 0$.
   4. $t = t + 1$.
Output: v.

Learning the RSSDs by solving Eq. (7) and classifying query viral RT sequences using the adaptive decision boundary method [29], our proposed RSSD method can be used for identifying HIV-1 drug resistance, as well as general multi-label classification problems.

3   An Efficient Solution Algorithm

Our new objective in Eq. (7) maximizes the ratio of the summations of a number of ℓ_1-norm distances, which is obviously not smooth and therefore difficult to solve in general. To solve this challenging optimization problem, we use the optimization method proposed in our previous works in [6,15].

We first turn to solve the following generalized the objective:

$$v_{\text{opt}} = \arg\max_{v \in \Omega} \frac{h(v)}{m(v)}, \qquad \forall v \in \Omega \quad \begin{cases} C_2 \geq m(v) \geq C_1 > 0, \\ C_4 \geq h(v) \geq C_3 > 0, \end{cases} \tag{8}$$

where Ω is the feasible domain. Next, we propose a simple, yet efficient, iterative framework in Algorithm 1 to solve the objective in Eq. (8). The convergence of Algorithm 1 is rigorously guaranteed by Theorem 1. Due to space limit, the proofs of all the theorems in this paper are provided in the extended journal version of this paper.

Theorem 1. *In Algorithm 1, for each iteration we have* $\frac{h(v^t)}{m(v^t)} \geq \frac{h(v^{t-1})}{m(v^{t-1})}$ *and* $\forall \delta$, *there must exist a $\hat{t}$ such that* $\forall t > \hat{t} \; \frac{h(v^t)}{m(v^t)} - \frac{h(v^{t-1})}{m(v^{t-1})} < \delta$.

3.1   Fixing s_{ik} to Solve $\mathbf{W}_k$

According to Step 3 in Algorithm 1, we can easily write the corresponding inequality of our objective in Eq. (7) as:

$$F(\mathbf{W}_k) = H(\mathbf{W}_k) - \lambda^t M(\mathbf{W}_k) \geq 0, \tag{9}$$

where λ^t is computed by

$$\lambda^t = \frac{\sum_{\mathbf{x}'_i \notin C_k} \sum_{\mathbf{x}_i \in C_k} \left\| (\mathbf{W}_k^{t-1})^T (\mathbf{x}_i - \mathbf{x}_{i'}) s_{ik} \right\|_1}{\sum_{\mathbf{x}'_i \in C_k} \sum_{\mathbf{x}_i \in C_k} \left\| (\mathbf{W}_k^{t-1})^T (\mathbf{x}_i - \mathbf{x}_{i'}) s_{ik} \right\|_1}. \tag{10}$$

In Eq. (10), $\mathbf{W}_k^{t-1}$ denotes the projection matrix in the $(t-1)$-th iteration. Here, we define the following:

$$H(\mathbf{W}_k) = \sum_{\mathbf{x}'_i \notin C_k} \sum_{\mathbf{x}_i \in C_k} \left\| \mathbf{W}_k^T (\mathbf{x}_i - \mathbf{x}_{i'}) s_{ik} \right\|_1,$$

$$M(\mathbf{W}_k) = \sum_{\mathbf{x}'_i \in C_k} \sum_{\mathbf{x}_i \in C_k} \left\| \mathbf{W}_k^T (\mathbf{x}_i - \mathbf{x}_{i'}) s_{ik} \right\|_1. \tag{11}$$

Now we need solve the problem in Eq. (9), for which we first introduce the following two lemmas:

Lemma 1. *[16, Theorem 1]. For any vector* $\boldsymbol{\xi} = [\xi_1, \cdots, \xi_m]^T \in \Re^m$, *we have* $\|\boldsymbol{\xi}\|_1 = \max_{\boldsymbol{\eta} \in \Re^m} (\text{sign}(\boldsymbol{\eta}))^T \boldsymbol{\xi}$, *where the maximum value is attained if and only if* $\boldsymbol{\eta} = a \times \boldsymbol{\xi}$, *where* $a > 0$ *is a scalar.*

Lemma 2. *[10, Lemma 3.1] For any vector* $\boldsymbol{\xi} = [\xi_1, \cdots, \xi_m]^T \in \Re^m$, *we have* $\|\boldsymbol{\xi}\|_1 = \min_{\boldsymbol{\eta} \in \Re_+^m} \frac{1}{2} \sum_{i=1}^m \frac{\xi_i^2}{\eta_i} + \frac{1}{2}\|\boldsymbol{\eta}\|_1$, *where the minimum value is attained if and only if* $\eta_j = |\xi_j|, j \in \{1, 2, \cdots, m\}$.

Motivated by Lemmas 1 and 2, we construct the following objective:

$$L(\mathbf{W}_k, \mathbf{W}_k^{t-1}) = K(\mathbf{W}_k) - \lambda^t N(\mathbf{W}_k), \tag{12}$$

where $K(\mathbf{W}_k)$ and $N(\mathbf{W}_k)$ are defined as:

$$K(\mathbf{W}_k) = \sum_{g=1}^r \mathbf{w}_g^T \mathbf{B} \, \text{sign} \left(\mathbf{B}^T \mathbf{w}_g^{t-1} \right),$$

$$N(\mathbf{W}_k) = \frac{1}{2} \sum_{g=1}^r \mathbf{w}_g^T \mathbf{A}_g \mathbf{w}_g + \left(\mathbf{w}_g^{t-1} \right)^T \mathbf{A}_g \mathbf{w}_g^{t-1}. \tag{13}$$

Here $\mathbf{w}_g$ and $\mathbf{w}_g^{t-1}$ denote the g-th column of matrices $\mathbf{W}_k$ and $\mathbf{W}_k^{t-1}$, respectively; $\mathbf{B}$ and $\mathbf{A}_g$ for $g = 1, 2, \cdots, r$ are defined as follows:

$$\mathbf{B} = [\bar{\mathbf{x}}_1 - \bar{\mathbf{x}}, \bar{\mathbf{x}}_2 - \bar{\mathbf{x}}, \cdots, \bar{\mathbf{x}}_n - \bar{\mathbf{x}}],$$

$$\mathbf{A}_g = \sum_{i=1}^n \sum_{\mathbf{x}_j \in \{\mathcal{N}_i \cup \{\mathbf{x}_i\}\}} \frac{(\mathbf{x}_j - \bar{\mathbf{x}}_i)(\mathbf{x}_j - \bar{\mathbf{x}}_i)^T}{\left| \left(\mathbf{w}_g^{t-1} \right)^T (\mathbf{x}_j - \bar{\mathbf{x}}_i) \right|}, \tag{14}$$

and $\text{sign}(x)$ is the sign function.

Then, using the definition of $L(\mathbf{W}_k, \mathbf{W}_k^{t-1})$ in Eq. (12) and Lemmas 1 and 2, we can prove the following theorem:

Theorem 2. *For any* $\mathbf{W}_k \in \Re^{d \times r}$, *we have:*

$$L(\mathbf{W}_k, \mathbf{W}_k^{t-1}) \leq F(\mathbf{W}_k). \tag{15}$$

The equality holds if and only if $\mathbf{W}_k = \mathbf{W}_k^{t-1}$.

Algorithm 2. Algorithm to maximize $F(\mathbf{W}_k)$.

Input: $\mathbf{W}_k^{t-1}$ and Armijo parameter $0 < \beta < 1$.
1. Calculate λ^k by Eq. (10).
2. Calculate the subgradient
$\mathbf{G}^{k-1} = \partial L(\mathbf{W}_k^{t-1}, \mathbf{W}_k^{t-1}) = \mathbf{B} \operatorname{sign}(\mathbf{B}^T \mathbf{W}_k^{t-1}) - \lambda^k [\mathbf{A}_1 \mathbf{w}_1, \mathbf{A}_2 \mathbf{w}_2, \cdots, \mathbf{A}_r \mathbf{w}_r]$.
3. Set $t = 1$.
while *not* $F(\mathbf{W}_k^t) > F(\mathbf{W}_k^{t-1}) = 0$ **do**
 4. Calculate $\mathbf{W}_k^t = P(\mathbf{W}_k^{t-1} + \beta^m \mathbf{G}^{t-1})$.
 5. Calculate $F(\mathbf{W}_k^t)$ by Eq. (9).
 6. $t = t + 1$.

Output: $\mathbf{W}_k^k$.

Algorithm 3. Algorithm for non-greedy ratio maximization of the ℓ_1-norm distances.

---

1. Randomly initialize $\mathbf{W}_k^0$ satisfying $(\mathbf{W}_k^0)^T \mathbf{W}_k^0 = \mathbf{I}$ and set $t = 1$.
while *not converge* **do**
 2. Calculate λ^t by Eq. (10).
 3. Find a $\mathbf{W}_k^t$ satisfying $F(\mathbf{W}_k^t) > F(\mathbf{W}_k^{t-1}) = 0$ by Algorithm 2.
 4. $t = t + 1$.

Output: $\mathbf{W}$.

Now we continue to solve our objective. Let $\mathbf{W}_k = \mathbf{W}_k^{t-1}$, by substituting it into the objective, we have $L(\mathbf{W}_k, \mathbf{W}_k^{k-1}) = F(\mathbf{W}_k^{t-1}) = 0$. In the k-th iteration in solving the objective in Eq. (7), $\mathbf{W}_k^{\star}$ satisfies $L(\mathbf{W}_k^{\star}, \mathbf{W}_k^{t-1}) \geq L(\mathbf{W}_k^{t-1}, \mathbf{W}_k^{t-1}) = 0$. Then, we have:

$$F(\mathbf{W}_k^{\star}) \geq L(\mathbf{W}_k^{\star}, \mathbf{W}_k^{t-1}) \geq L(\mathbf{W}_k^{t-1}, \mathbf{W}_k^{t-1}) = F(\mathbf{W}_k^{t-1}) = 0. \qquad (16)$$

Lemma 1 and Eq. (16) indicate that the solution of the objective function in Eq. (9) can be transformed to solve the objective function $L(\mathbf{W}_k, \mathbf{W}_k^{t-1}) \geq 0$, which can be easily solved by the projected subgradient method with Armijo line search [25]. Note that, for any matrix $\mathbf{W}_k$ the operator $P(\mathbf{W}_k) = \mathbf{W}_k (\mathbf{W}_k^T \mathbf{W}_k)^{-\frac{1}{2}}$ can project it onto an orthogonal cone. This guarantees the orthogonality constraint of the projection matrix, *i.e.* $(\mathbf{W}_k^t)^T (\mathbf{W}_k^t) = \mathbf{I}$. Algorithm 2 summarizes the algorithm to solve the objective in Eq. (9).

Finally, based on Algorithm 2, we can derive a simple yet efficient iterative algorithm as summarized in Algorithm 3 to solve our objective in Eq. (7) when s_{ik} is fixed. In addition, Theorem 3 indicates that our proposed Algorithm 3 monotonically increase the objective function value in each iteration. Theorem 4 indicates that the objective function is upper bounded, which, together with Theorem 3, indicates that Algorithm 3 converges to a local optimum.

Theorem 3. *If $\mathbf{W}_k^t$ is the solution of the objective function in Eq. (9) and satisfies $(\mathbf{W}_k^t)^T (\mathbf{W}_k^t) = \mathbf{I}$, then we have $\mathcal{J}(\mathbf{W}_k^t) \geq \mathcal{J}(\mathbf{W}_k^{t-1})$.*

Theorem 4. *The objective in Eq. (7) is upper bounded.*

3.2 Fixing $\mathbf{W}_k$ to Solve s_{ik}

When $\mathbf{W}_k$ is fixed, we define a scalar $d_{ii'k} = \left\| \mathbf{W}_k^T (\mathbf{x}_i - \mathbf{x}_{i'}) \right\|_1$. Then we write Eq. (7) as:

$$\max \frac{\sum_{\mathbf{x}_i' \notin C_k} \sum_{\mathbf{x}_i \in C_k} s_{ik} d_{ii'k}}{\sum_{\mathbf{x}_i' \in C_k} \sum_{\mathbf{x}_i \in C_k} s_{ik} d_{ii'k}}, \quad s.t. \ s_{ik} \geq 0. \tag{17}$$

Defining that $d_{ik}^w = \sum_{i' \in \pi_k} d_{ii'k}$ and $d_{ik}^b = \sum_{i' \notin \pi_k} d_{ii'k}$, we can further rewrite the objective as:

$$\max \frac{\sum_{\mathbf{x}_i \notin C_k} s_{ik} d_{ik}^w}{\sum_{\mathbf{x}_i \in C_k} s_{ik} d_{ik}^b}, \quad s.t. \ s_{ik} \geq 0. \tag{18}$$

Again, to solve Eq. (18), to Step 3 in Algorithm 1, we solve the following optimization problem:

$$\max \sum_{\mathbf{x}_i \in C_k} s_{ik} d_{ik}^w - \lambda \sum_{\mathbf{x}_i \in C_k} s_{ik} d_{ik}^b, \quad s.t. \ s_{ik} \geq 0, \tag{19}$$

where λ is computed as Eq. (10) in the t-th iteration.

Define that $d_{ik} = d_{ik}^w - \lambda d_{ik}^b$, we can rewrite the optimization problem in Eq. (19) as:

$$\max \sum_{\mathbf{x}_i \in C_k} s_{ik} d_{ik}, \quad s.t. \ s_{ik} \geq 0, \tag{20}$$

The problem in Eq. (20) can be decoupled to solve the following subproblems separately for each $\mathbf{x}_i \in C_k$:

$$\max \ s_{ik} d_{ik}, \quad s.t. \ s_{ik} \geq 0, \tag{21}$$

which is a convex linear programming problem [41] and can be solved efficiently by many off-the-shelf solution algorithms [41]. By inserting the solution to Eq. (21) after Step 3 of Algorithm 3, we can finally solve our objective in Eq. (7), which is equivalent to perform alternative optimization. Therefore, the algorithm is guaranteed to converge to a local optimum.

4 Experimental Results

We evaluate the proposed RSSD method using a publicly available HIV drug resistance database [22], which contains HIV-1 RT sequences with associated resistance factors measured by IC_{50} ratios. We analyze the drug resistance of these RT sequences against five nucleoside analogues: Lamivudine (3TC), Abacavir (ABC), Zidovudine (AZT), Stavudine ($d4T$) and Didanasine (ddI). Following [8], although the Tenofovir (TDF) nucleoside analogue is included in this database, it is not used in our study, because the number of the RT sequences resistant to this nucleoside analogue is very low. As a result, we end up with 623 RT sequences for our experiments.

Drug resistance of a particular HIV strain is measured by the IC_{50} ratio [7]. We label the viral RT sequences as "resistant" using the same drug-specific IC_{50} ratio cutoff thresholds as in [7], which are set to 3.0 for 3TC and *AZT*, 2.0 for *ABC*, and 1.5 for *ddI* and *d4T*. We use hydrophobicity characteristics [13] to represent the RT sequences, which has demonstrated good prediction performance in many protein classification studies [8]. For each RT sequence, we extract a hydrophobicity vector, which is obtained from the amino acid sequence and smoothed within a window. The length of the original hydrophobicity vectors may be different due to the different lengths of the RT sequences. In this study, following [7] we set a fixed window size of 11 and interpolate all hydrophobicity vectors to length 230 using the spline interpolation method [13].

4.1 Comparative Studies

Predicting drug resistance for HIV-1 RT sequences is a multi-label classification problem. Therefore, we evaluate the proposed method by two broadly used multi-label performance metrics [14]: Hamming loss and average precision. The Hamming loss is computed over all instances over all classes. The average precision is calculated for both the micro and macro averages. In multi-label classification, the macro average is computed as the average of the precision values over all the classes, thus it attributes equal weights to every individual class. In contrast, the micro average is obtained from the summation of contingency matrices for all binary classifiers, thus it gives equal weight to all classifiers and emphasizes the accuracy of categories with more positive samples.

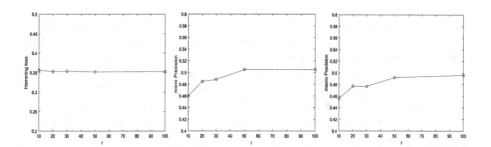

Fig. 2. Multi-label classification performance of the proposed method on the HIV-1 drug resistance data with respect to r (the dimensionality of $\mathbf{W}_k$).

Parameter Selection. The proposed RSSD has only one parameter: the dimensionality r of $\mathbf{W}_k$. Ideally, each class can have its own fine tuned parameter. Although, to reduce the experimental effort, we fix the parameter r across all classes in our studies. We evaluate the impacts of the parameter in a standard 5-fold cross-validation experiment, where we select r in the range from 10 to 100. The classification performance measured by the three aforementioned performance metrics are reported in Fig. 2, when we vary r. The results in these experiments show that the classification performance of the proposed method is reasonably stable when we vary r in a considerably large selection range. This illustrates that tuning parameters in our proposed method is not a difficult task; this property adds to the practical value of our method to solve real-world problems. Based on these observations, we fix $r = 50$ in all our future experiments for simplicity.

Comparative Studies. We use a standard 5-fold cross-validation to evaluate the predictive capability of the proposed RSSD method. We implement two versions of our proposed method, one version that defines $D(C_k, \mathbf{x}_{i'})$ using the ℓ_1-norm distances as in Eq. (6) (denoted as "Ours-ℓ_1") and another that defines $D(C_k, \mathbf{x}_{i'})$ using the squared ℓ_2-norm distances as in Eq. (5) (denoted as "Ours-ℓ_2^2"). We compare our new method to two broadly evaluated multi-label classification methods in literature: the Green's Function method [29] and the Sylvester Equation (SMSE) method [1]. We also compare the proposed method against two multi-label classification methods designed to study drug resistance in HIV-1: the Classifier Chain (CC) method and its ensemble version [7,20] (denoted as the ECC method). Finally, we also compare our method to two recent multi-instance classification methods: the multi-task learning (MTL) method [42] designed to study general drug resistance study and the deep MIML method [3] designed to study general multi-instance data. The Green's Function method and the Sylvester Equation methods are implemented following their original papers in [29] and [1] respectively, where the parameters are set to the suggested values. The CC method is implemented with logistic regression, where the chaining order for the CC method is $3TC \rightarrow ABC \rightarrow AZT \rightarrow d4T \rightarrow ddI$ as suggested in [7]. Following [7,23], we implement the ECC method by using both random forests and logistic regression as base classifiers, which are denoted as "ECC-RF" and "ECC-LR" respectively. The MTL method and the deep MIML method are implemented using the code published by the respective authors. The resistance prediction performances of the compared methods are reported in Table 1.

The comparison results in Table 1 show that the ℓ_1-norm version of the proposed method consistently outperforms all competing methods in terms of all the three performance metrics, sometime very significantly. The squared ℓ_2-norm version of our new method is, as expected, not as effective as its counterpart using the ℓ_1-norm distance, but it still provides adequate performance when compared to the other methods in Table 1.

4.2   A Case Study

We explore the learned distances by our method between RT sequence pairs and compared them with the Euclidean distances for the same RT sequence

Table 1. Performance of the compared methods by standard 5-fold cross validations, where "↓" means that smaller is better and "↑" means that bigger is better.

Compared methods	Hamming loss (↓)	Micro precision (↑)	Macro precision (↑)
Green's	0.450 ± 0.040	0.319 ± 0.046	0.241 ± 0.033
SMSE	0.385 ± 0.020	0.402 ± 0.032	0.241 ± 0.020
CC	0.302 ± 0.028	0.467 ± 0.046	0.434 ± 0.037
ECC-LR	0.313 ± 0.014	0.481 ± 0.011	0.442 ± 0.012
ECC-RF	0.301 ± 0.005	0.476 ± 0.020	0.461 ± 0.021
MTL	0.382 ± 0.010	0.475 ± 0.021	0.461 ± 0.010
Deep MIML	0.315 ± 0.010	0.478 ± 0.042	0.474 ± 0.022
Ours-ℓ_2^2	0.322 ± 0.015	0.505 ± 0.040	0.492 ± 0.050
Ours-ℓ_1	**0.282 ± 0.007**	**0.518 ± 0.012**	**0.527 ± 0.013**

pairs. The distance between two RT sequences by our method is defined as the sum of the two learned RSSDs: for the k-th class, the pairwise distance between sequence $\mathbf{x}_i$ and $\mathbf{x}_{i'}$ is the sum of $D(C_k, \mathbf{x}_i)$ and $D(C_k, \mathbf{x}_{i'})$. Because we learn a distance metric and significance coefficients for each class, this distance is class-dependent. Under this definition, the distances given by our method between sample pairs that belong to the same class are expected to be small and those between sample pairs not belonging to the same class are expected to be large. Using the learned class specific metrics and significance coefficients, we compute the pairwise distances between the RT sequences for every class (nucleoside analogue), which are plotted in Fig. 3. The Euclidean distances are also plotted for comparison.

To demonstrate the effectiveness of the proposed method, we study the distances between two example RT sequences, which are listed at the top of Fig. 3. These two RT sequences are known to be resistant to all five nucleoside analogues. As a result, the pairwise distance between these two RT sequences are expected to be small. However, as can be seen in top left panel of Fig. 3, the Euclidean distance between these two RT sequences is ranked at the 1855-th smallest distance among all pairwise Euclidean distances, which is not in accordance with the clinical evidences. In contrast, we can see that the pairwise distances between these RT sequences computed by our learned RSSDs for the five classes are small, which are at the 138-th smallest distance for *3CT*, the 525-th smallest distance for *ABC*, the 574-th smallest distance for *AZT*, the 406-th smallest distance for *d4T*, and 678-th smallest distance for *ddI*, respectively. This observation clearly demonstrates that the learned distances by our new methods, can better capture the relationships between data samples in terms of class membership.

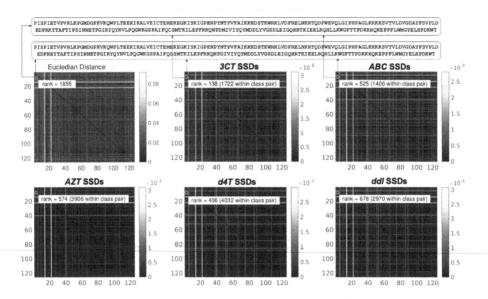

Fig. 3. Exploration of the learned sample-to-sample distance between RT sequence pairs for each class. **Top panel:** The two RT sequences (with known drug resistance) we are comparing; **Top Left Heatmap:** the Euclidean distances between RT sequence pairs. **Remaining Heatmaps:** the learned sample-to-sample distances between RT sequence pairs for each of the five classes. We can see that the sample-to-sample distance between the two RT sequences in the top panel for 3CT nucleoside analogue is ranked as the 138-th smallest pairwise distance among all 1722 RT sequence pairs. Compared to the Euclidean distance, which is ranked as 1855-th smallest distance, the pairwise distance computed by the projection and significance coefficients learned for this class is more clinically meaningful.

5   Conclusions

In this paper, we proposed a novel RSSD method for multi-label classification. To learn the parameters of the proposed RSSDs, we formulated a learning objective that maximizes the ratio of the summations of a number of ℓ_1-norm distances; this problem is difficult to solve in general. To solve this problem we derived a new efficient iterative algorithm with rigorously proved convergence. The promising experimental results have demonstrated the effectiveness of our new method for identifying HIV-1 drug resistances.

Acknowledgments. This work was partially supported by National Science Foundation under Grant NSF-IIS 1652943. This research was also partially supported by Army Research Office (ARO) under Grant W911NF-17-1-0447, U.S. Air Force Academy (USAFA) under Grant FA7000-18-2-0016, and the Distributed and Collaborative Intelligent Systems and Technology (DCIST) CRA under Grant W911NF-17-2-0181.

References

1. Chen, G., Song, Y., Wang, F., Zhang, C.: Semi-supervised multi-label learning by solving a sylvester equation. In: SDM, pp. 410–419. SIAM (2008)
2. Ding, C., Zhou, D., He, X., Zha, H.: R1-PCA: rotational invariant L1-norm principal component analysis for robust subspace factorization. In: ICML, pp. 281–288 (2006)
3. Feng, J., Zhou, Z.H.: Deep MIML network. In: AAAI (2017)
4. Fukunaga, K.: Introduction to Statistical Pattern Recognition. Elsevier, Amsterdam (2013)
5. Gönen, M., Margolin, A.A.: Drug susceptibility prediction against a panel of drugs using kernelized Bayesian multitask learning. Bioinformatics **30**(17), i556–i563 (2014)
6. Han, F., Wang, H., Zhang, H.: Learning of integrated holism-landmark representations for long-term loop closure detection. In: AAAI Conference on Artificial Intelligence (2018)
7. Heider, D., Senge, R., Cheng, W., Hüllermeier, E.: Multilabel classification for exploiting cross-resistance information in HIV-1 drug resistance prediction. Bioinformatics **29**(16), 1946–1952 (2013)
8. Heider, D., Verheyen, J., Hoffmann, D.: Predicting bevirimat resistance of HIV-1 from genotype. BMC Bioinform. **11**(1), 37 (2010)
9. Hepler, N.L., et al.: IDEPI: rapid prediction of HIV-1 antibody epitopes and other phenotypic features from sequence data using a flexible machine learning platform. PLOS Comput. Biol. **10**(9), e1003842 (2014)
10. Jenatton, R., Obozinski, G., Bach, F.: Structured sparse principal component analysis. In: International Conference on Artificial Intelligence and Statistics (2010)
11. Ke, Q., Kanade, T.: Robust L/sub 1/norm factorization in the presence of outliers and missing data by alternative convex programming. In: IEEE Computer Society Conference on Computer Vision and Pattern Recognition, CVPR 2005, vol. 1, pp. 739–746. IEEE (2005)
12. Kwak, N.: Principal component analysis based on L1-norm maximization. IEEE Trans. Pattern Anal. Mach. Intell. **30**, 1672–1680 (2008)
13. Kyte, J., Doolittle, R.F.: A simple method for displaying the hydropathic character of a protein. J. Mol. Biol. **157**(1), 105–132 (1982)
14. Lewis, D.D., Yang, Y., Rose, T.G., Li, F.: Rcv1: a new benchmark collection for text categorization research. J. Mach. Learn. Res. **5**, 361–397 (2004)
15. Liu, K., Wang, H., Nie, F., Zhang, H.: Learning multi-instance enriched image representations via non-greedy ratio maximization of the L1-norm distances. In: Proceedings of the IEEE Conference on Computer Vision and Pattern Recognition, pp. 7727–7735 (2018)
16. Liu, Y., Gao, Q., Miao, S., Gao, X., Nie, F., Li, Y.: A non-greedy algorithm for L1-norm LDA. IEEE Trans. Image Process. **26**(2), 684–695 (2017)
17. Nie, F., et al.: New L1-norm relaxations and optimizations for graph clustering. In: AAAI, pp. 1962–1968 (2016)
18. Nie, F., Wang, H., Huang, H., Ding, C.: Unsupervised and semi-supervised learning via ℓ_1-norm graph. In: 2011 IEEE International Conference on Computer Vision (ICCV), pp. 2268–2273. IEEE (2011)
19. Pennings, P.S.: Standing genetic variation and the evolution of drug resistance in HIV. PLoS Comput. Biol. **8**(6), e1002527 (2012)

20. Read, J., Pfahringer, B., Holmes, G., Frank, E.: Classifier chains for multi-label classification. Mach. Learn. **85**(3), 333–359 (2011)
21. Rhee, S.Y., Gonzales, M.J., Kantor, R., Betts, B.J., Ravela, J., Shafer, R.W.: Human immunodeficiency virus reverse transcriptase and protease sequence database. Nucleic Acids Res. **31**(1), 298–303 (2003)
22. Rhee, S.Y., Taylor, J., Wadhera, G., Ben-Hur, A., Brutlag, D.L., Shafer, R.W.: Genotypic predictors of human immunodeficiency virus type 1 drug resistance. Proc. Natl. Acad. Sci. **103**(46), 17355–17360 (2006)
23. Riemenschneider, M., Senge, R., Neumann, U., Hüllermeier, E., Heider, D.: Exploiting HIV-1 protease and reverse transcriptase cross-resistance information for improved drug resistance prediction by means of multi-label classification. Bio-Data Min. **9**(1), 10 (2016)
24. Smyth, R.P., Davenport, M.P., Mak, J.: The origin of genetic diversity in HIV-1. Virus Res. **169**(2), 415–429 (2012)
25. Sun, W., Yuan, Y.X.: Optimization Theory and Methods: Nonlinear Programming, vol. 1. Springer, Heidelberg (2006). https://doi.org/10.1007/b106451
26. Wang, H., Deng, C., Zhang, H., Gao, X., Huang, H.: Drosophila gene expression pattern annotations via multi-instance biological relevance learning. In: AAAI, pp. 1324–1330 (2016)
27. Wang, H., Ding, C., Huang, H.: Multi-label linear discriminant analysis. In: Daniilidis, K., Maragos, P., Paragios, N. (eds.) ECCV 2010. LNCS, vol. 6316, pp. 126–139. Springer, Heidelberg (2010). https://doi.org/10.1007/978-3-642-15567-3_10
28. Wang, H., Ding, C.H., Huang, H.: Multi-label classification: inconsistency and class balanced k-nearest neighbor. In: AAAI (2010)
29. Wang, H., Huang, H., Ding, C.: Image annotation using multi-label correlated green's function. In: 2009 IEEE 12th International Conference on Computer Vision, pp. 2029–2034. IEEE (2009)
30. Wang, H., Huang, H., Ding, C.: Multi-label feature transform for image classifications. In: Daniilidis, K., Maragos, P., Paragios, N. (eds.) ECCV 2010. LNCS, vol. 6314, pp. 793–806. Springer, Heidelberg (2010). https://doi.org/10.1007/978-3-642-15561-1_57
31. Wang, H., Huang, H., Ding, C.: Function-function correlated multi-label protein function prediction over interaction networks. J. Comput. Biol. **20**(4), 322–343 (2013)
32. Wang, H., Huang, H., Ding, C.: Correlated protein function prediction via maximization of data-knowledge consistency. J. Comput. Biol. **22**(6), 546–562 (2015)
33. Wang, H., Huang, H., Kamangar, F., Nie, F., Ding, C.H.: Maximum margin multi-instance learning. In: Advances in Neural Information Processing Systems, pp. 1–9 (2011)
34. Wang, H., Nie, F., Huang, H.: Learning instance specific distance for multi-instance classification. In: AAAI, vol. 2, p. 6 (2011)
35. Wang, H., Nie, F., Huang, H.: Robust and discriminative distance for multi-instance learning. In: CVPR. IEEE (2012)
36. Wang, H., Nie, F., Huang, H.: Robust and discriminative self-taught learning. In: International Conference on Machine Learning, pp. 298–306 (2013)
37. Wang, H., Nie, F., Huang, H.: Robust distance metric learning via simultaneous ℓ_1-norm minimization and maximization. In: ICML, pp. 1836–1844 (2014)
38. Wang, H., Nie, F., Huang, H., Yang, Y.: Learning frame relevance for video classification. In: Proceedings of the 19th ACM International Conference on Multimedia, pp. 1345–1348. ACM (2011)

39. Wang, H., Yan, L., Huang, H., Ding, C.: From protein sequence to protein function via multi-label linear discriminant analysis. IEEE/ACM Trans. Comput. Biol. Bioinform. (TCBB) **14**(3), 503–513 (2017)
40. Wright, J., Ganesh, A., Rao, S., Peng, Y., Ma, Y.: Robust principal component analysis: exact recovery of corrupted. In: NIPS, p. 116 (2009)
41. Wright, S.J., Nocedal, J.: Numerical optimization. Springer Sci. **35**(67–68), 7 (1999)
42. Yuan, H., Paskov, I., Paskov, H., González, A.J., Leslie, C.S.: Multitask learning improves prediction of cancer drug sensitivity. Sci. Rep. **6**, 31619 (2016)

MethCP: Differentially Methylated Region Detection with Change Point Models

Boying Gong[1] and Elizabeth Purdom[2(✉)]

[1] Division of Biostatistics, University of California, Berkeley, Berkeley, USA
jorothy_gong@berkeley.edu
[2] Department of Statistics, University of California, Berkeley, Berkeley, USA
epurdom@stat.berkeley.edu

Abstract. Whole-genome bisulfite sequencing (WGBS) provides a precise measure of methylation across the genome, yet presents a challenge in identifying regions that are differentially methylated (DMRs) between different conditions. Many methods have been developed, which focus primarily on the setting of two-group comparison. We develop a DMR detecting method MethCP for WGBS data, which is applicable for a wide range of experimental designs beyond the two-group comparisons, such as time-course data. MethCP identifies DMRs based on change point detection, which naturally segments the genome and provides region-level differential analysis. For simple two-group comparison, we show that our method outperforms developed methods in accurately detecting the complete DM region on a simulated dataset and an Arabidopsis dataset. Moreover, we show that MethCP is capable of detecting wide regions with small effect sizes, which can be common in some settings but existing techniques are poor in detecting such DMRs. We also demonstrate the use of MethCP for time-course data on another dataset following methylation throughout seed germination in Arabidopsis.

Availability: The package MethCP has been submitted to Bioconductor, and is currently available at https://github.com/boyinggong/MethCP.

Keywords: Differential methylation · Bisulfite sequencing · Change point detection

1 Introduction

DNA methylation is an important epigenetic mechanism for regulation of gene expression. Methylation is a process by which methyl groups are added to DNA cytosine (C) molecules. The methylation of promoter sites, in particular, is negatively correlated with gene expression while methylation in gene bodies is positively correlated with gene expression. Whole-genome bisulfite sequencing (WGBS) allows for precise measurement of DNA methylation across the genome. Briefly, when DNA is treated with bisulfite, the *unmethylated* cytosines

© Springer Nature Switzerland AG 2019
L. J. Cowen (Ed.): RECOMB 2019, LNBI 11467, pp. 68–84, 2019.
https://doi.org/10.1007/978-3-030-17083-7_5

are converted to uracil (U) leaving methylated cytosines unchanged. Sequencing of bisulfite-treated DNA and mapping of the sequenced reads to a reference genome then provides a quantification of the level of methylation at each cytosine. Methylation occurs in three different contexts: CG, CHG and CHH (where H corresponds to A, T or C). We will just refer to methylation of individual cytosine nucleotides.

In the analysis of BS-Seq data, a common interest is to identify regions of the genome where methylation patterns differ across populations of interest. Such regions are called differentially methylated regions (DMRs). Identifying DMRs are generally considered preferable than detection of individually differentially methylated cytosines (DMCs) from both statistical and biological perspective [22]. DNA methylation shows strong local patterns, and it is believed that region-level differences are more biologically important. Because of the low coverage and the fact that nearby cytosines usually have similar levels, combining them into regions substantially improves statistical power and lowers the false discovery rate. For the downstream analysis, reporting regions also reduces the redundancy.

A number of methods have been developed to identify regions from BS-Seq data that show differential methylation between two groups of samples (see [19] for a detailed review). One common strategy is to perform a test at every cytosine that appropriately accounts for the proportions and then use these significant results to determine the DMRs. For example, methylKit [1] performs either a logistic regression test or Fisher's exact test per cytosine; RADMeth [6] uses a beta-binomial regression, and a log-likelihood ratio test; DSS [8,17,24] uses a Bayesian hierarchical model with beta-binomial distribution to model the proportions and tests for per-cytosine significance with a Wald test. Other methods use the local dependency between neighboring cytosines to improve their per-cytosine test. BSmooth [10] and BiSeq [11] both use local likelihood regression to estimate an underlying methylation curve, and then test for differences in the smoothed methylation ratios between populations. HMM-DM [25,26] and HMM-Fisher [21,25] both use Hidden Markov Models along the genome to account for the dependency between cytosines. For many of these methods, the region is often either predefined or determined by merging adjacent DMCs based on specific criteria such as distance.

Another approach is to directly segment the methylation levels to find DMRs. The method metilene [13] uses a modified circular binary segmentation algorithm with statistics based on the mean differences in methylation ratios between two groups. The segments are tested for significance using Kolmogorov-Smirnov or Mann-Whitney U tests until the test results do not improve or the number of cytosines is too small.

We, too, propose a segmentation approach, MethCP, for finding DMRs from BS-Seq data. MethCP uses as input the results of a per-cytosine test statistic, like one of the methods described above, uses this input to segment the genome into regions, and then identifies which of those regions are DMRs. Our method, therefore, takes into account the coverages and biological variance between samples. Furthermore, all of the previously mentioned methods, including existing segmentation methods, are developed for simple two group comparisons, and are not straightforward to extend to more general experimental designs. MethCP, on the other hand, can be used in a wide variety of experimental designs.

We show via simulations that our method more accurately identifies regions differentially methylated between groups, as compared to competing methods. We illustrate the performance of MethCP on experimental data and show that its behavior on experimental data mirrors that of the simulations. We further demonstrate the flexibility of MethCP for use beyond the two-group setting by applying it to a time-course study.

2   Methods

MethCP assumes as input the results of a per-cytosine test of significance, such as those mentioned previously in the introduction. The main steps of MethCP are to (1) segment the test statistics into regions of similar values, and then (2) assign a p-value per region as to whether the region is a DMR.

Let T_k, $k = 1, \cdots, K$, be per-cytosine statistics for each of K cytosines, ordered by the location of the cytosines. We assume for now that the test statistics are independent (asymptotically) normally distributed, such as z-statistics or Wald statistics for testing equality of a proportion between two populations, and in Sect. 2.1 we extend this approach for other test statistics. We segment the T_k into regions of similar levels of significance based on the Circular Binary Segmentation (CBS) algorithm of [16], which was originally developed for segmentation of DNA copy number data. Note that MethCP applies the segmentation to test-statistics T_k, which is a summary per cytosine as to the difference of interest across the samples, so that it finds regions of similar population significance.

Briefly, binary segmentation methods involve testing over all of the possible breakpoints (cytosines) for whether there is a change in the mean of T at location $i \in [K]$; in the case of genomic data, the segmentation is applied per chromosome. The CBS algorithm performs a binary segmentation and adapts the algorithm so as to view the data from a chromosome as if it lies on a circle, segmenting the circle into two arcs. The segmentation procedure of CBS is then as follows: for each possible arc defined by, $1 \leq i < j \leq K$, the likelihood ratio test statistic Z_{ij} is calculated by comparing the mean value of T_k found in the arc from $i + 1$ to j with that found in the remaining circle. To find a significant breakpoint, CBS determines whether the statistic $Z = \max_{1 \leq i < j \leq K} |Z_{ij}|$ is significantly larger than 0. If so, this implies a detection that the arc $(i + 1, j)$ has a significantly different mean than the remaining arc and the two arcs are declared to be separate segments. The procedure is then applied recursively on each resulting segment until no more significant segments are detected.

The number of computation required for the segmentation is $O(K^2)$. However, due to the uneven distribution of methylation cytosine across the genome, "gaps" where nearby methylation cytosines are far from each other and almost uncorrelated naturally presegment the genome. Like other methods [13], MethCP can optionally presegment the genome and apply the algorithm separately to these highly separated segments. This reduces the computation to $O(KM)$, where $M \ll K$ is the maximum number of cytosines in these segments.

The underlying purpose of the segmentation step is to segment the differential region from the undifferentiated regions. After the completion of the segmentation, it remains to determine which of these segmented regions correspond to significant DMRs, rather than their surrounding undifferentiated regions. To classify these regions, we use meta-analysis principles to aggregate the per-cytosine statistics and obtain one single statistic per region, to which we apply significance tests.

We assume that for each cytosine the calculated statistic T_k can be written as $T_k = \frac{e_k}{\hat{\sigma}(e_k)}$, where e_k is the effect size that has approximate normal distribution with estimated variance $\hat{\sigma}^2(e_k)$. For a region i with a set of cytosines S_i, the weighted effect size is given by

$$e_i^* = \frac{\sum_{k \in S_i} w_k e_k}{\sum_{k \in S_i} w_k}, \tag{1}$$

where the weights w_k signify the contribution of cytosine k. Typically in meta-analysis applications, w_k is set to be $\hat{\sigma}(e_k)^{-1}$ [2]. In the case of WGBS, assuming that appropriate methods which account for the variability in the counts are used to calculate T_k, $\hat{\sigma}(e_k)^{-1}$ will be closely related to the coverage of the cytosine, which we designate as C_k. Alternatively, for example, when $\hat{\sigma}(e_k)$ is not available, we can use $w_k = C_k$, explicitly giving larger weights for high coverage cytosines.

A test statistic for a region found after segmentation is therefore calculated by $T_i^* = \frac{e_i^*}{\hat{\sigma}(e_i^*)}$, where $\hat{\sigma}(e_i^*)$ is the estimated variance of e_i^*. Based on our Gaussian distribution assumptions on the individual e_k, we call the region significant if $|T_i^*| > z_{\alpha/2}$, where α is the significance level.

In the standard meta-analysis, the individual T_k are often assumed to be independent so that the estimated variance of e_i^* is given by

$$\hat{\sigma}^2(e_i^*) = \frac{\sum_{k \in S_i} w_k^2 \hat{\sigma}^2(e_k)}{(\sum_{k \in S_i} w_k)^2}. \tag{2}$$

In the setting of methylation analysis, we have noted that the individual loci statistics are not independent. Even so, we show via simulation study (Sect. 3.1) that in using this estimate of $\hat{\sigma}^2(e_i^*)$, we still control the false discovery rate per region.

2.1   Generalizing Beyond z-statistics

The above approach relies on input statistics that are Gaussian. This can be limiting, since methods often produce other types of statistics, such as Fisher's exact test implemented by methylKit and log-likelihood ratio test from RADMeth. For this reason, we give a further adaptation in MethCP so as to be applicable for any cytosine-based parametric statistics that result in valid p-values. Let $p_1, p_2, \cdots, p_K$ be the p-values indexed by the location of cytosines.

For segmenting the genome into regions, we use the standard transform of the p-values to Z-scores,

$$z_k = \left[2\mathbb{1}(e_k \geq 0) - 1\right]\Phi^{-1}(1 - p_k/2), \tag{3}$$

where Φ is the cumulative distribution function of standard Gaussian. MethCP then performs CBS on the z_i's to segment the genome.

Region-level statistics can be obtained by aggregating p-values using Fisher's combined probability test [9] or Stouffer's weighted Z-method [20,23]. Namely, for a region i with a set of cytosines S_i, let

$$T_i^{\text{Fisher}} = -2 \sum_{k \in S_i} \log p_k, \tag{4}$$

$$T_i^{\text{Stouffer}} = \frac{\sum_{k \in S_i} w_k \Phi^{-1}(1 - p_k)}{\sqrt{\sum_{k \in S_i} w_k^2}}, \tag{5}$$

where w_k can be chosen to be constant or given by coverage C_k. We test T_i^{Fisher} against $\chi^2_{2|r|}$, and T_i^{Stouffer} against standard Gaussian for significance.

2.2   Quantifying Region Alignment

To quantify the performance of the methods or similarity of DMR sets detected by different methods, we need to define some measures for whether a region was successfully detected. One simple solution is just to calculate measures of specificity and sensitivity based on the percentage of individual cytosines were correctly called to be in a DMR or not. However, since our goal is to detect *regions*, this is unsatisfactory. Thus, we extend the specificity and sensitivity to region detection problem. Our framework of evaluation is closely related to supervised measures such as directional Hamming distance and segmentation covering in the image segmentation literature [12,18]. Specifically, to determine whether a detected region is considered a true or false positive, we set a parameter $\alpha \in (0, 1]$ that is the percentage of overlap required in order to be considered as having successfully detected a region. We then calculate true positive rates (TPR) and false positive rates (FPR) that vary with α.

Furthermore, we will see that some DMR methods are biased toward longer or shorter regions (Sect. 3), which can make comparing methods difficult. In order to account for different biases of regions found (in the following, we refer to number of cytosines in a region as the length of the region), we calculate the percent overlap between a detected region and a true region using three different denominators: that of the detected region, that of the true region and that of the union of the detected and true ones. The three measures can be interpreted as the local measures of precision, recall and Jaccard index. The first two allowed us to distinguish as to whether methods detected a high percentage of the true regions, versus if a high proportion of the detected regions were truly DMRs. The local Jaccard index allows us to measure the similarity between detected and true regions symmetrically.

We demonstrate our definitions using the local measure of precision – i.e., the overlap is determined by the proportion of the detected region that intersects the truth. Denote the detected and true region set as $\mathcal{R}^d$ and $\mathcal{S}^t$, respectively. To determine whether a detected region $R_i^d \in \mathcal{R}^d$ was a true positive (TP), we calculate the true positive indicator for R_i^d:

$$TP_i^d(\alpha) = I\left\{\max_{S_j^t \in \mathcal{S}^t} \frac{|R_i^d \cap S_j^t|}{|R_i^d|} \geq \alpha\right\}, \tag{6}$$

where $|R_i^d|$ is the length of the detected region R_i^d, and $\max_{S_j^t \in \mathcal{S}^t} |R_i^d \cap S_j^t|$ is the maximum overlapping cytosines of R_i^d with a true region. Note that we take the maximum over all true regions to account for the fact that a detected region may overlap multiple true regions (and vice versa). From the TP_i^d definitions, we calculate the total true positive (TP) as a function of α: $TP^d(\alpha) = \sum_{R_i^d \in \mathcal{R}^d} TP_i^d(\alpha)$.

The above formulas for TP can be extended to local measure of recall or Jaccard index by adjusting the denominator in Eq. (6), from $|R_i^d|$ to $|R_i^t|$ and $|R_i^d \cup S_j^t|$, respectively, for calculating overlap. Similarly, we can calculate the false positive (FP), false negative (FN) and true negatives (TN). Please refer to Appendix C.1 for detailed formulations.

3   Results

3.1   Simulation Study

In this section, we applied our method as well as five representative methods [19] BSmooth, HMM-Fisher, DSS, methylKit and metilene on simulated data with two population groups. Our method, MethCP, was run using the statistics of both DSS and methylKit as input (hereinafter referred to as MethCP-DSS and MethCP-methylKit). We evaluated their performances with the measures described in Sect. 2.2. The data simulation procedure and details of applying these methods can be found in Appendix B.

A Simulation for Comparing Two Population Groups. Figure 1 shows the summary of the length of the DMRs detected by the six methods using the default significance level (or test statistic thresholds); we also show the distribution of the lengths of true regions. MethCP and metilene gives the closest length distribution to that of the true regions. Although we shortened the smoothing window compared to the default, BSmooth and DSS detect much larger regions. In contrast, HMM-Fisher and methylKit both detect small, fragmented regions.

To evaluate the accuracy of the methods on the simulated data, we plot the ROC curve for both the local precision (Fig. 2a) and the local recall (Fig. 2b). The local precision requires that a large percent of the detected region overlap a true DMR (easier for shorter detected regions and conservative methods), while

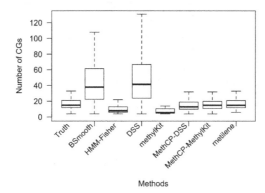

Fig. 1. Boxplot of number of CpGs in the DMRs. Number of CpGs in the true DMRs and in the DMRs detected by the seven methods compared for the simulated data.

the local recall requires that a large percent of the true region be overlapped by a detected region (easier for longer detected regions).

MethCP-DSS and MethCP-methylKit detect highly similar regions despite using different test statistics as input. For small α, MethCP and metilene achieve the highest true positive rate across all of the methods. And both TPR^d and TPR^t are close to 1, which suggests that most true and detected regions match in pairs, with slight disagreement in the region border. This is not the case for other four methods, which for a given FPR are usually strong in either TPR^d or TPR^t, but not both, which is evidence that either a proportion of the ground-truth regions are not detected (TPR^d high but TPR^t low), or a proportion of detected regions are not overlapping the ground-truth (TPR^t high but TPR^d low). DSS and BSmooth behave similarly in that TPR^t varies little with α while TPR^d decreases dramatically with the increase of α. This is an indication that both methods detect larger regions than the truth, which has been shown in Fig. 1. Despite detecting regions wider than the true regions, BSmooth misses at least 20% of the true regions, as indicated by the values of TPR^t, while DSS misses a smaller proportion.

HMM-Fisher and methylKit exhibit the opposite behavior, calling fewer and smaller regions significant, but as a result not obtaining good coverage of the true regions. The DMRs identified by these two methods are generally a subset of the true regions as indicated by the high values of TPR^d regardless of the significance level α. However, they also miss a good number of regions as shown in their lower TPR^t (Fig. 2b). metilene and MethCP both rely on segmentation procedures, and metilene achieves better performance than the other competing methods, with high levels of both FPR^d and FPR^t, though metilene still gives smaller TPR than MethCP.

However, in addition to assessing their overall sensitivity and specificity, we can consider whether the methods actually control the false positive rate at the desired level, and here we can see an even stronger difference between metilene and MethCP. In particular, for Fig. 2a and b, the sold portions of the

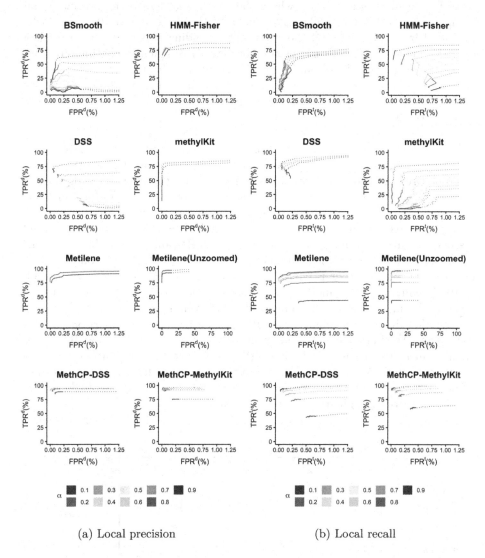

(a) Local precision   (b) Local recall

Fig. 2. ROC curve: method comparison on the simulated data. False positive rate (FPR) versus true positive rate (TPR) as we change the parameter α for (a) local precision measure and (b) local recall measure. The solid lines indicate significance levels smaller than 0.05 for HMM-Fisher, DSS, metilene, MethCP-DSS and MethCP-methylKit, statistics threshold larger than 4 for BSmooth (the author recommendation is 4.6), and q-values cutoffs less than 0.05 for methylKit. Thus, in real applications, we only focus on the regions of solid curves. For the completeness of the graph, we extend the curve to larger significance levels. methylKit uses an FDR correction procedure and only report q-values. Their false positive rate is reasonably controlled especially when we use the detected length as the denominator of our measure (FPR^d).

precision-recall curves indicate where the cutoff was due to the *p*-value cutoff being less than 0.05, meaning that control of the FPR at level 0.05 would be indicated by solid portion of the curve not extending beyond the true FPR of 0.05 given in the x-axis. For `metilene`, however, the solid curve continues beyond true FPR of 0.1 (see "Unzoomed" `metilene` plot), indicating that `metilene` is not accurately controlling the FPR. `MethCP`, on the other hand, does not have this property, and is quite conservative, despite having a higher TPR than the other methods.

A Simulation for Small Effect Size Regions. Recent studies [3,7,15] have shown that small-magnitude effect sizes are functionally important for DNA methylation and are associated with specific phenotypes. Furthermore, in plants, there exist other contexts of methylation other than CpG pairing (CHG and CHH methylation), where the baseline levels can be much lower, and hence their changes are also much smaller in scale, usually less than 10%. However, related discoveries have been hampered by a lack of DMR-calling tool addressing this issue [3,7]. We now show via a simulation study that `MethCP` is the only method of those we consider that is capable of accurately detecting regions with small (<10%) changes in DNA methylation. In fact, none of the DMR detection methods we consider here, other than `metilene`, even identify *any* regions with methylation differences smaller than 10%, so we focus our comparison only on `metilene`. We simulated data sets with DMRs that have 2.5%, 5%, 10%, 20% changes in methylation, and their lengths vary from 25 methylation cytosines to 400 cytosines. As shown in Fig. 3, `MethCP` achieves a high true positive rate while keeping the false positive rate low. `MethCP` consistently has higher TPR and lower FPR throughout the range of values.

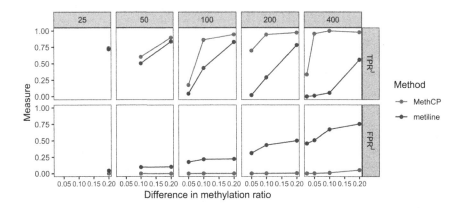

Fig. 3. A comparison between `metilene` and MethCP for the small effect size simulation. Upper panel shows TPR^J, and the lower panel shows FPR^J ($\alpha = 0.5$). On the x-axis are the simulated methylation differences between control and treatment group. The five columns show simulated DMR length from short DMRs (25 cytosines) to large DMRs (400 cytosines).

3.2 Arabidopsis Dataset

To further illustrate the performance of MethCP on real datasets, we consider their performance on actual WGBS dataset. We do not have WGBS with actual ground truth (i.e., knowledge of where DMR regions are and are not located), so we use a dataset with a relatively large number of replicates that allow us to use permutation methods to give some idea of the performance of the tests; we combine this with a consideration of whether the important features seen in our simulation results are mimicked in real data. We use a WGBS on *Arabidopsis Thaliana* ([4], GEO accession number GSE39045) which contained six biological replicates each of a wild-type control line and an H2A.Z mutant line.

Figure 4a shows the boxplot of the size of the DMRs detected (i.e., number of CpGs) under significance level 10^{-2} and FDR corrected level 10^{-2}. We see that the sizes of the regions are mostly fairly consistent across the different significance levels, and that the relative sizes between methods resemble that of the simulated data. Namely, DSS and BSmooth detect large regions, HMM-Fisher and methylKit identified small ones, and MethCP-DSS and MethCP-methylKit identified highly similar regions despite different test statistics they use. metilene, stands out as sensitive to the significance level, with smaller required significance levels resulting in more small regions being reported (and HMM-Fisher finds no DMRs at more stringent significance levels).

To compare the false positives produced by six methods, we create a null dataset, i.e., one with no region of cytosines different between the two groups, by randomly assigning the six controls into two groups of three. Individual differences between the mean methylation levels, combined with the fact that nearby sytosines have similar methylation levels in a sample, meaning that by chance there are regions of cytosines that are different between the two random groups. Therefore, we further permute the data in two different ways to create null data sets that have no DMRs between the two groups. The first method permutes the sequencing counts (methylated and unmethylated count pairs) across the samples for *each* CpG position (Permutation 1). The second method permutes the cytosine positions but keeps the sample labels, thus breaking any residual spatial signal between neighboring CpGs (Permutation 2). Figure 4 shows the number of DMRs and proportion of CpGs detected by each method. In both permutations, MethCP has a small number of DMRs relative to the other methods. Indeed for the second permutation, MethCP detects 0 DMRs. For the first permutation, only HMM-Fisher finds fewer, but HMM-Fisher also appears to have less power in detecting the real differences, Fig. 4b. If we further consider how well MethCP assigns individual cytosines correctly to DMRs (i.e. its ability to detect of DMCs), then MethCP also results in fewer false assignments of individual cytosines than the other methods, which is surprising since it is not a DMC method.

We would note that despite detecting fewer false positive DMRs in our permutation analysis, MethCP still remains competitive, as compared to the other methods, in terms of the number of DMRs it finds on the real data (Fig. 4b), indicating that it is not suffering from a lack of power. DSS and methylKit both find more regions, but we see from our permutation analysis that DSS and methylKit tend to find many more false positives than `MethCP`. Furthermore, the regions found by `methylKit` are small (Fig. 4a), suggesting that like in the simulations `methylKit` may be missing or fragmenting large parts of the true DMRs.

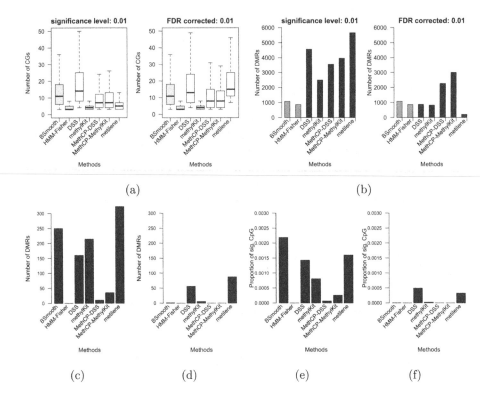

Fig. 4. (a, b) Summary of DMRs detected for Arabidopsis dataset. (a) Boxplot of the number of CpGs in the DMRs. (b) Number of DMRs detected by different methods. We summarize the results under significance level 10^{-2} and FDR corrected level 10^{-2}. `BSmooth` and `HMM-Fisher` are colored because we use the author recommended test statistic cutoff (4.6) for `BSmooth` and significance level 0.05 for `HMM-Fisher` on three plots (Small significance level for `HMM-Fisher` returns no DMR). (c,d, e, f) Permutation results for Arabidopsis dataset. (c) Permutation 1, number of DMRs detected. (d) Permutation 2, number of DMRs detected. (e) Permutation 1, the proportion of significant CpGs. (f) Permutation 2, the proportion of significant CpGs. Methods other than `BSmooth` uses significance level 0.01, and `BSmooth` uses the author recommended test statistic cutoff (4.6).

3.3   Time-Course Dataset

A strength of `MethCP` is that it is flexible for a wide variety of experimental designs, because it only requires an appropriate per-cytosine test-statistic, which can be calculated by standard tests on a per-cytosine basis. We demonstrate this utility on a seed germination dataset from *Arabidopsis thaliana* ([14], GEO accession number GSE94712). The data is generated from tracking over time germinating seeds after the dry seeds are given water. Two experimental conditions are considered: wild-type plants (*Col-0*) and *ros1*, *dml2*, and *dml3* triple demethylase mutants plants (*rdd*). ROS1, DML2 and DML3 are closely related DNA demethylation enzymes that mainly act in vegetative tissues. Two replicates were collected at each of 0–4 days after introduction of water (DAI), resulting in six time points, including the dry seed.

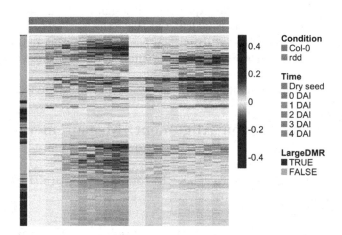

Fig. 5. Heatmap of DMRs for the seed germination dataset. Heatmap of average methylation levels in the DMRs detected. Here we show the results for CpG context. For each condition and each DMR, we subtract the ratios by the average dry seed methylation ratios, so that the heatmap better shows the changes with time. We annotate the DMRs with greater than 200 CpGs as large regions.

The original authors did not conduct a DMR analysis between these groups. Rather they did a two-group test of the overall methylation levels by grouping together the summarized methylation ratios and testing for the difference in distributions at each time point. They saw a modest overall increase between the two groups at each time point. We perform a DMR analysis with MethCP by fitting per CpG a linear model on the arcsine-transformed methylation ratios and choosing the statistic T_k to be the difference between the time coefficient of *Col-0* and *rdd* groups. Figure 6 shows the histogram of DMR length. Unlike the length distribution in Sect. 3.2, we detect some very large DMRs with small changes of methylation over time. Our ability to pick up both small and long regions gives us the ability to see multiple effects. We see the overall general increase, reported by the authors, represented by the long regions of small increase (Fig. 5, regions colored black). But we also detect some smaller regions with an opposite pattern of decrease in methylation in the mutant samples (Fig. 5, regions colored grey).

4   Conclusion

We proposed a method `MethCP` for identifying differentially methylated regions. We presented the results of `MethCP-methylKit` and `MethCP-DSS` on simulated and real datasets. And we showed that `MethCP` gives better accuracy and a lower number of false positives, as compared to existing methods. We also show that `MethCP` is the only method that detects large DMRs with small effect sizes, which is prevalent in DMR methylation data. Other than comparing two groups, we presented an example of a time-course study with `MethCP`. Our framework is in principle flexible for general experimental design assuming an appropriate single-cytosine test statistic can be calculated. Thus the method can be expanded immediately to more complicated situations, such as comparing multiple groups or measurements of methylation status over time or developmental progression.

Funding. This work has been supported in part by a DOE BER grant, DE-SC0014081.

A   Quantifying Region Alignment: Measurement Formulations

Similar to TP_i^d definitions, we calculate the total false positive (FP), and false negative (FN) as a function of α:

$$FP_i^d(\alpha) = I\left\{ \max_{S_j^t \in \mathcal{S}^t} \frac{|R_i^d \cap S_j^t|}{|R_i^d|} < \alpha \right\}, \qquad FP^d(\alpha) = \sum_{R_i^d \in \mathcal{R}^d} FP_i^d(\alpha), \qquad (7)$$

$$FN_j^d(\alpha) = I\left\{ \max_{R_i^d \in \mathcal{R}^d} \frac{|R_i^d \cap S_j^t|}{|R_i^d|} < \alpha \right\}, \qquad FN^d(\alpha) = \sum_{S_j^t \in \mathcal{S}^t} FN_j^d(\alpha), \qquad (8)$$

where the false negative is interpreted as the number of true regions that do have overlap greater than α with any detected positives.

The above formula can be extended to use local measure of recall or Jaccard index by adjusting the denominator in Eq. 6, from $|R_i^d|$ to $|R_i^t|$ and $|R_i^d \cup S_j^t|$, respectively, for calculating overlap.

For example, measuring the number of true positives, we have:

$$TP_j^t(\alpha) = I\left\{ \max_{R_i^d \in \mathcal{R}^d} \frac{|R_i^d \cap S_j^t|}{|S_j^t|} \geq \alpha \right\}, \qquad TP^t(\alpha) = \sum_{S_j^t \in \mathcal{S}^t} TP_j^t(\alpha), \qquad (9)$$

$$TP_i^J(\alpha) = I\left\{ \max_{S_j^t \in \mathcal{S}^t} \frac{|R_i^d \cap S_j^t|}{|R_i^d \cup S_j^t|} \geq \alpha \right\}, \qquad TP^J(\alpha) = \sum_{R_i^d \in \mathcal{R}^d} TP_i^J(\alpha). \qquad (10)$$

Calculating True Negatives. Calculating the total number of true negatives would require a calculation of the number of detected regions that were truly not significant (i.e. equally methylated). However, an equally methylated region is a more nebulous quantity for a region (unlike for cytosines). Unlike the different DMRs, all equally methylated regions are equivalent from the point of view of all of these methods: arbitrarily defining separate regions within a large block of equally methylated regions could not be detected by any method. Instead, we use the following formula to get an estimation of the total number of true negatives:

$$\frac{\# \text{ CpGs in Total} - \# \text{ CpGs in (TP + FP + FN)}}{\text{Average True DMR Length}} \tag{11}$$

B   Simulation Study

How We Applied Other Methods. We applied our method as well as five representative methods [19] BSmooth, HMM-Fisher, DSS, methylKit and metilene on both real and simulated data. Our method, MethCP, was run using the statistics of both DSS and methylKit as input. To be fair between methods, we remove the coverage filter for individual cytosines, as it varies by methods. Furthermore, the reads in symmetric CpG sites are collapsed. We set the same length filter (3 cytosines) and absolute mean methylation level difference filter (0.1) for DMRs, where the numbers are the default of the majority of methods. We shorten the smoothing window of BSmooth from default 1000 bps to 500 bps, which gives better results for our simulated dataset. For DSS, we use moving average smoothing, which is recommended in the documentation. For methylKit, the output is DMCs rather than DMRs. We combine adjacent DMCs to DMRs. Resulting DMRs that are smaller than 3 cytosines are discarded. All other parameters other than the significance level (test statistics cutoffs) were left at the default values.

Generation of Simulated Data. We generate simulated BS-Seq data by the following procedure adapted from [25]. We assume there are K cytosines in the simulated genome and two groups of samples ("treatment" and "control"), each of size $n = 3$ to compare; this is similar to the level of replication that is often seen in WGBS. We designate regions within this genome to be classified as DMR by generating region size (number of CpGs) from a negative binomial distribution $NB(r = 6, p = 0.25)$. We further require that the number of CpGs to be greater than 3 in each region. The starting positions of the DMR were chosen by random sampling. This divided the genome into differentially methylated and equally methylated regions.

To mimic the read coverage and the methylation ratio in real datasets, the actual sequencing counts were generated based on a human senescent cells dataset [5] available from the Gene Expression Omnibus (GEO) with accession number GSE48580. To determine the total read coverage, we randomly sampled from the observed coverage distribution of each cytosine in the human dataset. The number of reads determined to be methylated per cytosine was based on a binomial distribution, with the probability of proportion depending on what treatment group the sample was in and if the cytosine was in a DMR or not. For samples in the control group or for cytosines in the equally methylated regions, the binomial probability parameter was chosen from the observed distribution of the per-cytosine average methylation ratio in the senescent cells dataset. For DM regions, each DM region was randomly assigned one of five beta distributions from which the methylation probability of the treatment samples would follow; in addition, we require that the absolute difference between the mean of binomial distribution in treatment and in the corresponding control group is at least 0.2, which eliminated some of the five beta distributions from consideration. These beta distributions represent five different methylation levels, from poorly methylated to highly methylated (specific parameters of the beta distribution are given in Table 1). Then the cytosine methylation probability for samples in the treatment group was generated according to the beta distribution chosen for that DMR region. To take into account the high correlation of methylation levels between neighboring CpGs, we simulate smoothing DMR boundaries. For a DMR of length l, a region of length $w \sim \text{Unif}(0.1l, 0.3l)$ is added to each side of the DMR where the methylation probability is given by a mixture of treatment and control. The weights of the treatment group decrease as we move to the edge of the DMR. In this paper, we show only results with the smoothing boundary, but simulation without smoothing boundaries give similar results.

Table 1. Parameters of beta distributions for simulating the methylated counts in the treatment group.

Distribution	(a)	(b)	(c)	(d)	(e)
α	2	6	10	14	18
β	18	14	10	6	2
Mean probability	0.1	0.3	0.5	0.7	0.9

C   Figures

C.1   DMR Lengths in the Time-Course Dataset

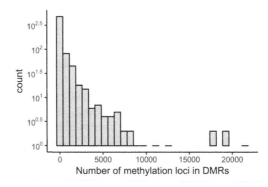

Fig. 6. Lengths (number of methylation cytosine) of DMRs detected in the seed germination dataset (CpG context).

References

1. Akalin, A., et al.: MethylKit: a comprehensive R package for the analysis of genome-wide DNA methylation profiles. Genome Biol. **13**(10), R87 (2012)
2. Borenstein, M., Hedges, L.V., Higgins, J., Rothstein, H.R.: Introduction to Meta-Analysis. Wiley, Hoboken (2009)
3. Breton, C.V., et al.: Small-magnitude effect sizes in epigenetic end points are important in children's environmental health studies: the children's environmental health and disease prevention research centers epigenetics working group. Environ. Health Perspect. **125**(4), 511 (2017)
4. Coleman-Derr, D., Zilberman, D.: Deposition of histone variant H2a. Z within gene bodies regulates responsive genes. PLoS Genet. **8**(10), e1002988 (2012)
5. Cruickshanks, H.A., et al.: Senescent cells harbour features of the cancer epigenome. Nat. Cell Biol. **15**(12), 1495 (2013)
6. Dolzhenko, E., Smith, A.D.: Using beta-binomial regression for high-precision differential methylation analysis in multifactor whole-genome bisulfite sequencing experiments. BMC Bioinform. **15**(1), 215 (2014)
7. Eichten, S.R., Springer, N.M.: Minimal evidence for consistent changes in maize DNA methylation patterns following environmental stress. Front. Plant Sci. **6**, 308 (2015)
8. Feng, H., Conneely, K.N., Wu, H.: A Bayesian hierarchical model to detect differentially methylated loci from single nucleotide resolution sequencing data. Nucleic Acids Res. **42**(8), e69 (2014)
9. Fisher, R.A.: Statistical methods for research workers (1934)
10. Hansen, K.D., Langmead, B., Irizarry, R.A.: Bsmooth: from whole genome bisulfite sequencing reads to differentially methylated regions. Genome Biol. **13**(10), R83 (2012)
11. Hebestreit, K., Dugas, M., Klein, H.-U.: Detection of significantly differentially methylated regions in targeted bisulfite sequencing data. Bioinformatics **29**(13), 1647–1653 (2013)
12. Huang, Q., Dom, B.: Quantitative methods of evaluating image segmentation. In: Proceedings of International Conference on Image Processing, vol. 3, pp. 53–56. IEEE (1995)
13. Jühling, F., Kretzmer, H., Bernhart, S.H., Otto, C., Stadler, P.F., Hoffmann, S.: metilene: fast and sensitive calling of differentially methylated regions from bisulfite sequencing data. Genome Res. **26**(2), 256–262 (2016)
14. Kawakatsu, T., Nery, J.R., Castanon, R., Ecker, J.R.: Dynamic DNA methylation reconfiguration during seed development and germination. Genome Biol. **18**(1), 171 (2017)
15. Leenen, F.A.D., Muller, C.P., Turner, J.D.: DNA methylation: conducting the orchestra from exposure to phenotype? Clin. Epigenetics **8**(1), 92 (2016)
16. Olshen, A.B., Venkatraman, E.S., Lucito, R., Wigler, M.: Circular binary segmentation for the analysis of array-based DNA copy number data. Biostatistics **5**(4), 557–572 (2004)
17. Park, Y., Hao, W.: Differential methylation analysis for BS-seq data under general experimental design. Bioinformatics **32**(10), 1446–1453 (2016)
18. Pont-Tuset, J., Marques, F.: Supervised evaluation of image segmentation and object proposal techniques. IEEE Trans. Pattern Anal. Mach. Intell. **38**(7), 1465–1478 (2016)

19. Shafi, A., Mitrea, C., Nguyen, T., Draghici, S.: A survey of the approaches for identifying differential methylation using bisulfite sequencing data. Briefings Bioinform. **19**, 737–753 (2017)

20. Stouffer, S.A., Suchman, E.A., DeVinney, L.C., Star, S.A., Williams Jr., R.M.: The American Soldier: Adjustment During Army Life. (Studies in Social Psychology in World War II), vol. 1 (1949)

21. Sun, S., Yu, X.: HMM-Fisher: identifying differential methylation using a hidden Markov model and fisher's exact test. Stat. Appl. Genet. Mol. Biol. **15**(1), 55–67 (2016)

22. Teschendorff, A.E., Relton, C.L.: Statistical and integrative system-level analysis of DNA methylation data. Nat. Rev. Genet. **19**(3), 129 (2018)

23. Whitlock, M.C.: Combining probability from independent tests: the weighted Z-method is superior to fisher's approach. J. Evol. Biol. **18**(5), 1368–1373 (2005)

24. Wu, H., et al.: Detection of differentially methylated regions from whole-genome bisulfite sequencing data without replicates. Nucleic Acids Res. **43**(21), e141 (2015)

25. Xiaoqing, Y., Sun, S.: Comparing five statistical methods of differential methylation identification using bisulfite sequencing data. Stat. Appl. Genet. Mol. Biol. **15**(2), 173–191 (2016)

26. Xiaoqing, Y., Sun, S.: HMM-DM: identifying differentially methylated regions using a hidden Markov model. Stat. Appl. Genet. Mol. Biol. **15**(1), 69–81 (2016)

On the Complexity of Sequence to Graph Alignment

Chirag Jain, Haowen Zhang, Yu Gao, and Srinivas Aluru[⊠]

College of Computing, Georgia Institute of Technology, Atlanta, USA
aluru@cc.gatech.edu

Abstract. Availability of extensive genetics data across multiple individuals and populations is driving the growing importance of graph based reference representations. Aligning sequences to graphs is a fundamental operation on several types of sequence graphs (variation graphs, assembly graphs, pan-genomes, etc.) and their biological applications. Though research on sequence to graph alignments is nascent, it can draw from related work on pattern matching in hypertext. In this paper, we study sequence to graph alignment problems under Hamming and edit distance models, and linear and affine gap penalty functions, for multiple variants of the problem that allow changes in query alone, graph alone, or in both. We prove that when changes are permitted in graphs either standalone or in conjunction with changes in the query, the sequence to graph alignment problem is $\mathcal{NP}$-complete under both Hamming and edit distance models for alphabets of size ≥ 2. For the case where only changes to the sequence are permitted, we present an $O(|V| + m|E|)$ time algorithm, where m denotes the query size, and V and E denote the vertex and edge sets of the graph, respectively. Our result is generalizable to both linear and affine gap penalty functions, and improves upon the run-time complexity of existing algorithms.

1 Introduction

Aligning sequences to graphs is becoming increasingly important in the context of several applications in computational biology, including variant calling [6,7,9,22], genome assembly [2,8,33], long read error-correction [28,32,34], RNA-seq data analysis [4,13], and more recently, antimicrobial resistance profiling [27]. Much of this has been driven by the growing ease and ubiquity of sequencing at personal, population, and environmental-scale, leading to significant growth in availability of datasets. Graph based representations provide a natural mechanism for compact representation of related sequences and variations among them. Some of the most useful graph based data structures are de-Bruijn graphs [25], variation graphs [23], string graphs [19], and partial order graphs [14].

C. Jain and H. Zhang—Should be regarded as joint first-authors.

© Springer Nature Switzerland AG 2019
L. J. Cowen (Ed.): RECOMB 2019, LNBI 11467, pp. 85–100, 2019.
https://doi.org/10.1007/978-3-030-17083-7_6

Decades of progress made towards designing provably good algorithms for the classic sequence to sequence alignment problems serves as the foundation for mapping tools currently used in genomics, and similar efforts are necessary for sequence to graph alignment. To address the growing list of biological applications that require aligning sequences to a graph, several heuristics [9,11,12,15,16] and a few provably good algorithms [26,29,31] have been developed in recent years. In addition, sequence to graph alignment has been studied much earlier in the string literature through its counterpart, approximate pattern matching to hypertext [17]. Since then, important complexity results and algorithms have been obtained for different variants of this problem [1,20,30].

Many versions of the classic sequence to sequence alignment problem were considered in the literature, e.g., different alignment modes – local/global, scoring functions – linear/affine/arbitrary gap penalty, and so on [21]. The list further proliferates when considering a graph-based reference. This is because the nature of the problem changes depending on whether the input graphs are cyclic or acyclic [20], and whether edits are allowed in the graph, or query, or both [1].

In this paper, we present new complexity results and improved algorithms for multiple variants of the sequence to graph alignment problem. Consider a query sequence of length m and a directed graph $G(V, E)$ with string-labeled vertices, over the alphabet Σ. We make the following contributions:

- The problem variants that allow changes to the graph labels are known to be $\mathcal{NP}$-complete [1], via proofs that assume $|\Sigma| \geq |V|$. To date, tractability of these problems remains unknown for the case of constant sized alphabets, which is an important consideration when aligning DNA, RNA, or protein sequences to corresponding graphs. We close this knowledge gap by proving that four variants of the problem, characterized by changes to graph alone or both graph and query, under the Hamming or edit distance models, remain $\mathcal{NP}$-complete for $|\Sigma| \geq 2$.
- Allowing changes to the query sequence alone makes the problem polynomially solvable. For graphs with character-labeled vertices, we propose an algorithm that achieves $O(|V| + m|E|)$ time bound for both linear and affine gap

Table 1. Comparison of run-time complexity achieved by different algorithms for the sequence to graph alignment problem when changes are allowed in the query sequence alone.

	Linear gap penalty		Affine gap penalty
	Edit distance	Arbitrary costs	
Amir et al. [1]	$O(m(|V| \log |V| + |E|))$	$O(m(|V| \log |V| + |E|))$	-
Navarro [20]	$O(m(|V| + |E|))$	-	-
HybridSpades [2]	$O(m(|V| \log(m|V|) + |E|))$	$O(m(|V| \log(m|V|) + |E|))$	-
V-ALIGN [31]	$O(m|V||E|)$	$O(m|V||E|)$	$O(m|V||E|)$
Rautiainen and Marschall [26]	$O(|V| + m|E|)$	$O(m(|V| \log |V| + |E|))$	$O(m(|V| \log |V| + |E|))$
This work	$O(|V| + m|E|)$	$O(|V| + m|E|)$	$O(|V| + m|E|)$

penalty cases, superior to the best existing algorithms (Table 1). An important attribute of the proposed algorithm is that it achieves the same time and space complexity as required for the easier problem of sequence alignment to acyclic graphs [17, 20], under both scoring models.

2 Preliminaries

Let Σ denote an alphabet, and x and y be two strings over Σ. We use $x[i]$ to denote the ith character of x, and $|x|$ to denote its length. Let $x[i, j]$ ($1 \leq i \leq j \leq |x|$) denote $x[i]x[i+1]\ldots x[j]$, the substring of x beginning at the ith position and ending at the jth position. Concatenation of x and y is denoted as xy. Let x^k denote string x concatenated with itself k times.

Definition 1. *Sequence Graph: A sequence graph $G(V, E, \sigma)$ is a directed graph with vertices V and edges E. Function $\sigma : V \rightarrow \Sigma^+$ labels each vertex $v \in V$ with string $\sigma(v)$ over the alphabet Σ.*

Naturally, path $p = v_i, v_{i+1}, \ldots, v_j$ in $G(V, E, \sigma)$ spells the sequence $\sigma(v_i)$ $\sigma(v_{i+1})\ldots\sigma(v_j)$. Given a query sequence q, we seek its best matching path sequence in the graph. Alignment problems are formulated such that distance between the computed path and the query sequence is minimized, subject to a specified distance metric such as Hamming or edit distance. Typically, an alignment is scored using either a linear or an affine gap penalty function. The cost of a gap is proportional to its length, when using a linear gap penalty function. An affine gap penalty function imposes an additional constant cost to initiate a gap.

3 Complexity Analysis

3.1 Asymmetry of Edit Locations

An alignment between two sequences also specifies possible changes to the sequences (e.g. substitutions, insertions, deletions) to make them identical, with alignment distance specifying the cumulative penalty for the changes. The changes can be individually applied either to the first or the second sequence, or any combination thereof. Such a symmetry is no longer valid when aligning sequences to graphs [1]. This is because alignments can occur along cyclic paths in the graph. If the label of a vertex in the graph is changed, then an alignment path visiting that vertex k times reflects the same change at k different positions in the alignment. On the other hand, a change in one position of the sequence only reflects that change in the corresponding position in the alignment. As such, optimal alignment scores vary depending on whether changes are permitted in just the sequence, just the graph, or both (see Fig. 1 for an illustration). This characteristic leads to *three different problems*, with each potentially resulting in a different optimal distance.

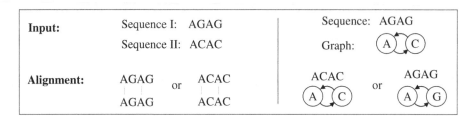

Input:	Sequence I: AGAG	Sequence: AGAG
	Sequence II: ACAC	Graph:
Alignment:	AGAG or ACAC	ACAC or AGAG
	AGAG ACAC	

Fig. 1. Asymmetry w.r.t. the location of changes in sequence to graph alignment illustrated using Hamming distance. Two substitutions are required in the sequence, whereas just one is sufficient if made in the graph.

Consider the sequence to graph alignment problem under the Hamming or edit distance metrics. For each distance metric, there are three versions of the problem depending on whether changes are allowed in query alone, graph alone, or both in the query and graph. Consider the decision versions of these problems, which ask whether there exists an alignment with $\leq d$ modifications (substitutions or edits), as per the distance metric. Restricting substitutions or edits to the query sequence alone admits polynomial time solutions [1, 20, 26]. In the pioneering work of Amir *et al.* [1] in the domain of string to hypertext matching, it has been proved that the other problem variants which permit changes to graph are $\mathcal{NP}$-complete. The proofs provided in their work assume an alphabet size $\geq |V|$. To date, tractability of these problems remains unknown for the case of constant sized alphabets (e.g., for DNA, RNA, or protein sequences). In what follows, we close this knowledge gap by showing that the problems remain $\mathcal{NP}$-complete for any alphabet of size at least 2.

3.2   Alignment Using Hamming Distance

Theorem 1. *The problem "Can we substitute a total of $\leq d$ characters in graph G and query q such that q will have a matching path in G?" is $\mathcal{NP}$-complete for $|\Sigma| \geq 2$.*

Proof. The problem is in $\mathcal{NP}$. Given a solution, the set of substitutions can be used to obtain the corrected graph and query. Next, we can leverage any polynomial time algorithm [1, 20, 24] to verify if the corrected query matches a path in the corrected graph.

To show that the problem is $\mathcal{NP}$-hard, we perform a reduction using the directed Hamiltonian cycle problem. Suppose $G'(V, E)$ is a directed graph in which we seek a Hamiltonian cycle. Let $n = |V|$. We transform it into a sequence graph $G(V, E, \sigma)$ over the alphabet $\Sigma = \{\alpha, \beta\}$ by simply labeling each vertex $v \in V$ with α^n (Fig. 2). Note that the graph structure remains unchanged. Next, we construct query sequence q. Let token t_i be the sequence of n characters $\alpha^{n-i-1}\beta\alpha^i$. We choose query q to be the $n^2(2n+2)$ long sequence: $(t_0 t_1 \dots t_{n-1})^{2n+2}$. We claim that a Hamiltonian cycle exists in $G'(V, E)$ if and only if q can be matched after substituting a total of $\leq n$ characters in $G(V, E, \sigma)$ and q.

Suppose there is a Hamiltonian cycle in $G'(V, E)$. We can follow the corresponding loop in $G(V, E, \sigma)$ from the first character of any vertex label. To match each token in the query q, we require one $\alpha \to \beta$ substitution per vertex. Thus, the query q matches $G(V, E, \sigma)$ after making exactly n substitutions in the graph.

Conversely, suppose the query q matches the graph $G(V, E, \sigma)$ after making $\leq n$ substitutions in the query and the graph. Consider the following substring q_{sub} of q: $t_0 t_1 \ldots t_{n-1} t_0 t_1$. Note that there are $n + 1$ non-overlapping instances of q_{sub} in q. Even if all the n substitutions occur in the query, at least one instance of q_{sub} must remain unchanged. As a result, q_{sub} must match to a path in the corrected $G(V, E, \sigma)$.

Case 1: q_{sub} starts matching from the first character of a vertex label. Note that the first n tokens $q_{sub}[1, n] = t_0$, $q_{sub}[n+1, 2n] = t_1, \ldots, q_{sub}[n^2 - n + 1, n^2] = t_{n-1}$ are all unique followed by $q_{sub}[n^2 + 1, n^2 + n] = t_0$. Therefore, this requires a Hamiltonian cycle in $G(V, E, \sigma)$. Accordingly, there is a Hamiltonian cycle in $G'(V, E)$.

Case 2: q_{sub} starts somewhere other than the starting position within a vertex label. Let $q_{sub}[k]$ $(1 < k \leq n)$ be the first character that matches at the beginning of the next vertex on the path matching q. Similar to the previous case, the following n sequences $q_{sub}[k, n + k - 1]$, $q_{sub}[n + k, 2n + k - 1], \ldots, q_{sub}[n^2 - n + k, n^2 + k - 1]$ are unique due to the spacing between β characters in q_{sub}. Therefore, the matching path must yield a Hamiltonian cycle.

Corollary 1. *The problem "Can we substitute $\leq d$ characters in graph G such that q will have a matching path in G?" is $\mathcal{NP}$-complete for $|\Sigma| \geq 2$.*

Proof. The setup used in the proof of Theorem 1 can be trivially extended to prove the above claim. Alternatively, we can simplify the proof by using the query sequence $q = (t_0 t_1 \ldots t_{n-1})^2$ since only one instance of the substring q_{sub} in q is needed for the subsequent arguments. This is because substitutions in the query sequence are not permitted.

Using the above two results, we conclude that Hamming-distance based decision formulations of sequence to graph alignment problems are $\mathcal{NP}$-complete when substitutions are allowed in graph labels, for $|\Sigma| \geq 2$. In fact, it can be easily shown that $|\Sigma| \geq 2$ reflects a tight bound. Using $|\Sigma| = 1$, all the problem instances can be decided in polynomial time using straightforward application of standard graph algorithms.

3.3   Alignment Using Edit Distance

We next show that edit distance based decision problems that permit changes in graph labels are $\mathcal{NP}$-complete if $|\Sigma| \geq 2$. Similar to our previous claims, allowing edits in the graph makes the sequence to graph alignment problem intractable. Proofs used for Hamming distance do not apply here as edits also permit insertions and deletions. Length of vertex labels can grow or shrink using insertion and deletion edits respectively.

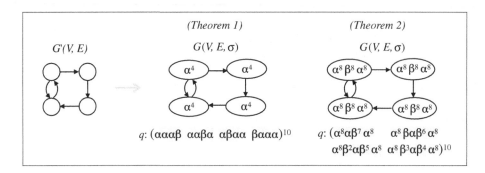

Fig. 2. The constructs used for reductions in proofs of Theorems 1 and 2.

Theorem 2. *The problem "Can we perform a total of $\leq d$ edits in graph G and query q so that q will match in G?" is $\mathcal{NP}$-complete for $|\Sigma| \geq 2$.*

Proof. Clearly the problem is in $\mathcal{NP}$. We again use the directed Hamiltonian cycle problem for reduction. Given an instance $G'(V, E)$ of the directed Hamiltonian cycle problem, we design an instance $G(V, E, \sigma)$ using $\Sigma = \{\alpha, \beta\}$. Let $n = |V|$. Label each vertex v in V using a sequence of $6n$ characters $\alpha^{2n}\beta^{2n}\alpha^{2n}$ (Fig. 2). Let token t_i be a sequence of length $6n$: $\alpha^{2n} \beta^i \alpha \beta^{2n-1-i} \alpha^{2n}$. Using such tokens, we build a query sequence q of length $6n^2(2n + 2)$ as $(t_0 t_1 \ldots t_{n-1})^{2n+2}$. We claim that a Hamiltonian cycle exists in $G'(V, E)$ if and only if we can match the sequence q to the graph $G(V, E, \sigma)$ using $\leq n$ total edits.

If there is a Hamiltonian cycle in $G'(V, E)$, we can follow the same loop in $G(V, E, \sigma)$ to align q. The alignment requires one substitution per vertex. To prove the converse, suppose query q matches graph $G(V, E, \sigma)$ after making a total of $\leq n$ edits in q and $G(V, E, \sigma)$. Consider the substring q_{sub} of q: $t_0 t_1 \ldots t_{n-1} t_0$. Note that there are $n + 1$ non-overlapping instances of q_{sub} in q, at least one of which must remain unchanged. Accordingly, the substring q_{sub} must match corrected $G(V, E, \sigma)$.

For the token t_i, let $k_i = \beta^i \alpha \beta^{2n-1-i}$ be its *kernel* sequence of length $2n$. It follows that $t_i = \alpha^{2n} k_i \alpha^{2n}$. We show that a kernel must be matched entirely within a vertex in $G(V, E, \sigma)$ using the following two arguments. First, since any vertex label cannot shrink from length $6n$ to $< 5n$, a kernel cannot be matched to an entire vertex after the edits. It implies that a kernel must match to ≤ 2 vertices. Second, if a kernel aligns across two vertices, $(2n - 1)$ β's must be required in place of α's at the two vertex ends, thus requiring $> n$ edits. Therefore, a kernel can only be matched within a single vertex label. Finally, it is easy to observe that any vertex label after $\leq n$ edits cannot be matched to more than one kernel. When combining these arguments with the fact that all n consecutive kernels in q_{sub} are unique, we establish that the alignment path of q_{sub} must follow a Hamiltonian cycle in $G(V, E, \sigma)$. Accordingly, there is a Hamiltonian cycle in $G'(V, E)$.

Corollary 2. *The problem "Can we perform $\leq d$ edits in graph G so that q will match in G?" is $\mathcal{NP}$-complete for $|\Sigma| \geq 2$.*

Proof. The setup used to prove Theorem 2 can be trivially extended to prove the above claim.

It is straightforward to prove that other problem variants, e.g., with linear gap penalty or affine gap penalty scoring functions are at least as hard as the edit-distance based formulations. Therefore, the sequence to graph alignment problem remains $\mathcal{NP}$-complete even on constant sized alphabets for these classes of scoring functions also if changes are permitted in the graph.

4   Sequence-to-Graph Alignment with Edits in Sequence

The sequence to graph alignment problem is polynomially solvable when changes are allowed on the query sequence alone [1,20]. Here, we improve upon the state-of-the-art by presenting an algorithm with $O(|V| + m|E|)$ run-time. Our algorithm matches the run-time complexity achieved previously by Rautiainen and Marschall [26] for edit distance, while improving that for linear and affine gap penalty functions. In addition, it is simpler to implement because it only uses elementary queue data structures. Note that edit distance is a special case of linear gap penalty when cost per unit length of the gap is 1, and substitution penalty is also 1. We first present our algorithm for the case of a linear gap penalty function, and subsequently show its generalization to affine gap penalty. From hereon, we assume that the sequence graph $G(V, E, \sigma)$ is a character labeled graph, i.e., $\sigma(v) \in \Sigma, v \in V$. This assumption simplifies the description of the algorithm. Note that it is straightforward to transform a graph from string-labeled form to character-labeled form, and vice versa.

4.1   Linear Gap Penalty

Alignment Graph. In the literature on the classic sequence to sequence alignment problem, the problem is either formulated as a dynamic programming problem or an equivalent graph shortest-path problem in an appropriately constructed edge-weighted *edit graph* or *alignment graph* [18]. However, formulating the sequence to graph alignment problem as a dynamic programming recursion, while easy for directed acyclic graphs through the use of topological ordering, is difficult for general graphs due to the possibility of cycles. As it turns out, formulation as a shortest-path problem in an alignment graph is still rather convenient, even for graphs with cycles [1,26]. The alignment graph, described below, is constructed using the given query sequence, the sequence graph and the scoring parameters.

The alignment graph is a weighted directed graph which is constructed such that each valid alignment of the query sequence to the sequence graph corresponds to a path from source vertex s to sink vertex t in the alignment graph, and vice versa (Fig. 3). The alignment cost is equal to the corresponding path

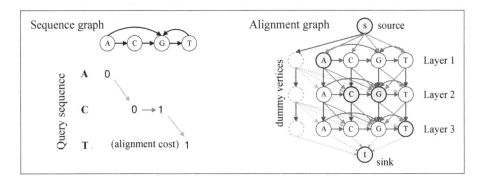

Fig. 3. An example to illustrate the construction of an alignment graph from a given sequence graph and a query sequence. Multiple colors are used to show weighted edges of different categories in the alignment graph. The red, blue and green edges are weighted as insertion, deletion and substitution costs respectively. (Color figure online)

distance from the source to the sink. Note that the alignment graph is a multi-layer graph containing m 'copies' of the sequence graph, one in each layer. A column of dummy vertices is required in addition to accommodate the possibility of deleting a prefix of the query sequence. Edges that emanate from a vertex are equivalent to the choices available while solving the alignment problem. A formal definition of the alignment graph follows:

Definition 2. *Alignment graph: Given a query sequence q, a sequence graph $G(V, E, \sigma)$, linear gap penalty parameters $\Delta_{del}, \Delta_{ins}$, and a substitution cost parameter Δ_{sub}, the corresponding alignment graph is a weighted directed graph $G_a(V_a, E_a, \omega_a)$, where $V_a = (\{1, \ldots, m\} \times (V \cup \{\delta\})) \cup \{s, t\}$ is the vertex set, and $\omega_a : E_a \to \mathbb{R}_{\geq 0}$ is the weight function defined as*

$$
\omega_a(x, y) = \begin{cases}
\Delta_{i,v} & x = (i-1, u), y = (i, v) & 1 < i \leq m \ \& \ (u, v) \in E \\
\Delta_{ins} & x = (i, u), y = (i, v) & 1 \leq i \leq m \ \& \ (u, v) \in E \\
\Delta_{del} & x = (i-1, v), y = (i, v) & 1 < i \leq m \ \& \ (v, v) \notin E \\
\min(\Delta_{del}, \Delta_{i,v}) & x = (i-1, v), y = (i, v) & 1 < i \leq m \ \& \ (v, v) \in E \\
\textit{for source and sink vertices:} \\
\Delta_{1,v} & x = s, y = (1, v) & v \in V \\
\Delta_{del} & x = s, y = (1, \delta) \\
0 & x = (m, v), y = t & v \in V \cup \{\delta\} \\
\textit{for dummy vertices:} \\
\Delta_{del} & x = (i-1, \delta), y = (i, \delta) & 1 < i \leq m \\
\Delta_{i,v} & x = (i-1, \delta), y = (i, v) & 1 < i \leq m \ \& \ v \in V
\end{cases}
$$

Edges $(x, y) \in E_a$ are defined implicitly, as those pairs (x, y) for which ω_a is defined above. $\Delta_{i,v} = \Delta_{sub}$ if $q[i] \neq \sigma(v), v \in V$, and 0 otherwise. Δ_{sub} denotes the cost of substituting $q[i]$ with $\sigma(v)$.

Existing definitions of the alignment graph [1, 26] did not include the dummy vertices, and were incomplete. Using the alignment graph, we reformulate the problem of computing an optimal alignment to finding the shortest path in the alignment graph. Even though the alignment graph defined by Amir *et al.* [1] has minor differences, proof in their work can be easily adapted to state the following claim:

Lemma 1 (Amir *et al.* [1]). *Shortest distance from the source vertex s to the sink vertex t in the alignment graph $G_a(V_a, E_a, \omega_a)$ equals cost of optimal alignment between the query q and the sequence graph $G(V, E, \sigma)$.*

One way of solving the above shortest path problem is to directly apply Dijkstra's algorithm [1, 2]. However, it results in an $O\big(m|V|\log(m|V|) + m|E|\big)$ time algorithm. We next show how to solve the problem in $O(|V| + m|E|)$ time.

Proposed Algorithm. While searching for a shortest path from the source to the sink vertex, we compute the shortest distances from the source to intermediate vertices $V_a \backslash \{s, t\}$ in the alignment graph. An edge from a vertex in layer i is either directed to a vertex in the same layer or to a vertex in the next layer. As a result, the shortest distances to nodes in a layer can be computed once the distances for the previous layer are known. This also makes it feasible to solve for the layers 1 to m, one by one [20]. We use a two-stage strategy to achieve linear $O(|V| + |E|)$ run-time per layer. Before describing the details, we give an outline of the algorithm and its two stages.

Any path from the source vertex to a vertex v in a layer must extend a path ending in the previous layer using either a deletion or a substitution cost weighted edge. Afterwards, a path that ends in the same layer but not at v can be further extended to v using the insertion cost weighted edges if it results in the shortest path to the source. Roughly speaking, the first stage executes the former task, while the second takes care of the latter. The two stages together are invoked m times during the algorithm until the optimal distances are known for the last layer (Algorithm 1). Input to the first stage *InitializeDistance* is an array of the shortest distances of the vertices in previous layer sorted in non-decreasing order. This stage computes the 'tentative' distances of all vertices in the current layer because it ignores the insertion cost weighted edges during the computation. It outputs the sorted tentative distances as an input to the second stage *PropagateInsertion*. The PropagateInsertion stage returns the optimal distances of all vertices in the current layer while maintaining the sorted order for a subsequent iteration.

The following are two important aspects of our algorithm. First, we are able to maintain the sorted order of vertices by spending $O(|V|)$ time per layer during the first stage (Lemma 2). Secondly, we propagate insertion costs through the edges in $O(|V| + |E|)$ time per layer during the second stage by eluding the need for standard priority queue implementations (Lemmas 3–5). Both of these features exploit characteristics specific to the alignment graphs.

Algorithm 1. Algorithm for sequence to graph alignment

Result: The length of shortest path from s to t

1 $PreviousLayer = [s]$;
2 $s.distance = 0$;
3 **for** $i = 1$ **to** m **do**   /* Do the computation layer by layer */
4 | $CurrentLayer = [(i, v_1), (i, v_2), \ldots, (i, v_n), (i, k)]$;
5 | $x.distance = \infty \; \forall x \in CurrentLayer$;
6 | InitializeDistance($PreviousLayer, CurrentLayer$);
7 | PropagateInsertion($CurrentLayer$);
8 | $PreviousLayer = CurrentLayer$;
9 **return** Min($PreviousLayer.distance$);

Algorithm 2. Algorithm to initialize and sort layer before insertion propagation

---

Result: A sorted layer $CurrentLayer$ with distances initialized using $PreviousLayer$

1 **Function** InitializeDistance($PreviousLayer, CurrentLayer$)
2 | **foreach** $x \in PreviousLayer$ **do**
3 | | **foreach** $y \in x.neighbor$ & $y \in CurrentLayer$ **do**
4 | | | **if** $y.distance > x.distance + \omega_a(x, y)$ **then**
5 | | | | $y.distance = x.distance + \omega_a(x, y)$;
6 | Sort($CurrentLayer$);

The InitializeDistance Stage. We compute tentative distances for each vertex in the current layer by using shortest distances computed for the previous layer (Algorithm 2). Because all deletion and substitution cost weighted edges are directed from the previous layer towards the current, this only requires a straightforward linear $O(|V| + |E|)$ time traversal (lines 2–5). In addition, we are required to maintain the current layer as per sorted order of distances. Note that vertices in the previous layer are already available in sorted order of their shortest distances from s. A vertex v in the previous layer can assign only three possible distance values ($v.distance$, $v.distance + \Delta_{sub}$, or $v.distance + \Delta_{del}$) to a neighbor in the current layer. By maintaining three separate lists for each of the three possibilities, we can create the three lists in sorted order and merge them in $O(|V|)$ time. The relative order of vertices in the current layer can be easily determined in linear time by tracking the positions of their distance values in the merged list. As a result, the current layer can be obtained in sorted form in $O(|V|)$ time and $O(|V|)$ space, leading to the following claim.

Lemma 2. *Time and space complexity of the sorting procedure in Algorithm 2 is $O(|V|)$.*

The PropagateInsertion Stage. Note that the tentative distance computed for a vertex is sub-optimal if its shortest path from the source vertex traverses any insertion cost weighted edge in the current layer. One approach to compute

---

Algorithm 3. Algorithm to propagate insertions in the same layer

---

Result: A sorted layer $CurrentLayer$ with optimal distance values

1 **Function** PropagateInsertion($CurrentLayer$)
2 $x.resolved = false \; \forall x \in CurrentLayer$;
3 Queue $q_1 = \emptyset$, $q_2 = \emptyset$;
4 q_1.Enqueue($CurrentLayer$);
5 $CurrentLayer = [\;]$;
6 **while** $q_1 \neq \emptyset$ **or** $q_2 \neq \emptyset$ **do**
7   $q_{min} = q_1$.Front() $< q_2$.Front() ? $q_1 : q_2$;
8   $x = q_{min}$.Dequeue();
9   **if** $x.resolved = false$ **then**
10    $x.resolved = true$;
11    $CurrentLayer$.Append(x);
12   **foreach** $y \in x.neighbor$ & $y.layer = x.layer$ **do**
13   **if** $y.distance > x.distance + \Delta_{ins}$ **then**
14      $y.distance = x.distance + \Delta_{ins}$;
15   q_2.Enqueue(y);

---

optimal distance values is to process vertices in their sorted distance order (minimum first) and update the neighbor vertices, similar to Dijkstra's algorithm. When processing vertex v, the distance of its neighbor should be adjusted such that it is no more than $v.distance + \Delta_{ins}$. Selecting vertices with minimum scores can be achieved using a standard priority queue implementation (e.g., Fibonacci heap); however, it would require $O(|E| + |V| \log |V|)$ time per layer. A key property that can be leveraged here is that all edges being considered in this stage have uniform weights (Δ_{ins}). Therefore, we propose a simpler and faster algorithm using two First-In-First-Out queues (Algorithm 3). The first queue q_1 is initialized with sorted vertices in the current layer, and the second queue q_2 is initialized as empty (line 4). The minimum distance vertex is always dequeued from either of the two queues (line 8). As and when distance of a vertex is updated by its neighbor, it is enqueued to q_2 (line 15). Following lemmas establish the correctness and an $O(|E| + |V|)$ time bound for the PropagateInsertion stage in the algorithm.

Lemma 3. *In each iteration at line 8, Algorithm 3 dequeues a vertex with the minimum overall distance in q_1 and q_2.*

Proof. The queue q_1 always maintains its non-decreasing sorted order at the beginning of each loop iteration (line 6) in Algorithm 3 as we never enqueue new elements into q_1. We prove by contradiction that q_2 also maintains the order. Maintaining this invariant would immediately imply the above claim. Let i be the first iteration where q_2 lost the order. Clearly $i > 1$. Because i is the first such iteration, new vertices (say $y_1, y_2, \ldots, y_k$) must have been enqueued to q_2 in the previous iteration (line 15), and the vertex (say x) which caused these additions must have been dequeued (line 8). Note that the distance of all the new vertices, the y_i's, equals $x.distance + \Delta_{ins}$. Therefore, the vertex prior to y_1 (say

y_{pre}) must have a distance higher than y_1. However, this leads to a contradiction because if we consider the iteration when y_{pre} was enqueued to q_2, the distance of the vertex that caused addition of y_{pre} could not be higher than the distance of the vertex x.

Lemma 4. *Once a vertex is dequeued in Algorithm 3, its computed distance equals the shortest distance from the source vertex.*

Proof. Lemma 3 establishes that Algorithm 3 processes all vertices that belong to the current layer in sorted order. Therefore, it mimics the choices made by Dijkstra's algorithm [5].

Lemma 5. *Algorithm 3 uses $O(|V| + |E|)$ time and $O(|V|)$ space to compute shortest distances in a layer.*

Proof. Each vertex in the current layer enqueues its updated neighbor vertices into q_2 at most once. Note that distance of a vertex can be updated at most once, therefore the maximum number of enqueue operations into q_2 is $|V|$. In addition, enqueue operations are never performed in q_1. Accordingly, the number of outer loop iterations (line 6) is bounded by $O(|V|)$. The inner loop (line 12) is executed at most once per vertex, therefore the amortized run-time of the inner loop is $O(|V| + |E|)$.

The above claims yield an $O(m(|V| + |E|))$ time algorithm for aligning the query sequence to sequence graph. Assuming a constant alphabet, we can further tighten the bound to $O(|V| + m|E|)$ by using a simple preprocessing step suggested in [26]. This step transforms the sequence graph by merging all vertices with 0 in-degree into $\leq |\Sigma|$ vertices. As a result, the preprocessing ensures that the count of vertices in the new graph is no more than $|E| + |\Sigma|$ without affecting the correctness. Summary of the above claims is presented as a following theorem:

Theorem 3. *Algorithm 1 computes the optimal cost of aligning a query sequence of length m to graph $G(V, E, \sigma)$ in $O(|V| + m|E|)$ time and $O(|V|)$ space using a linear gap penalty cost function.*

It is natural to wonder whether there exist faster algorithms for solving the sequence to graph alignment problem. As noted in [26], the sequence to sequence alignment problem is a special case of the sequence to graph alignment problem because a sequence can be represented as a directed chain graph with character labels. As a result, existence of either $O(m^{1-\epsilon}|E|)$ or $O(m|E|^{1-\epsilon}), \epsilon > 0$ time algorithm for solving the sequence to graph alignment problem is unlikely because it would also yield a strongly sub-quadratic algorithm for solving the sequence to sequence alignment problem, further contradicting SETH [3].

4.2 Affine Gap Penalty

Supporting affine gap penalty functions in the dynamic programming algorithm for sequence to sequence alignment is typically done by using three rather than one scoring matrix [10]. Similarly, the alignment graph can be extended to contain three sub-graphs with substitution, deletion, and insertion cost weighted edges respectively [26]. The edge weights are adjusted for the affine gap penalty model such that a cost for opening a gap is penalized whenever a path leaves the match sub-graph to either the insertion or the deletion sub-graph (Appendix Fig. 4). The properties that were leveraged to design faster algorithm for linear gap penalty functions continue to hold in the new alignment graph. In particular, the sorting still requires linear time during the InitializeDistance stage, and insertion propagation is still executed over uniformly weighted edges in the insertion sub-graph. As a result, the two-stage algorithm can be extended to operate using affine gap penalty function in the same time and space complexity as with the linear gap penalty function.

5 Conclusions and Open Problems

In this paper, we show that the sequence to graph alignment problem is $\mathcal{NP}$-complete when changes are allowed in the sequence graph, for any alphabet of size ≥ 2. When changes are allowed in the query sequence alone, we provide a faster polynomial time algorithm that generalizes to linear gap penalty and affine gap penalty functions. The proposed algorithms use elementary data structures, therefore are simple to implement. Overall, the theoretical results presented in this work enhance the fundamental understanding of the problem, and will aid the development of faster tools for mapping to graphs. The alignment problem for sequence graphs is a rich area with several unsolved problems. For the intractable problem variants, development of faster exact and approximate algorithms are fertile grounds for future research. In addition, working towards robust indexing schemes and heuristics that scale to large input graphs and different sequencing technologies remains an important direction.

Acknowledgements. This work is supported in part by US National Science Foundation grant CCF-1816027. Yu Gao was supported by the ACO Program at Georgia Institute of Technology.

A   Appendix

See Fig. 4.

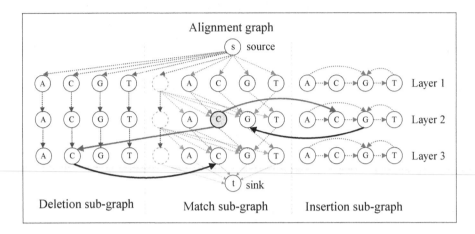

Fig. 4. An example to illustrate the construction of an alignment graph for sequence to graph alignment using affine gap penalty. The alignment graph now contains three sub-graphs separated by the gray dash lines. The deletion and insertion weighted edges in the alignment graph for linear gap penalty are shifted to deletion sub-graph and insertion sub-graph, respectively. Their weights are also changed to the gap extension penalty. Besides, more edges are added to connect the sub-graphs with each other. For simplicity, we use the highlighted vertex as an example to illustrate how to open a gap and extend it. The weight of magenta colored edges is the sum of gap open penalty and gap extension penalty, and the weight of the black colored edges is 0. (Color figure online)

References

1. Amir, A., Lewenstein, M., Lewenstein, N.: Pattern matching in hypertext. J. Algorithms **35**(1), 82–99 (2000)
2. Antipov, D., Korobeynikov, A., McLean, J.S., Pevzner, P.A.: hybridSPAdes: an algorithm for hybrid assembly of short and long reads. Bioinformatics **32**(7), 1009–1015 (2015)
3. Backurs, A., Indyk, P.: Edit distance cannot be computed in strongly subquadratic time (unless SETH is false). In: Proceedings of the Forty-Seventh Annual ACM Symposium on Theory of Computing, pp. 51–58. ACM (2015)
4. Beretta, S., Bonizzoni, P., Denti, L., Previtali, M., Rizzi, R.: Mapping RNA-seq data to a transcript graph via approximate pattern matching to a hypertext. In: Figueiredo, D., Martín-Vide, C., Pratas, D., Vega-Rodríguez, M.A. (eds.) AlCoB 2017. LNCS, vol. 10252, pp. 49–61. Springer, Cham (2017). https://doi.org/10.1007/978-3-319-58163-7_3
5. Cormen, T.H., Leiserson, C.E., Rivest, R.L., Stein, C.: Introduction to Algorithms. MIT Press, Cambridge (2009)

6. Dilthey, A., Cox, C., Iqbal, Z., Nelson, M.R., McVean, G.: Improved genome inference in the MHC using a population reference graph. Nat. Genet. **47**(6), 682 (2015)
7. Eggertsson, H.P., et al.: Graphtyper enables population-scale genotyping using pangenome graphs. Nat. Genet. **49**(11), 1654 (2017)
8. Garg, S., Rautiainen, M., Novak, A.M., Garrison, E., Durbin, R., Marschall, T.: A graph-based approach to diploid genome assembly. Bioinformatics **34**(13), i105–i114 (2018)
9. Garrison, E., et al.: Variation graph toolkit improves read mapping by representing genetic variation in the reference. Nat. Biotechnol. **36**, 875–879 (2018)
10. Gotoh, O.: An improved algorithm for matching biological sequences. J. Mol. Biol. **162**(3), 705–708 (1982)
11. Heydari, M., Miclotte, G., Van de Peer, Y., Fostier, J.: BrownieAligner: accurate alignment of illumina sequencing data to de Bruijn graphs. BMC Bioinform. **19**(1), 311 (2018)
12. Huang, L., Popic, V., Batzoglou, S.: Short read alignment with populations of genomes. Bioinformatics **29**(13), i361–i370 (2013)
13. Kuosmanen, A., Paavilainen, T., Gagie, T., Chikhi, R., Tomescu, A., Mäkinen, V.: Using minimum path cover to boost dynamic programming on DAGs: co-linear chaining extended. In: Raphael, B.J. (ed.) RECOMB 2018. LNCS, vol. 10812, pp. 105–121. Springer, Cham (2018). https://doi.org/10.1007/978-3-319-89929-9_7
14. Lee, C., Grasso, C., Sharlow, M.F.: Multiple sequence alignment using partial order graphs. Bioinformatics **18**(3), 452–464 (2002)
15. Limasset, A., Cazaux, B., Rivals, E., Peterlongo, P.: Read mapping on de Bruijn graphs. BMC Bioinform. **17**(1), 237 (2016)
16. Liu, B., Guo, H., Brudno, M., Wang, Y.: deBGA: read alignment with de Bruijn graph-based seed and extension. Bioinformatics **32**(21), 3224–3232 (2016)
17. Manber, U., Wu, S.: Approximate string matching with arbitrary costs for text and hypertext. In: Advances in Structural and Syntactic Pattern Recognition, pp. 22–33. World Scientific (1992)
18. Myers, E.W.: An overview of sequence comparison algorithms in molecular biology. University of Arizona, Department of Computer Science (1991)
19. Myers, E.W.: The fragment assembly string graph. Bioinformatics **21**(Suppl_2), ii79–ii85 (2005)
20. Navarro, G.: Improved approximate pattern matching on hypertext. Theoret. Comput. Sci. **237**(1–2), 455–463 (2000)
21. Navarro, G.: A guided tour to approximate string matching. ACM Comput. Surv. (CSUR) **33**(1), 31–88 (2001)
22. Nguyen, N., et al.: Building a pan-genome reference for a population. J. Comput. Biol. **22**(5), 387–401 (2015)
23. Novak, A.M., et al.: Genome graphs. Preprint at bioRxiv (2017). https://doi.org/10.1101/101378
24. Park, K., Kim, D.K.: String matching in hypertext. In: Galil, Z., Ukkonen, E. (eds.) CPM 1995. LNCS, vol. 937, pp. 318–329. Springer, Heidelberg (1995). https://doi.org/10.1007/3-540-60044-2_51
25. Pevzner, P.A., Tang, H., Waterman, M.S.: An Eulerian path approach to DNA fragment assembly. Proc. Natl. Acad. Sci. **98**(17), 9748–9753 (2001)
26. Rautiainen, M., Marschall, T.: Aligning sequences to general graphs in O(V + mE) time. Preprint at bioRxiv (2017). https://doi.org/10.1101/216127
27. Rowe, W.P., Winn, M.D.: Indexed variation graphs for efficient and accurate resistome profiling. Bioinformatics **1**, 8 (2018)

28. Salmela, L., Rivals, E.: LoRDEC: accurate and efficient long read error correction. Bioinformatics **30**(24), 3506–3514 (2014)
29. Sirén, J., Välimäki, N., Mäkinen, V.: Indexing graphs for path queries with applications in genome research. IEEE/ACM Trans. Comput. Biol. Bioinform. (TCBB) **11**(2), 375–388 (2014)
30. Thachuk, C.: Indexing hypertext. J. Discrete Algorithms **18**, 113–122 (2013)
31. Vaddadi, K., Tayal, K., Srinivasan, R., Sivadasan, N.: Sequence alignment on directed graphs. J. Comput. Biol. **26**(1), 53–67 (2018)
32. Wang, J.R., Holt, J., McMillan, L., Jones, C.D.: FMLRC: hybrid long read error correction using an FM-index. BMC Bioinform. **19**(1), 50 (2018)
33. Wick, R.R., Judd, L.M., Gorrie, C.L., Holt, K.E.: Unicycler: resolving bacterial genome assemblies from short and long sequencing reads. PLoS Comput. Biol. **13**(6), e1005595 (2017)
34. Zhang, H., Jain, C., Aluru, S.: A comprehensive evaluation of long read error correction methods. Preprint at bioRxiv (2019). https://doi.org/10.1101/519330

Minimization-Aware Recursive K^* ($MARK^*$): A Novel, Provable Algorithm that Accelerates Ensemble-Based Protein Design and Provably Approximates the Energy Landscape

Jonathan D. Jou[1], Graham T. Holt[1,2], Anna U. Lowegard[1,2], and Bruce R. Donald[1,3,4(✉)]

[1] Department of Computer Science, Duke University, Durham, NC, USA
brd+recomb19@cs.duke.edu
[2] Computational Biology and Bioinformatics Program, Duke University, Durham, NC, USA
[3] Department of Biochemistry, Duke University Medical Center, Durham, NC, USA
[4] Department of Chemistry, Duke University, Durham, NC, USA

Abstract. Protein design algorithms that model continuous sidechain flexibility and conformational ensembles better approximate the *in vitro* and *in vivo* behavior of proteins. The previous state of the art, iMinDEE-A^*-K^*, computes provable ε-approximations to partition functions of protein states (e.g., bound vs. unbound) by computing provable, admissible pairwise-minimized energy lower bounds on protein conformations and using the A^* enumeration algorithm to return a gap-free list of lowest-energy conformations. iMinDEE-A^*-K^* runs in time sublinear in the number of conformations, but can be trapped in loosely-bounded, low-energy conformational wells containing many conformations with highly similar energies. That is, iMinDEE-A^*-K^* is unable to exploit the correlation between protein conformation and energy: *similar conformations often have similar energy*. We introduce two new concepts that exploit this correlation: Minimization-Aware Enumeration and Recursive K^*. We combine these two insights into a novel algorithm, Minimization-Aware Recursive K^* ($MARK^*$), that tightens bounds not on single conformations, but instead on *distinct regions of the conformation space*. We compare the performance of iMinDEE-A^*-K^* vs. $MARK^*$ by running the BBK^* algorithm, which provably returns sequences in order of decreasing K^* score, using either iMinDEE-A^*-K^* or $MARK^*$ to approximate partition functions. We show on 200 design problems that $MARK^*$ not only enumerates and minimizes vastly fewer conformations than the previous state of the art, but also runs up to two orders of magnitude faster. Finally, we show that $MARK^*$ not only efficiently approximates the partition function, but also *provably* approximates the

J. D. Jou and G. T. Holt—These authors contributed equally to the work.

L. J. Cowen (Ed.): RECOMB 2019, LNBI 11467, pp. 101–119, 2019.
https://doi.org/10.1007/978-3-030-17083-7_7

energy landscape. To our knowledge, $MARK^*$ is the first algorithm to do so. We use $MARK^*$ to analyze the change in energy landscape of the bound and unbound states of the HIV-1 capsid protein C-terminal domain in complex with camelid $V_H H$, and measure the change in conformational entropy induced by binding. Thus, $MARK^*$ both accelerates existing designs and offers new capabilities not possible with previous algorithms.

1   Introduction

The objectives of computational structure-based protein design algorithms are (1) to accurately calculate properties of a protein or protein complex (e.g., stability, binding affinity, etc.), and (2) to efficiently search for optimal sequences given an objective function defined on these properties. These algorithms search over a space defined by a user-specified input model (i.e., a structural model, allowed sidechain and backbone flexibility, allowed mutations, energy function, etc.). Designs for ensemble-average, macromolecular properties like binding affinity and stability are more biophysically accurate when modeling thermodynamic, conformational ensembles [3,13,15,18,33,44,48,49,53]. However, accurately modeling these ensembles can be challenging: the space of possible conformations available *in vitro* and *in vivo* to a protein can be massive, and furthermore grows exponentially with the number of residues.

Various simplifications to the input model and the search methodology have been used in order to reduce the complexity of this problem, of which we will discuss three: (1) discretized, rigid sidechain and backbone flexibility; (2) design to a single, static global minimum energy conformation (GMEC); and (3) nonprovable search over possible conformations and sequences. (1) Although amino acid sidechains are continuously flexible, sidechains are often modeled as discrete, frequently observed low-energy states called *rotational isomers*, or *rotamers* [36]. Furthermore, protein backbone flexibility is frequently modeled as fixed, or restricted to a small set of discrete alternate conformations [30,50,52,55]. Designs made with these simplifications do not model small, commonly observed sidechain and backbone movements, much less larger structural rearrangements. Even under these simplifications, calculating the partition function for a protein remains #P-hard [37,54,55]. Moreover, the conformation space grows exponentially with the number of residues. (2) As a result, many design algorithms optimize the energy of a single, static GMEC structure [2,4,14,21,22,29,52]. GMEC-based algorithms do not model conformational entropy, which can contribute significantly to protein structure and function [7,8], and as a result can overlook thermodynamically favorable sequences [44]. (3) Finally, some algorithms attempt to estimate the partition function by stochastically sampling the conformation space for low-energy microstates [28,30,31]. These algorithms provide no guarantees on the quality of the lowest-energy conformation returned, much less on the quality of the approximation of the overall partition function. Indeed, [50] demonstrated that as the size of the search space increases, the probability that stochastic methods find even the GMEC falls rapidly to zero.

Furthermore, because these methods are non-deterministic, it is profoundly difficult to deconvolve methodological error (i.e., undersampling) from input model error [6,10].

Algorithms distributed in the OSPREY [24] package efficiently solve protein design problems without the above simplifications, provably returning the optimal sequences and conformations without sacrificing accuracy. OSPREY models not only continuous sidechain flexibility [11,15,21,22], but also discrete and continuous backbone flexibility [12,13,20,23]. Additionally, the *BBK** algorithm [40] provably returns protein sequences in order of decreasing binding affinity and runs in time sublinear in the number of sequences. These algorithms have been used to prospectively predict drug resistance [9,39, 43] and design enzymes [3,12, 15,33,51], new drugs [19], peptide inhibitors of protein-protein interactions [44], epitope-specific antibody probes [16], and broadly-neutralizing antibodies [17,47]. These designs have been experi-

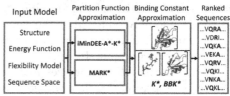

Fig. 1. Provably computing the best-binding sequences with respect to the input model. When designing for macromolecular, ensemble-average properties such as binding affinity K_a, provable algorithms such as K^* [15,33,44] and *BBK** [40] take as input an input structure, energy function, allowed backbone and sidechain flexibility, and allowed mutations which define the sequence space. Both algorithms used the previous state of the art, iMinDEE-A^*-K^*, to provably approximate partition functions (with respect to the input model) and combined these partition function approximations into a K^* score [33], which approximates K_a. By approximating K_a, designers can rank candidate sequences in order of binding affinity, and identify the best-binding sequence (green) with respect to the input model. In this design protocol, *MARK** replaces iMinDEE-A^*-K^* as a provable partition function approximation module. (Color figure online)

mentally validated *in vitro*, some *in vivo*, and one designed anti-HIV broadly-neutralizing antibody, VRC07-523LS, is currently in 6 clinical trials [1].

The K^* algorithm in OSPREY estimates binding affinity with the K^* score [33], a ratio of ε-approximate, Boltzmann-weighted partition functions for bound and unbound states. These partition functions are computed by combining an admissible lower bound on conformational energy with the A^* search algorithm to quickly and provably enumerate a gap-free list of the lowest energy conformations [25,29,46]. We will refer to algorithms that compute K^* scores using A^* as A^*-K^* *algorithms*. While significantly more efficient than exhaustive enumeration, A^*-K^* algorithms are guaranteed to return the GMEC first, and therefore focus on efficiently finding low-energy *conformations*. However, a GMEC-first enumeration strategy may not efficiently approximate the full partition function. Modeling continuous flexibility further compounds the difficulties of partition function approximation. Previous A^*-K^* algorithms [11,15] that incorporate continuous flexibility, such as iMinDEE-A^*-K^* [11], enumerate conformations in order of energy lower bounds on the minimized energy. However, when these bounds are loose, iMinDEE-A^*-K^* must perform many computationally expensive *full minimizations* (wherein all mutable and flexible residues minimize simultaneously) to provably approximate the partition function. In the worst case, A^*-K^* algorithms must minimize a combinatorial number of conformations that are loosely bounded at the same residues.

To overcome the limitations of A^*-K^*, we present a novel algorithm that combines two new concepts: Recursive K^* (RK^*) and Minimization-Aware Enumeration (MAE). RK^* prioritizes *low-entropy regions* of the energy landscape, instead of prioritizing *low-energy conformations* (Fig. 2D vs. C), and MAE tightens bounds on a combinatorial number of conformationally similar loosely bounded conformations (Fig. 2E). This combination, Minimization-Aware Recursive K^* ($MARK^*$), achieves significant efficiency and runtime improvements for large protein design problems that confound previous A^*-K^* algorithms, as well as algorithms that call A^*-K^* algorithms as a subroutine, such as BBK^* [40] (Fig. 1). Because $MARK^*$ replaces iMinDEE-A^*-K^*, we ran BBK^* with iMinDEE-A^*-K^* as a control, and compared it to the performance of BBK^* with $MARK^*$ on 200 protein design problems. We found that $MARK^*$ accelerates BBK^* by up to 2 orders of magnitude, efficiently completing designs an order of magnitude larger than was possible using BBK^* with iMinDEE-A^*-K^*. Finally, we show that $MARK^*$ not only outperforms the previous state of the art in speed, but also offers new design capabilities. Because $MARK^*$ tightly bounds low-entropy regions of the conformation space instead of low-energy conformations, it computes a provable approximation of the *energy landscape*, which bounds the energy of every conformation in the conformation space. $MARK^*$ is, to our knowledge, the first provable algorithm to do so. In contrast, previous algorithms (provably or non-provably) returned only low-energy conformations, and do not tightly and provably approximate the energy landscape. This energy landscape approximation provides additional insight into the higher-energy regions between tightly-bounded low-energy conformational wells (Fig. 2D). Using $MARK^*$ to compute the partition function and energy landscape for the design problem of an HIV-related protein-protein interface, we demonstrate the ability of $MARK^*$ to reveal components of binding thermodynamics. That is, we show that $MARK^*$ not only approximates the partition function more efficiently, but also computes an entire energy landscape that enables insight into thermodynamics.

By presenting this algorithm, our paper makes the following contributions:

1. A novel algorithm that more quickly and efficiently predicts binding affinity using partition functions over molecular ensembles.
2. Proofs of correctness and admissibility of the bounds used in the branch and bound strategy by $MARK^*$, as well as the optimality of $MARK^*$ for a given energy bounding function.
3. 200 designs showing that BBK^* with $MARK^*$ returned the five best sequences up to two orders of magnitude faster, minimized 685-fold fewer conformations, and completed designs up to an order of magnitude larger than was possible using BBK^* with iMinDEE-A^*-K^*.

4. An application of $MARK^*$ to compute a provable approximation of the energy landscape for an HIV-related protein-protein interface, revealing components of binding thermodynamics.
5. An implementation of $MARK^*$ in our lab's open-source protein design software, OSPREY [24].

2 Background

To accurately model macroscopic properties like binding affinity, design algorithms must approximate the Boltzmann-weighted partition function over bound and unbound states. For a protein design with n residues, let the sequence s be a set of n ordered pairs (i, a), each containing the residue index i and an amino acid a. For a sequence s, we can define the conformation space $Q(s)$ to be the set of conformations defined by s. Additionally, we denote the maximum number of rotamers (at any one residue) to be q. Let $E_X(c)$ be the minimized energy of a conformation c in state X (e.g., bound or unbound). Under this formulation, the partition function $Z_X(s)$ for a protein with sequence s in state X can be defined as

$$Z_X(s) = \sum_{c \in Q(s)} \exp(-E_X(c)/RT). \tag{1}$$

Notably, the set of all conformations $Q(s)$ grows exponentially with the number of residues n, and therefore the exact value of the partition function becomes intractable to compute as n increases. As a result, many protein design algorithms instead approximate Z_X with stochastic [26,32,35,38] or provable [15,33,34,39,44,49,55] methods. Provable algorithms have mathematical guarantees on their computed approximation of Z_X, and thus obviate any need for deconvolution of error in the output.

One class of provable algorithms computes an ε-approximation of the partition function by using the A^* search algorithm to enumerate a gap-free list of conformations in order of increasing energy [11,15,33,40,44]. By enumerating a gap-free list of low-energy conformations, A^*-K^* algorithms compute both upper and lower bounds on the overall partition function, and return a partition function approximation that is guaranteed to be within a $(1 - \varepsilon)$ factor of the true partition function. When incorporating continuous flexibility, A^*-K^*-based enumeration proceeds in order of a provable lower bound $E^\ominus$ on the full-minimized energy of a conformation. By minimizing the enumerated conformations, A^*-K^* algorithms tighten the upper and lower bounds on the partition function.

In practice, A^*-K^* algorithms have been shown to run in time sublinear in the number of conformations [33]. However, in their focus on returning the lowest-energy conformations, these algorithms can tightly bound the energy of a large number of low-energy conformations while still achieving only a loose energy lower bound on the unenumerated conformations, and thus the overall partition function upper bound. In the worst case, A^*-K^* algorithms must enumerate a large number of conformations to compute an ε-approximate partition function.

This issue is especially common when a design problem contains many low-energy conformations with similar energies. Furthermore, when the energy lower bounds are loose, the difference between the partition function upper and lower bounds can remain large, even after enumerating and minimizing a large number of conformations. As a result, A^*-K^* algorithms can be trapped in loosely-bounded low-energy wells, enumerating and minimizing a combinatorial number of low-energy conformations without efficiently tightening the partition function bounds. To overcome the limitations of A^*-K^*-based methods, we introduce two concepts: Minimization-Aware Enumeration and Recursive K^*, both of which exploit the correlation between protein structure and energy to efficiently bound a combinatorial number of conformations.

3   Algorithm

3.1   Recursive K^* (RK^*): Enumerating in Order of Z-error

It may at first seem counter-intuitive to tightly bound the partition function of a protein conformation space without computing the energies of any one conformation. Indeed, previous provable algorithms have efficiently approximated the partition function by computing a gap-free list of the lowest-energy conformations [15,33,40,44]. The key insight is that *structurally similar conformations are often energetically similar*: although a set of low-energy conformations may constitute the vast majority of the partition function, these conformations may in fact be both structurally and energetically similar (Fig. 2A). Therefore, computing one upper and one lower bound on a set of similar conformations can efficiently bound the partition function contribution of the *entire set*. More formally, when the energy upper and lower bounds on a set C of structurally similar conformations are very close, the statistical weight of the set may be tightly approximated by simply scaling the upper and lower bounds by $|C|$. The following definitions of these new bounds are sufficient for the theorems provided in the main paper – the precise definitions involve some subtleties, which are deferred to Section A of the **Supplementary Information (SI)** [27]. For a set of conformations C that all share the partial conformation c', we can define partition function upper and lower bounds as follows:

$$U(c') = \exp(-E^{\ominus}(c')/RT)\tau(c') \tag{2}$$

$$L(c') = \exp(-E^{\oplus}(c')/RT)\tau(c') \tag{3}$$

where $E^{\ominus}$ and $E^{\oplus}$ are lower and upper energy bounds on the best and worst energies of any conformation in C, respectively, and $\tau(c') = |C|$.

Fundamentally, computing K^* scores for a sequence can be formulated as computing energy upper and lower bounds on the conformation spaces (one for each state, e.g., bound and unbound) in order to reduce the difference between partition function upper and lower bounds. We will refer to this difference as Z-error. To directly explore the conformation space *in order of Z-error*, RK^*

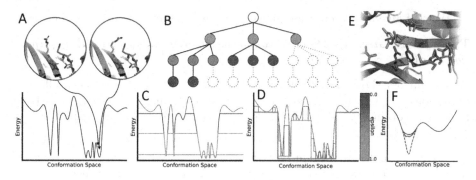

Fig. 2. Recursive K^* and Minimization-Aware Enumeration exploit positive correlation between conformation and energy to efficiently bound the partition function. (A) Structurally similar conformations within the same energy well often have similar energies, shown as two points in the black energy landscape. (B) When the conformation space is represented as a conformation tree, some conformations (white leaf nodes) may be tightly bounded by computing bounds on their parent nodes (colored internal nodes). (C) Previous provable partition function approximation algorithms tightly bounded all conformations within some energy window of the GMEC. To decrease the error bound ε (colored by the scale beside D), these algorithms incrementally increased the energy window, computing exact energies for more and more conformations (colored curves). (D) Recursive K^* instead exploits the correspondence between conformation and energy to more efficiently bound similar conformations with bounds on regions of the energy landscape. As the error bound ε decreases and the approximation becomes more accurate (colored step curves), Recursive K^* iteratively tightens bounds on loosely bounded (and often *low-entropy*) regions of the landscape, rather than tightly bounding *low-energy conformations*. (E) Loosely-bounded pairwise-minimized bounds can affect a combinatorial number of conformations, shown as an ensemble of conformations that share the same sidechain assignments at the blue residues. Although the blue residues have favorable pairwise-minimized lower bounds, when all three are minimized in concert, their post-minimized energy is higher. (F) By computing a tighter bound on the three blue residues, Minimization-Aware Enumeration tightens the bounds on the combinatorial number of conformations containing the sidechain assignments at the blue residues. Thus, a loosely-bounded energy well (black curve vs. dotted blue curve) may be bounded more tightly (solid blue curve) without minimizing all conformations in the well. (Color figure online)

calculates the Z-error of a full or partial conformation c by computing the difference between the upper and lower bounds on its partition function contribution:

$$\gamma(c) = U(c) - L(c). \tag{4}$$

In effect, RK^* divides the conformation space into smaller subspaces along the allowed rotamers at a residue, and bounds *regions of the conformation space* rather than tightly bounding the next best unenumerated conformation. By using $\gamma(c)$ to explore the conformation space in order of Z-error, RK^* approximates the partition function by branching and bounding the most loosely-bounded regions first, rather than enumerating the next-lowest energy conformation as A^* does.

We now give a theorem that shows $\gamma(c)$ never underestimates the Z-error:

Theorem 1. *Given a set C of all conformations containing a partial conformation c', $\gamma(c') \geq \gamma(c)$ for all $c \in C$.*

We can further show that, when the energy lower and upper bounds on all conformations are tight (e.g., when using a pairwise-decomposable energy function for a rigid-rotamer, rigid-backbone design), the number of nodes expanded by RK^* is *optimal*.

Theorem 2. *Let T be a conformation tree where each conformation in the conformation space corresponds to a leaf node in T. For any given upper bounding function $U(c')$ and lower bounding function $L(c')$ over T, RK^* expands the minimum number of nodes required to compute a provable ε-approximation of the partition function Z using those bounding functions.*

For details on these bounds and the full proofs of Theorems 1 and 2, see Section A in the **SI** [27].

Figure 2C and D illustrate this strategy, showing how RK^* can use lower bounds on partial conformations (shown as colored step curves) to efficiently and incrementally bound regions of the energy landscape (black curve). For clarity, the figure omits upper bounds, but the same strategy can be applied. $MARK^*$ computes one upper bound and one lower bound on a combinatorial number of energetically similar conformations that contain the same partial conformation c', whereas in the worst case A^*-K^* must enumerate all $q^{n-|c'|}$ conformations, where $|c'|$ is the number of residues whose sidechain conformations are assigned by c'.

3.2   Minimization-Aware Enumeration (MAE): Tightening Loose Bounds During Enumeration

Minimization-Aware Enumeration exploits the conformational similarity between loosely-bounded conformations to *tighten* loosely-bounded pairwise-minimized lower bounds as they are encountered during conformation enumeration. Upon encountering a loosely-bounded pair, a tighter bound is computed by minimizing the pair in the presence of a third *witness* residue (Fig. 2E). For a pairwise-decomposable energy function, a loosely bounded residue pair is only overly optimistic when the presence of the other flexible residues changes the post-minimization conformation and energy of that pair. As has been shown previously, these higher-order interactions are often represented with high accuracy by simply modeling three-residue interactions as well, as is done by the LUTE [22] algorithm. Indeed, the HOT/PartCR [45] algorithm has identified both *overly flexible* residues whose pairwise-minimized conformation varies widely depending on the conformation of nearby residues, and *higher-order clashing* tuples whose pairwise-minimized conformation are clash-free, but cannot be achieved when all residues in the tuple are minimized together. While these algorithms both successfully tighten pairwise-minimized lower bounds, both are effectively *preprocessing algorithms*. HOT/PartCR changes the conformation space each iteration, repeatedly restarting A^* search, whereas LUTE is run before A^*-K^* enumeration. In effect, both HOT/PartCR and LUTE must be run to satisfactory energy bound tightness before any approximation of the partition function can be computed. MAE instead computes LUTE-like energy *corrections* when a conformation with a loose lower bound is encountered, and applies all computed corrections to unexplored regions of the conformation space. Notably, MAE combines LUTE-like corrections with HOT/PartCR-like lowest lower bound-first tightening, thus correcting only loosely-bounded conformations that also share rotamer assignments with other conformations with low lower bounds. Unlike either algorithm, MAE can then incorporate these corrections into its partition function computation *without restarting A^* search*: that is, it corrects the energy of a combinatorial number of conformations *online*. Thus, MAE provides an efficient way to tighten conformational lower bounds during partition function approximation, further reducing computational cost.

3.3   Minimization-Aware Recursive K^* ($MARK^*$)

In combination, the improvements of MAE and RK^* are further enhanced. RK^* not only prioritizes low-entropy regions of the conformation space, it is able to also weigh the potential benefits of full minimization vs. branching and bounding. MAE converts the tighter bounds on each full minimization into tighter bounds on a *region of the conformation space* (i.e., a combinatorial number of conformations). Thus, $MARK^*$ chooses the most effective of both possible bound-tightening strategies: recursively bounding one region of the conformation space, or minimizing another. In doing so, $MARK^*$ distinguishes itself from the GMEC-first, A^*-K^*-based previous state of the art: rather than enumerating or minimizing one conformation at a time, it bounds and minimizes *regions of the conformation space*. For a full description of the algorithm, see Section A of the SI [27].

4   Computational Experiments

We implemented $MARK^*$ in our laboratory's open source OSPREY [24] protein design package and compared our algorithm to the previous state-of-the-art, iMinDEE-A^*-K^*. To do so, we first measured performance of the BBK^* [40] algorithm with either iMinDEE-A^*-K^* (A^*-BBK^*) or $MARK^*$ ($MARK^*$-BBK^*) as its partition function approximation subroutine. Using A^*-BBK^* and $MARK^*$-BBK^*, we computed the 5 best-binding sequences for 200 different protein design problems from 38 different protein-ligand complexes used in [40]. This was a head-to-head comparison: for both A^*-BBK^* and $MARK^*$-BBK^*, we measured performance using the BBK^* implementation from [24]. The size of the resulting design problems ranged from 18 to 400 sequences, and the number of conformations over all sequences (which is the total size of a design problem) ranged from 1.62×10^3 to 3.26×10^{17} conformations. In all cases, we modeled continuous sidechain flexibility using continuous rotamers [11,45]. As in [11,15,40], rotamers from the Penultimate Rotamer Library [36] were allowed to minimize to any conformation plus or minus 9° of their modal χ-angles (18° of dihedral angle flexibility). Next, to investigate the comparative advantage of RK^* over A^*-K^*, we performed additional computational experiments designed to deconvolve the challenge of minimizing conformations from the challenge of exploring the conformation space. We computed the wildtype K^* scores for 344 rigid rotamer, rigid backbone design problems, created from 38 protein structures used in [40]. For each rigid design problem, we selected up to 29 residues at a protein-protein interface to be flexible. The size of the resulting design problems ranged from 3.46×10^3 to 6.76×10^{25} conformations.

For all design problems, each algorithm computed ε-approximate bounds to an accuracy of $\varepsilon < 0.683$ (as was derived in [40]) or was terminated after 7 days for the continuous design problems and 6 days for the rigid design problems. All continuous designs were run on 40–48 core Intel Xeon nodes with up to 200 GB of memory, and rigid designs were run on the same machines with 60 GB of memory. A detailed description of the 544 total protein design problems, the 38 protein-ligand systems they are based on, and our continuous and rigid sidechain flexibility experimental protocols is in Section B of the **SI** [27].

5   Results

We first compared overall runtime and demonstrated that, for large designs, $MARK^*$-BBK^* completed designs faster than A^*-BBK^* (Fig. 3A). Notably, $MARK^*$-BBK^* completes designs that were previously too large or memory-intensive for the previous state of the art. Out of 200 total designs, iMinDEE-A^*-K^* computed an ε-approximation to the partition function within 7 days for only 185. For 10 design problems, iMinDEE-A^*-K^* ran for more than 7 days and was terminated, and for 5 other cases iMinDEE-A^*-K^* ran out of 200 GB of memory. In particular, iMinDEE-A^*-K^* was unable to complete any of the largest designs which contained more than 10^{17} conformations. In contrast, $MARK^*$ provably returned

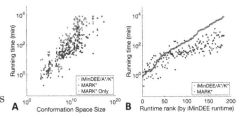

Fig. 3. Speed: $MARK^*$ is up to 135 times faster than iMinDEE- A^* -K^*, and its speedups increase as iMinDEE- A^* -K^* takes longer. (A,B) Times to return the 5 best sequences for $MARK^*$ (blue, red) and iMinDEE-A^*-K^* (green) are shown. (A) Times for all 200 continuous design problems are shown, plotted against conformation space size. $MARK^*$ completes 15 challenging design problems (size larger than 10^{10} conformations, red triangles) that iMinDEE-A^*-K^* cannot. (B) Runtimes for the 185 designs completed by iMinDEE-A^*-K^*, sorted along the x-axis by iMinDEE-A^*-K^* runtime (ranks for designs in Table 1 of the **SI** [27]) are shown. For all designs that required longer than 146 min, $MARK^*$ required less time (up to 135 times faster). (Color figure online)

the 5 best sequences for all 200 in under 6 days, including the 15 for which iMinDEE-A^*-K^* could not (Fig. 3A). The largest design, a 17-residue design of llama antibody in complex with the *C. Botulinum* neurotoxin serotype A catalytic domain (PDB id: 3k3q), contained 3.26×10^{17} conformations, which is an order of magnitude larger than the largest design completed by iMinDEE-A^*-K^*. Whereas iMinDEE-A^*-K^* ran out of memory after 5 days, $MARK^*$ returned the 5 best-binding sequences in 55 h. Furthermore, the advantage of $MARK^*$ over iMinDEE-A^*-K^* grew as designs became more complex. As can be seen in Fig. 3A, although the performance of iMinDEE-A^*-K^* varied as conformation space size increased, the design problems for which iMinDEE-A^*-K^* performed slowly are the very designs where $MARK^*$ demonstrated the largest improvements (Fig. 3B). For design problems for which iMinDEE-A^*-K^* required longer than 146 min, $MARK^*$ required less time (completing up to two orders of magnitude faster) to calculate an ε-approximation to the K^* score for the best 5 sequences. In one design at the binding interface between HIV-1 capsid protein C-terminal domain in complex with a camelid $V_H H$ (PDB id: 2xxm), the conformation space was 1.14×10^{12} conformations, and iMinDEE-A^*-K^* computed provable ε-approximate K^* scores for the 5 best sequences in 4.5×10^3 min. In contrast, $MARK^*$ completed in 33 min, 135 times faster than iMinDEE-A^*-K^*. To further elucidate the improvements of $MARK^*$, we measure the effects of RK^* and MAE separately. While, for reasons of accuracy, we recommend always using continuous flexibility, we show that the speed improvements for rigid rotamer, rigid backbone wildtype K^* score computation is even more dramatic. Results from these simplified design problems suggest that design with continuous flexibility considerably increases the challenge of the design problem. In particular, $MARK^*$, with its RK^* bounding strategy, is able to efficiently bound the conformation space when the conformational energy upper and lower bounds are tight, but cannot avoid minimizing conformations with loose energy bounds.

5.1 RK^* Is Orders of Magnitude More Efficient and Faster than A^*-K^*

For the 344 rigid rotamer, wildtype-only design problems, we compared overall runtime and the number of conformations enumerated, shown in Fig. 4. Notably, RK^* computed the K^* score for all 344 design problems, whereas A^*-K^* was only able to do so for 321. Of the 30 largest design problems (conformation space size of 4.5×10^{22} or more conformations), A^*-K^* completed only 8. In fact, A^*-K^* was unable to compute any K^* scores for design problems containing more than 10^{25} conformations, showing that RK^* is able to complete designs larger than was possible with the previous state of the art. For the largest design problem, a 24-residue design of the Llama $V_H H$-02 binder of ORF49 (PDB id: 4hem), the conformation space was 6.76×10^{25} conformations, and A^*-K^* timed out after 6 days, whereas RK^* finished in merely 4.7 min. Furthermore, RK^* finishes up to 3 orders of magnitude faster than A^*-K^*. In the case of the 24-residue design of HIV-1 capsid protein bound to camelid $V_H H$ (PDB id: 2xxm), A^*-K^*

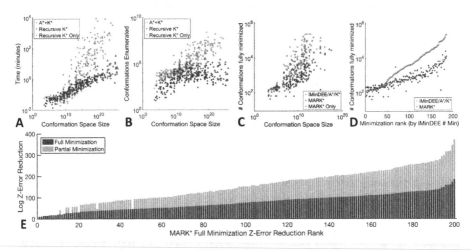

Fig. 4. RK^* **and MAE both significantly improve efficiency over iMinDEE-A^*-K^*.** (A,B) Times and conformations enumerated by $MARK^*$ (blue, red) and iMinDEE-A^*-K^* (green) plotted against conformation space size. When computing wildtype K^* scores with rigid backbones and rigid rotamers, $MARK^*$ not only computes ε-approximate scores 3 orders of magnitude faster and 5 orders of magnitude more efficiently, it also completes design problems larger than was possible with iMinDEE-A^*-K^* (red triangles). (C,D) Number of conformations minimized by $MARK^*$ (blue, red) and iMinDEE-A^*-K^* (green) are shown, plotted against conformation space size (C) or full minimizations computed by iMinDEE-A^*-K^* (D, ranks for designs in Table 1 of the **SI** [27]). $MARK^*$ completes 15 challenging design problems (size larger than 10^{10} conformations, red triangles) that iMinDEE-A^*-K^* cannot. $MARK^*$ minimizes up to 685-fold fewer conformations than iMinDEE-A^*-K^*, and is more efficient on the problems for which iMinDEE-A^*-K^* minimizes the most conformations. (E) Log Z-error reduction achieved by full minimization (blue) and corrections from partial minimizations (yellow) are shown as stacked bar charts. x-axis shows design problems, sorted by Z-error reduction attributable to full minimization (ranks for designs in Table 1 of the **SI** [27]). As the number of full minimizations increases, MAE efficiency increases, reducing Z-error by up to three orders of magnitude more than full minimization does. (Color figure online)

enumerated more than 162 *million* conformations, taking 5.4 days, whereas RK^* enumerated merely 11,699 conformations in under 75 *seconds*, finishing 6,230 times faster than the previous state of the art. RK^* also enumerated far fewer conformations than A^*-K^*. In the case of a 29-residue design of the TRF2 TRFH domain bound to Apollo peptide, A^*-K^* took 2.2 days to enumerate over 52 million conformations. In contrast, RK^* enumerated merely 576 conformations in 2.2 min, and was over 90,800 times more efficient.

5.2 MAE Is More Efficient and Effective than Full Minimization Alone

Using MAE, $MARK^*$ tightens bounds on a potentially exponential number (up to $\mathcal{O}(q^{n-3})$) of conformations by performing a merely polynomial number ($\mathcal{O}(\binom{n}{3}q^3)$) of minimizations. In contrast, iMinDEE-A^*-K^* must, in the worst case, minimize the same (potentially exponential) number of loosely-bounded conformations. In our experiments, the energy bounds were often very loose. The median energy difference between pairwise-minimized lower bounds and full-minimized energy was 4.9 kcal/mol, leading to overestimation of statistical weight by *orders of magnitude* for many conformations. To measure the efficiency of MAE, first we compared the number of full conformations minimized by $MARK^*$ and by iMinDEE-A^*-K^*. Then, to analyze the benefits of the partial minimizations performed by MAE, we measured the reduction of Z-error (*Z-error reduction*) from full minimizations and MAE corrections for each of the 200 continuous design problems. Figure 4 illustrates the improvement in efficiency of $MARK^*$ over iMinDEE-A^*-K^*: $MARK^*$ minimizes up to 685-fold fewer leaf nodes. As can be seen in Fig. 4D, $MARK^*$ minimizes fewer conformations than iMinDEE-A^*-K^* for all designs in which iMinDEE-A^*-K^* minimizes more than 1344 conformations. Additionally, the bound-correcting effect of MAE increases as the conformation space grows larger and more complex. For one design of a Scribble PDZ34 domain complexed with its target peptide (PDB id: 4wyu), total Z-error reduction from full minimizations was 4.12×10^{97}, and total Z-error reduction from partial minimizations was 8.36×10^{100}. Thus, for every full minimization computed by $MARK^*$, MAE achieved Z-error reduction equivalent to 2030 additional minimizations. The trend in Fig. 4E emphasizes the increasing number of loosely-bounded conformations as the conformation space grows, showing that for every conformation $MARK^*$ minimizes, it also tightens the bound on a combinatorial number of conformations.

6 Discussion

6.1 $MARK^*$ Reveals Components of Binding Thermodynamics

To test the ability of $MARK^*$ to approximate the energy landscape, we ran $MARK^*$ on a 10-residue design problem at the protein-protein interface of HIV-1 capsid protein C-terminal domain bound to camelid V_HH and compared the partition functions of the wildtype sequence for both proteins in the bound (PDB id: 2xxm), unbound camelid V_HH (PDB id: 2xxc), and unbound HIV-1 capsid protein C-terminal domain (PDB id: 3ds2) states with continuous sidechain flexibility, in a similar fashion as was done in [41] (Fig. 5). For these three states, we modeled bound and unbound backbones using the three separate bound and unbound structures described above. We computed a provable ε-approximation of $\varepsilon < 0.01$ for the partition functions of the bound and unbound states for both proteins, and computed the corresponding energy landscapes, where the energies of all conformations within 5 kcal/mol of the GMEC were computed exactly. Next, for each protein we computed bounds on its *binding-competent ensemble*, which is the ensemble consisting of all conformations that exist in the bound state, modeled with the energies of the unbound state.

As has been observed in [24,43], the correlation between K^* scores and K_a is not yet quantitative, although it is good enough for ranking. In particular, for current K^* computations only a subset of biologically available structural flexibility is allowed and waters are not explicitly modeled, both of which lead to underestimated entropy. Additionally, most physical effective energy functions are based on small-molecule energetics, which can overestimate van der Waals terms and thereby overestimate internal energy. Despite these input model limitations, in [24,43] significant changes in energy corresponded to large changes in K^* score, and correlated well with experimental measurements. Therefore, in our comparisons we expect K^* scores to (1) correctly predict if one state is more favorable than another, and (2) compute free energy terms that are comparable within an order of magnitude. For our analysis, scaling entropy up by a factor of 2 and internal energy down by a factor of 8 resulted in energies within the range of typical experimental measurements for 10 residues at a protein-protein interface. We report all computed energies after scaling.

Using the results from $MARK^*$, we computed the ensemble-weighted internal energy and entropy for both binding partners in their bound and unbound states. At the computed temperature of ~298 K, HIV-1 capsid protein undergoes a change in its conformational distribution upon binding. This entails a decrease in entropy, lowering $T\Delta S$ by 3.06 kcal/mol. $T\Delta S$ decreases by 1.74 kcal/mol for camelid $V_H H$, as well. To compensate, the complex internal energy decreases upon binding. Whereas the unbound protein and the ligand have a combined ensemble-weighted internal energy of -8.69 kcal/mol, the internal energy of the complex is -14.0 kcal/mol, which is 5.31 kcal/mol lower than the combined internal energy of unbound HIV capsid protein and camelid $V_H H$. The change in Helmholtz free energy ΔF is therefore -0.51 kcal/mol. Importantly, the internal energy of the binding-competent HIV-1 capsid protein ensemble is only 0.044 kcal/mol less than the internal energy of its free ensemble, which agrees with the unfavorable overall increase in Helmholtz free energy of 3.01 kcal/mol between the free ensemble and the binding-competent ensemble. Similarly, the internal energy of the binding-competent camelid $V_H H$ ensemble is 0.939 kcal/mol higher than the energy of the free ensemble, for a total increase in Helmholtz free energy of 2.68 kcal/mol. As this data shows, both binding partners incur an energy penalty when assuming the binding-competent ensemble, which is overcome by favorable interactions gained upon binding.

Thus, $MARK^*$ reveals the loss of entropy, and its commensurate increase in internal energy upon binding. Figure 5 shows the change in the conformational ensemble between the free and binding-competent ensemble, followed by internal energy change upon binding. As can be seen in Fig. 5, there are many low-energy states in the free ensemble of camelid $V_H H$, shown as numerous blue and purple arcs, and by the comparatively small green arc for the GMEC of the free ensemble. In contrast, both the binding-competent ensemble and the bound ensemble show significantly fewer low-energy states, and in both ensembles the

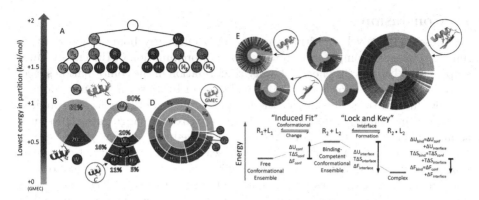

Fig. 5. *MARK** reveals components of binding thermodynamics. Upper bounds on the Boltzmann-weighted partition function for a 10-residue design at the protein-protein interface between HIV-1 capsid protein and camelid V_HH domain are shown as colored ring charts (explained in panels A–D). For context, secondary structure near the design problem is shown when identifying conformations (colored proteins), but the full protein is not shown. (A) The conformation space can be represented as a tree, where leaf nodes correspond to full conformations, and each node c in the tree is colored by the smallest energy difference $\min\limits_{c^* \in C} E(c^*) - \min\limits_{c' \in C'} E(c')$ between the GMEC c^* and the lowest-energy conformation c' contained in the subtree beneath c. (B,C,D) The partition function can be projected onto the tree, in the form of concentric ring charts. Each node of the tree in (A) corresponds to an arc in (D). Arc angle for each node is proportional to the partition function contribution of all nodes in a subtree (80% vs. 20% for rotamer W_1 vs. rotamer W_2). Notably, high-energy conformations are shown as white gaps in (D). (E) For HIV-1 capsid protein bound to camelid V_HH domain, the binding reaction is broken down into two parts: change in conformational ensembles of HIV-1 capsid protein and camelid V_HH from R_1 and L_1 (conformational ensembles when not bound) to R_2 and L_2 (conformational ensembles when bound), and the formation of the protein-protein interface. The order of events is arbitrary and the overall binding mechanism can be deconstructed without regard to mechanism. The left ring charts ($R_1 + L_1$) show the distribution of the free conformational ensembles for HIV-1 capsid protein and camelid V_HH. The middle charts ($R_2 + L_2$) show the change in conformational distribution, where non-binding low-energy states have reduced statistical weight, and the right ring chart ($R_2 \cdot L_2$) represents the the bound conformational ensemble, where the internal energy of the bound complex increases due to binding. When energetic contributions from the conformational change and interface formation reactions are added, they give the thermodynamics of the overall coupled binding reaction. As described in the text (Sect. 6.1), bounds on ΔU_{conf}, ΔS_{conf}, $\Delta U_{interface}$, and $\Delta S_{interface}$ can all be estimated from the energy landscape approximation returned by *MARK**. (Color figure online)

GMEC comprises a much larger fraction of the corresponding partition function. Accordingly, our novel ring charts for the energy landscapes of the free, binding-competent, and bound states show visually how the bound and unbound states differ, emphasizing the novel capabilities of *MARK**, and the significance of modeling more than just the lowest-energy conformations when designing for affinity.

7 Conclusion

We presented a novel algorithm that not only efficiently bounds the partition function, but also computes a provably good approximation of the energy landscape, which bounds the energy of every conformation in the conformation space. $MARK^*$ is, to our knowledge, the first algorithm to do so. Previously, designers were limited to optimizing for the lowest-energy conformations for a limited number of predefined states, and could only approximate aggregate values such as internal energy or K_a. With $MARK^*$, we showed that designers can directly measure changes to the entire energy landscape, such as conformational rearrangement upon binding. With this capability, it even becomes possible to compare the energy landscapes of different sequences. That is, $MARK^*$ empowers designers to evaluate sequences not by low-energy conformations, but instead *by energy landscape*. Thus, $MARK^*$ enables not only faster design, but also a new potential strategy to design for *conformational dynamics* [5,42]. We believe that $MARK^*$ will not only accelerate existing designs, it will enhance future designs, and enable a novel, dynamics-based strategy for computational structure-based protein design.

Acknowledgements. We thank Goke Ojewole, Mark Hallen, Jeffrey Martin, Marcel Frenkel, Terrence Oas, Jane and Dave Richardson, Hong Niu, and all members of the lab for helpful discussions; Jeffrey Martin for software optimizations; and the NIH (R01-GM078031 and R01-GM118543 to BRD) for funding.

References

1. ClinicalTrials.gov Identifier: NCT02840474. NIAID and National Institutes of Health Clinical Center, September 2018. https://clinicaltrials.gov/ct2/results?cond=&term=VRC07

2. Chazelle, B., Kingsford, C., Singh, M.: A semidefinite programming approach to side chain positioning with new rounding strategies. INFORMS J. Comput. **16**(4), 380–392 (2004). https://doi.org/10.1287/ijoc.1040.0096

3. Chen, C.Y., Georgiev, I., Anderson, A.C., Donald, B.R.: Computational structure-based redesign of enzyme activity. Proc. Natl. Acad. Sci. USA **106**(10), 3764–9 (2009). https://doi.org/10.1073/pnas.0900266106

4. Dahiyat, B.I., Mayo, S.L.: De novo protein design: fully automated sequence selection. Science **278**(5335), 82–87 (1997)

5. Davey, J.A., Damry, A.M., Goto, N.K., Chica, R.A.: Rational design of proteins that exchange on functional timescales. Nat. Chem. Biol. **13**(12), 1280–1285 (2017)

6. Donald, B.R.: Algorithms in Structural Molecular Biology. MIT Press, Cambridge (2011)

7. Fleishman, S.J., Khare, S.D., Koga, N., Baker, D.: Restricted sidechain plasticity in the structures of native proteins and complexes. Protein Sci. **20**(4), 753–757 (2011). https://doi.org/10.1002/pro.604

8. Frederick, K.K., Marlow, M.S., Valentine, K.G., Wand, A.J.: Conformational entropy in molecular recognition by proteins. Nature **448**(7151), 325–329 (2007). https://doi.org/10.1038/nature05959

9. Frey, K.M., Georgiev, I., Donald, B.R., Anderson, A.C.: Predicting resistance mutations using protein design algorithms. Proc. Natl. Acad. Sci. U.S.A. **107**(31), 13,707–13,712 (2010). https://doi.org/10.1073/pnas.1002162107

10. Gainza, P., Nisonoff, H.M., Donald, B.R.: Algorithms for protein design. Curr. Opin. Struct. Biol. **39**, 16–26 (2016)

11. Gainza, P., Roberts, K.E., Donald, B.R.: Protein design using continuous rotamers. PLoS Comput. Biol. **8**(1), e1002335 (2012). https://doi.org/10.1371/journal.pcbi. 1002335

12. Georgiev, I., Donald, B.R.: Dead-end elimination with backbone flexibility. Bioinformatics **23**(13), i185–i194 (2007). https://doi.org/10.1093/bioinformatics/ btm197

13. Georgiev, I., Keedy, D., Richardson, J.S., Richardson, D.C., Donald, B.R.: Algorithm for backrub motions in protein design. Bioinformatics **24**(13), i196–i204 (2008). https://doi.org/10.1093/bioinformatics/btn169

14. Georgiev, I., Lilien, R.H., Donald, B.R.: Improved pruning algorithms and divide-and-conquer strategies for dead-end elimination, with application to protein design. Bioinformatics **22**(14), e174–e183 (2006). https://doi.org/10.1093/bioinformatics/ btl220

15. Georgiev, I., Lilien, R.H., Donald, B.R.: The minimized dead-end elimination criterion and its application to protein redesign in a hybrid scoring and search algorithm for computing partition functions over molecular ensembles. J. Comput. Chem. **29**(10), 1527–1542 (2008). https://doi.org/10.1002/jcc.20909

16. Georgiev, I., et al.: Design of epitope-specific probes for sera analysis and antibody isolation. Retrovirology **9**, P50 (2012)

17. Georgiev, I.S., et al.: Antibodies VRC01 and 10E8 neutralize HIV-1 with high breadth and potency even with IG-framework regions substantially reverted to germline. J. Immunol. **192**(3), 1100–1106 (2014). https://doi.org/10.4049/ jimmunol.1302515

18. Gilson, M.K., Given, J.A., Bush, B.L., McCammon, J.A.: The statistical-thermodynamic basis for computation of binding affinities: a critical review. Biophys. J. **72**(3), 1047–1069 (1997). https://doi.org/10.1016/S0006-3495(97)78756-3

19. Gorczynski, M.J., et al.: Allosteric inhibition of the protein-protein interaction between the leukemia-associated proteins Runx1 and CBFbeta. Chem. Biol. **14**(10), 1186–1197 (2007). https://doi.org/10.1016/j.chembiol.2007.09.006

20. Hallen, M.A., Donald, B.R.: CATS (coordinates of atoms by taylor series): protein design with backbone flexibility in all locally feasible directions. Bioinformatics **33**(14), i5–i12 (2017). https://doi.org/10.1093/bioinformatics/btx277

21. Hallen, M.A., Gainza, P., Donald, B.R.: Compact representation of continuous energy surfaces for more efficient protein design. J. Chem. Theory Comput. **11**(5), 2292–2306 (2015). https://doi.org/10.1021/ct501031m

22. Hallen, M.A., Jou, J.D., Donald, B.R.: LUTE (local unpruned tuple expansion): accurate continuously flexible protein design with general energy functions and rigid rotamer-like efficiency. J. Comput. Biol. **24**(6), 536–546 (2017). https://doi. org/10.1089/cmb.2016.0136

23. Hallen, M.A., Keedy, D.A., Donald, B.R.: Dead-end elimination with perturbations (DEEPer): a provable protein design algorithm with continuous sidechain and backbone flexibility. Proteins **81**(1), 18–39 (2013). https://doi.org/10.1002/ prot.24150

24. Hallen, M.A., et al.: OSPREY 3.0: open-source protein redesign for you, with powerful new features. J. Comput. Chem. **39**(30), 2494–2507 (2018)

25. Hart, P., Nilsson, N.J., Raphael, B.: A formal basis for the heuristic determination of minimum cost paths. IEEE Trans. SSC **4**, 100–114 (1968)

26. Hastings, W.: Monte Carlo sampling methods using Markov chains and their applications. Biometrika **57**(1), 97–109 (1970). https://doi.org/10.1093/biomet/57.1.97

27. Jou, J.D., Holt, G.T., Lowegard, A.U., Donald, B.R.: Supplementary information: minimization-aware recursive: K^* ($MARK^*$): A novel, provable partition function approximation algorithm that accelerates ensemble-based protein design and provably approximates the energy landscape (2019). (Available at http://www.cs.duke.edu/donaldlab/Supplementary/recomb19/markstar)

28. Kuhlman, B., Baker, D.: Native protein sequences are close to optimal for their structures. Proc. Natl. Acad. Sci. U.S.A. **97**(19), 10,383–10,388 (2000)

29. Leach, A.R., Lemon, A.P.: Exploring the conformational space of protein side chains using dead-end elimination and the A* algorithm. Proteins **33**(2), 227–239 (1998)

30. Leaver-Fay, A., et al.: Rosetta3: an object-oriented software suite for the simulation and design of macromolecules. Methods Enzymol. **487**, 545–574 (2011). https://doi.org/10.1016/B978-0-12-381270-4.00019-6

31. Lee, C., Subbiah, S.: Prediction of protein side-chain conformation by packing optimization. J. Mol. Biol. **217**(2), 373–388 (1991)

32. Lee, J.: New Monte Carlo algorithm: entropic sampling. Phys. Rev. Lett. **71**(2), 211–214 (1993). https://doi.org/10.1103/PhysRevLett.71.211

33. Lilien, R.H., Stevens, B.W., Anderson, A.C., Donald, B.R.: A novel ensemble-based scoring and search algorithm for protein redesign and its application to modify the substrate specificity of the gramicidin synthetase a phenylalanine adenylation enzyme. J. Comput. Biol. **12**(6), 740–761 (2005). https://doi.org/10.1089/cmb.2005.12.740

34. Lou, Q., Dechter, R., Ihler, A.T.: Anytime anyspace and/or search for bounding the partition function. In: AAAI (2017)

35. Lou, Q., Dechter, R., Ihler, A.T.: Dynamic importance sampling for anytime bounds of the partition function. In: NIPS (2017)

36. Lovell, S.C., Word, J.M., Richardson, J.S., Richardson, D.C.: The penultimate rotamer library. Proteins **40**(3), 389–408 (2000)

37. Nisonoff, H.: Efficient partition function estimation in computational protein design: probabalistic guarantees and characterization of a novel algorithm. B.S. thesis. Department of Mathematics, Duke University (2015). http://hdl.handle.net/10161/9746

38. Nosé, S.: A molecular dynamics method for simulations in the canonical ensemble. Mol. Phys. **52**(2), 255–268 (2006). https://doi.org/10.1080/00268978400101201

39. Ojewole, A., et al.: OSPREY predicts resistance mutations using positive and negative computational protein design. Methods Mol. Biol. **1529**, 291–306 (2017)

40. Ojewole, A.A., Jou, J.D., Fowler, V.G., Donald, B.R.: BBK* (Branch and Bound over K*): a provable and efficient ensemble-based protein design algorithm to optimize stability and binding affinity over large sequence spaces. J. Comput. Biol. **25**(7), 726–739 (2018). https://doi.org/10.1089/cmb.2017.0267

41. Qi, Y., et al.: Continuous interdomain orientation distributions reveal components of binding thermodynamics. J. Mol. Biol. **430**(18 Pt B), 3412–3426 (2018)

42. Reardon, P.N., et al.: Structure of an HIV-1-neutralizing antibody target, the lipid-bound gp41 envelope membrane proximal region trimer. Proc. Natl. Acad. Sci. U.S.A. **111**(4), 1391–1396 (2014). https://doi.org/10.1073/pnas.1309842111

43. Reeve, S.M., Gainza, P., Frey, K.M., Georgiev, I., Donald, B.R., Anderson, A.C.: Protein design algorithms predict viable resistance to an experimental antifolate. Proc. Natl. Acad. Sci. U.S.A. **112**(3), 749–754 (2015). https://doi.org/10.1073/pnas.1411548112

44. Roberts, K.E., Cushing, P.R., Boisguerin, P., Madden, D.R., Donald, B.R.: Computational design of a PDZ domain peptide inhibitor that rescues CFTR activity. PLoS Comput. Biol. **8**(4), e1002477 (2012). https://doi.org/10.1371/journal.pcbi.1002477

45. Roberts, K.E., Donald, B.R.: Improved energy bound accuracy enhances the efficiency of continuous protein design. Proteins **83**(6), 1151–1164 (2015). https://doi.org/10.1002/prot.24808

46. Roberts, K.E., Gainza, P., Hallen, M.A., Donald, B.R.: Fast gap-free enumeration of conformations and sequences for protein design. Proteins **83**(10), 1859–1877 (2015). https://doi.org/10.1002/prot.24870

47. Rudicell, R.S., et al.: Enhanced potency of a broadly neutralizing HIV-1 antibody in vitro improves protection against lentiviral infection in vivo. J. Virol. **88**(21), 12,669–12,682 (2014). https://doi.org/10.1128/JVI.02213-14

48. Sciretti, D., Bruscolini, P., Pelizzola, A., Pretti, M., Jaramillo, A.: Computational protein design with side-chain conformational entropy. Proteins **74**(1), 176–191 (2009). https://doi.org/10.1002/prot.22145

49. Silver, N.W., et al.: Efficient computation of small-molecule configurational binding entropy and free energy changes by ensemble enumeration. J. Chem. Theory Comput. **9**(11), 5098–5115 (2013). https://doi.org/10.1021/ct400383v

50. Simoncini, D., Allouche, D., de Givry, S., Delmas, C., Barbe, S., Schiex, T.: Guaranteed discrete energy optimization on large protein design problems. J. Chem. Theory Comput. **11**(12), 5980–5989 (2015). https://doi.org/10.1021/acs.jctc.5b00594

51. Stevens, B.W., Lilien, R.H., Georgiev, I., Donald, B.R., Anderson, A.C.: Redesigning the PheA domain of gramicidin synthetase leads to a new understanding of the enzyme's mechanism and selectivity. Biochemistry **45**(51), 15,495–15,504 (2006). https://doi.org/10.1021/bi061788m

52. Traoré, S., et al.: A new framework for computational protein design through cost function network optimization. Bioinformatics **29**(17), 2129–2136 (2013). https://doi.org/10.1093/bioinformatics/btt374

53. Tzeng, S.R., Kalodimos, C.G.: Protein activity regulation by conformational entropy. Nature **488**(7410), 236–240 (2012). https://doi.org/10.1038/nature11271

54. Valiant, L.G.: The complexity of computing the permanent. Theoret. Comput. Sci. **8**(2), 189–201 (1979)

55. Viricel, C., Simoncini, D., Barbe, S., Schiex, T.: Guaranteed weighted counting for affinity computation: beyond determinism and structure. In: Rueher, M. (ed.) CP 2016. LNCS, vol. 9892, pp. 733–750. Springer, Cham (2016). https://doi.org/10.1007/978-3-319-44953-1_46

Sparse Binary Relation Representations for Genome Graph Annotation

Mikhail Karasikov[1,2,3] ⓘ, Harun Mustafa[1,2,3] ⓘ, Amir Joudaki[1,2] ⓘ,
Sara Javadzadeh-No[1] ⓘ, Gunnar Rätsch[1,2,3(✉)] ⓘ, and André Kahles[1,2,3(✉)] ⓘ

[1] Department of Computer Science, ETH Zurich, 8092 Zurich, Switzerland
{raetsch,andre.kahles}@inf.ethz.ch
[2] University Hospital Zurich, Biomedical Informatics Research,
8091 Zurich, Switzerland
[3] SIB Swiss Institute of Bioinformatics, 1015 Lausanne, Switzerland
http://bmi.inf.ethz.ch

Abstract. High-throughput DNA sequencing data is accumulating in public repositories, and efficient approaches for storing and indexing such data are in high demand. In recent research, several graph data structures have been proposed to represent large sets of sequencing data and to allow for efficient querying of sequences. In particular, the concept of labeled de Bruijn graphs has been explored by several groups. While there has been good progress towards representing the sequence graph in small space, methods for storing a set of labels on top of such graphs are still not sufficiently explored. It is also currently not clear how characteristics of the input data, such as the sparsity and correlations of labels, can help to inform the choice of method to compress the graph labeling. In this work, we present a new compression approach, *Multi-BRWT*, which is adaptive to different kinds of input data. We show an up to 29% improvement in compression performance over the basic BRWT method, and up to a 68% improvement over the current state-of-the-art for de Bruijn graph label compression. To put our results into perspective, we present a systematic analysis of five different state-of-the-art annotation compression schemes, evaluate key metrics on both artificial and real-world data and discuss how different data characteristics influence the compression performance. We show that the improvements of our new method can be robustly reproduced for different representative real-world datasets.

Keywords: Sparse binary matrices · Binary relations · Genome graph annotation · Compression

1 Introduction

Over the past decade, there has been an exponential growth in the global capacity for generating DNA sequencing data [23]. Various sequencing efforts

© Springer Nature Switzerland AG 2019
L. J. Cowen (Ed.): RECOMB 2019, LNBI 11467, pp. 120–135, 2019.
https://doi.org/10.1007/978-3-030-17083-7_8

have started to amass data from populations of humans [1] and other organisms [24,25]. For these well studied organisms, already assembled reference sequences are the common starting point for comparative and functional analyses. Unfortunately, a large proportion of DNA sequencing data, in particular data originating from non-model organisms or collected in metagenomics studies, are lacking a genome reference. Whereas general guidelines exist for the former case [9], genome assembly for metagenomics is much less well defined. Its vastness and the currently lacking standards for indexing such data make an integrated analysis daunting even for field experts.

To make this host of data efficiently searchable, it is necessary to employ a search index. However, when building an index on the sequence data alone, only presence or absence of a query can be tested. To support relating queries to information such as source genomes, haplotypes, or functional annotations, additional labels must be associated with the index. To facilitate this, approaches for storing additional data on an indexed graph have been suggested, such as the gPBWT [17] for storing haplotype information as genome graphs or succinct representations of *labeled* de Bruijn graphs [4,12,14] for the representation of sets of sequences. In this context, dynamic representations of such data have also recently received attention [15,20].

The problem of efficiently representing these types of relations is also addressed in other fields. Commonly referred to as *compressed binary relations*, a growing body of theoretical work addresses such approaches [7]. Successful applications of similar techniques include the efficient representation of large web-graphs [8] and RDF data sets [5]. We will provide a more detailed description of some of these approaches in Sect. 2.

In this work, we present a new method for compressing abstract binary relations. Providing as background a comprehensive benchmark of existing compression schemes, we show that our approach has superior performance on both artificial and real-world data sets.

Our paper has the following structure: After introducing our notation, we begin by defining the abstract graph, and the associated annotation structures that we wish to compress (Sect. 2.1). We then provide descriptions of our proposed compression technique and competing methods (Sect. 2.2). Finally, we compare the compression performance of these methods on different types of graph annotations (Sect. 3) and close with a brief discussion of our results and an outlook on future work (Sect. 4).

2   Methods

After introducing our notation, we will give an overview of all methods implemented for this work and provide a description of our methodological contributions.

2.1   Preliminaries

We will operate in the following setting. We are given a k-dimensional de Bruijn graph over a set of given input sequences S. The node set V shall be defined as the set of all consecutive sub-sequences of length k (k-mers) of sequences in S

$$V = \{s_{i:i+k-1} \mid s \in S, \ i = 1, \ldots, |s| - k + 1\}, \tag{1}$$

where $s_{i:j}$ denotes the sub-sequence of s from position i up to and including position j, and $|s|$ is the length of s. A directed edge exists from node u to node v, if $u_{2:k} = v_{1:k-1}$.

In order to represent relations between sources of the input sequences S and the nodes V, we now define the concept of a *labeled de Bruijn graph* and proceed by discussing the more general problem of representing a graph labeling.

Each node $v \in V$ that we refer to as an *object* is assigned a finite set of labels $\ell(v) \subset L$. We represent this *graph labeling* as a binary relation $\mathcal{R} \subset V \times L$. A trivial representation of $\mathcal{R}$ taking $|V| \cdot |L|$ bits of space is a binary matrix $A \in \{0,1\}^{|V| \times |L|}$. We will use A^i and A_j to denote its rows and columns, respectively.

In the following sections, we will discuss various methods described in the recent literature and present our improvements in efficiently representing $\mathcal{R}$. In addition to minimal space, we also require that the following set of operations can be carried out efficiently on the compressed representation of $\mathcal{R}$:

query_labels$(v) = \{l \in L \mid (v,l) \in \mathcal{R}\}$ Given an object $v \in V$ (a k-mer in the underlying de Bruijn graph), return the set of labels $\ell(v)$ assigned to it.

query_objects$(l) = \{v \in V \mid (v,l) \in \mathcal{R}\}$ Given a label $l \in L$ (e.g., a genome or sample ID), return the set of objects assigned to that label.

query_relation(v,l) Given an object $v \in V$ and a label $l \in L$, check whether (v,l) is in the relation $\mathcal{R}$, query_relation$(v,l) = 1_{\{(v,l) \in \mathcal{R}\}}$.

2.2   Binary Relation Representation Schemes

For compressing the binary relation $\mathcal{R}$, we consider the following representations suggested in recent literature. As an abstraction, we will use the representation of $\mathcal{R}$ as a binary matrix $A \in \{0,1\}^{|V| \times |L|}$ (referred to as the *binary relation matrix*) to illustrate the individual methods.

Column-Major Sparse Matrix Representation. As a simple baseline technique, we compress the positions of the non-zero indices in each column independently using Elias-Fano encoding [18]. While this method does not take into account correlations between columns for compression, this feature allows for a trivial parallel construction implementation in which each column is computed in a separate process. For our experiments, this serves as the initial representation of the binary matrix, which is then queried during the construction of all other matrix representations.

Flat Row-Major Representation. As a second baseline method, this representation concatenates all rows of A into a joint vector that is subsequently compressed using Elias-Fano encoding. This approach, for instance, is used by VARI [14] and its extensions [3].

Rainbowfish. The current state-of-the-art for genome graph labeling is a row-major representation of the binary relation matrix A in which an optimal coding is constructed for the set of rows in A [4]. More precisely, let $A^{i_1}, \ldots, A^{i_r} \in \{0,1\}^{|L|}$ denote the unique rows of A, sorted by their number of occurrences in A in non-increasing order, where $r \leq |V|$. To encode A, we start by forming a matrix $A' \in \{0,1\}^{r \times |L|}$ of sorted unique rows, $A'_j = A^{i_j}$. Then we compress A' with the flat row-major representation using an RRR vector (named after the initials of the three original authors [21]) as the underlying storage technique and construct a *coding vector* $(i(v) - 1)_{v \in V}$, where $i(v)$ maps each node $v \in V$ to the index of the row in A' corresponding to the labeling of v. The coding vector is represented in a variable-length packed binary coding with a delimiter vector [4] compressed into an RRR vector [21].

Binary Relation Compressed with Wavelet Trees (BinRel-WT). This method involves a translation of the $|\mathcal{R}|$ non-zero elements of A into a string, which is then represented using a conventional wavelet tree [7]. Given the binary relation matrix, its set bits are iterated in row-major order and their respective column indices are stored contiguously in a string over the alphabet $\{1, \ldots, |L|\}$ represented with a wavelet tree that enables efficient queries. The numbers of set bits in each row of A are stored in a delimiter vector using unary coding and compressed into an RRR vector.

Hierarchical Compressed Column-Major Representation (BRWT). Described as Binary Relation Wavelet Trees (BRWT) in the original literature [7], in contrast to BinRel-WT, this representation directly acts on binary matrices without translation into a sequence. First, an index vector I with elements $I_i = \bigvee_j A^i_j$ is computed by merging all matrix columns through bitwise-OR operations on the rows and stored to represent the root of the tree. Then, the rows composed entirely of 0s are discarded from A and two equal-sized submatrices A' and A'' (which may contain rows composed entirely of 0s) of the binary relation matrix A are constructed by splitting A and are passed to the left and right children of the root. The compression proceeds recursively. Construction terminates when a node is assigned a single column, which is stored as its index column (see Fig. 1(a)). For reconstruction of the matrix elements, it is sufficient to only store the index vectors associated with each node of the BRWT tree.

In the next section we consider the problem of topology optimization during BRWT tree construction and propose *Multi-BRWT*, an extension of BRWT that allows its nodes to have arbitrary numbers of children as well as to arbitrarily distribute the columns of a parent node to its children. Afterwards, we propose a two-step approach for Multi-BRWT construction along with two specific

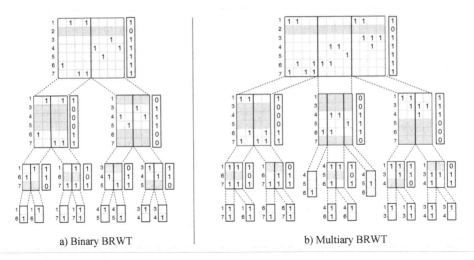

a) Binary BRWT   b) Multiary BRWT

Fig. 1. Schematic of hierarchical compressed column-major representations—(a) BRWT for the binary case. Grey rows correspond to all-zero rows, also indicated through the vector to the right of each matrix. Each child encodes only non-zero rows of the submatrix passed to it by its respective parent. Numbers to the left of each matrix are the respective row-indices in the initial matrix. (b) Multi-BRWT in the multiary case. Notation is as in the binary case. Stored vectors are shown in red (Color figure online).

algorithms as its implementation for improving the compression performance of Multi-BRWT.

2.3   Multiary, Topology-Optimized BRWTs

Our first extension to the BRWT scheme is the introduction of an n-ary tree topology, Multi-BRWT (Split n), allowing for matrices to be vertically split into more than two submatrices (see Fig. 1(b)). The construction and querying for Multi-BRWT (Split n) is analogous to the case of binary BRWT. In computational experiments on artificial and real-world data we show that in most cases, Multi-BRWT (Split n) with arity greater than two provides a higher compression ratio than the simple binary BRWT scheme (see Sect. 3). Note that Multi-BRWT (Split n) with the maximum allowed arity $n = |L|$ is equivalent to the baseline column-major sparse matrix representation as it keeps all columns of the input binary relation matrix unchanged except for the case when the input matrix has all-zero rows.

To proceed with our second extension, let us consider binary relations with far fewer labels than objects, $|L| \ll |V|$, a condition that is commonly met in annotated genome graphs from biological data. In these contexts, the number of k-mers is usually in the billions and the number of labels is on the order of thousands (see Sect. 3.2).

Our second extension consists in introducing arbitrary assignments of columns from the matrices encoded in the nodes of the Multi-BRWT to their children. These assignments are represented by dictionaries stored in the Multi-BRWT nodes, but the $|L| \ll |V|$ constraint makes the space overhead from storing these negligible compared to the space needed to encode the index vectors. Thus, we exclude the problem of representing these assignments from further consideration and leave that as a small technical detail.

We now focus on the problem of constructing a Multi-BRWT tree structure that satisfies certain local optimality conditions with respect to compression ratio.

Problem Setting. Let us set this problem formally. Given a binary matrix A, let $\mathcal{T}$ be the set of all Multi-BRWT trees representing A (i.e., the set of all rooted trees with $|L|$ labeled leaves). Let $Size(I)$ denote the size of a compressed binary vector I in bits. For instance, if I is of length n with m set bits, $Size(I) = n$ for an uncompressed bit vector, and $Size(I) \approx \lceil \log_2 \binom{n}{m} \rceil$ for RRR vectors [21]. We then neglect the space required for dictionaries defining the column assignments and we define the size of the Multi-BRWT tree $T \in \mathcal{T}$ as the space required to store all its index vectors including the vectors in leaves:

$$Size(T) := \sum_{i \in N} Size(I_i), \tag{2}$$

where N is the set of all nodes of the Multi-BRWT tree T and I_i corresponds to the index vector stored in node i. Thus, we wish to find an optimal Multi-BRWT tree by minimizing the storage space,

$$T^* = \arg\min_{T \in \mathcal{T}} Size(T). \tag{3}$$

We will refer to this as the MULTI-BRWT problem.

Optimized Multi-BRWT Construction. By analogy to the NoSQL table compaction problem [10], it can be shown that Multi-BRWT constrained on the space of binary trees with the uncompressed bit vector representation as the underlying structure for storing the index vectors is NP-hard. Thus, we propose a two-step approach for finding a good Multi-BRWT structure (see Fig. 2). First, we build a binary Multi-BRWT tree by hierarchical clustering of the index vectors according to their similarity, the number of shared set bits. Then, we optimize the arity of the chosen Multi-BRWT by selecting a node subset N' which includes the root and leaves of the base Multi-BRWT tree, $\{r, v_1, \ldots, v_{|L|}\} \subset N' \subset N$. To keep the resulting Multi-BRWT tree valid (allowing for reconstruction of the initial matrix A), we reassign all nodes in N' to their nearest common ancestors remaining in N'.

As a specific implementation of the proposed two-step construction approach, we consider two heuristic greedy optimization procedures. In the first step, we

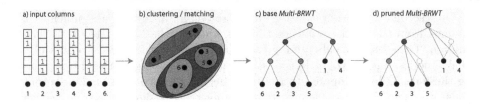

Fig. 2. Schematic describing the construction of *Multi-BRWT*—(a) The columns of the input binary matrix depicted as numbered black dots are considered independently. (b/c) Columns are hierarchically pair-matched based on number of shared entries, forming the base Multi-BRWT topology. (d) Pruning internal nodes of Multi-BRWT to optimize the tree structure for a smaller representation size.

perform greedy matching of the index vectors starting from the columns of the input binary relation matrix, and repeat recursively for the aggregated parent index vectors until we merge all into a single index vector placed in the root. In the second step of the construction approach, we consider another greedy algorithm for optimizing the size of the Multi-BRWT tree by removing some of its internal nodes and thereby increasing the arity of the tree.

Greedy Pairwise Matching for Finding a Base Multi-BRWT Approximation. To find an initial approximate solution to the MULTI-BRWT problem (base Multi-BRWT), we propose a greedy algorithm in which an initial greedy pairwise matching (GPM) step is performed on the columns of the input binary relation matrix A to optimize their initial order prior to construction (see Fig. 2(a–c)). Given the input columns $A_1, \ldots, A_{|L|}$ and their corresponding object queries $o_i = \texttt{query\_objects}(i)$, we first compute cardinalities of their pairwise intersections $s_{ij} = |o_i \cap o_j|$. Then, we sort all the computed similarities $\{s_{ij}\}$ in non-increasing order and match pairs of columns greedily. Afterwards, we compute the aggregated index columns by merging the matched columns through bitwise-OR operations to form the index vectors and repeat this algorithm recursively.

Efficient Pairwise Distance Estimation. The proposed greedy approximation method takes as input a matrix of pairwise column similarities $\{s_{ij}\}$. For m input columns of length n, computing each entry of this matrix costs $\mathcal{O}(n)$, and thus, the time complexity of computing the full similarity matrix is $\mathcal{O}(nm^2)$, which is a considerable overhead for datasets with a typical size of $m \sim 10^3$ and $n \sim 10^9$. To make the estimation of the pairwise similarities cheaper, we approximate these on a submatrix composed of rows sampled randomly from matrix A. Moreover, we prove the following lemma to show that using just $\mathcal{O}\left(\log(m)/\varepsilon^2\right)$ random rows is sufficient for approximating the pairwise similarities with a small relative error ε with high probability, if each column has a sufficiently large number of set bits.

Lemma 1 (Subsampling Lemma). *Suppose we are given subsets of a universe set, $o_1, \ldots, o_m \subset \{1, \ldots, n\}$, with the minimum cardinality $d = \min_{i=1}^{m} |o_i|$, $d > 0$. We sample the elements of $\{1, \ldots, n\}$ independently with the same probability p, to form a sampled set of objects $S \subset \{1, \ldots, n\}$ and define subsampled* sets as $\tilde{o}_i = o_i \cap S$. *Consider the union cardinalities* $u_{ij} = |o_i \cup o_j|$ *with their approximators* $\hat{u}_{ij} = \frac{1}{p}|\tilde{o}_i \cup \tilde{o}_j|$. *For all* $0 < \varepsilon < 1$, $0 < \delta < 1$, *and*

$$p \geq \min\{\frac{3\ln(\frac{m^2+m}{\delta})}{d\varepsilon^2}, 1\},$$

we claim

$$\Pr\left(\bigcap_{i,j=1}^{m} \left\{|\hat{u}_{ij} - u_{ij}| < \varepsilon u_{ij}\right\}\right) \geq 1 - \delta.$$

See Supplementary Section 2 for proof.

According to Lemma 1, with the subsampling technique we can approximate the union cardinalities up to an ε-fraction with high probability. Similar bounds can be obtained for sufficiently large intersection cardinalities, estimated in the proposed greedy pairwise matching algorithm.

Refining Multi-BRWT by Pruning. Starting the procedure in the leaves' parents and applying it to each node except for the root recursively, we estimate the cost of removing each current node by the following formula

$$\mathrm{CostRem}(v) = \sum_{c \in \mathrm{Children}(v)} Size(I'(c))$$

$$- \left[Size(I(v)) + \sum_{c \in \mathrm{Children}(v)} Size(I(c))\right], \quad (4)$$

where $I(v)$ denotes the index vector stored in the node v and $I'(c)$ denotes the updated index vector that would be stored in the node c if its parent v was removed and the node c was reassigned to its grandparent. Now we simplify the formula for estimating the cost of removing a node in Multi-BRWT by introducing an assumption that the size of bit vector I of length n with m set bits is fully defined by these two parameters, i.e. $Size(I) = Size(n, m)$. Now, it is easy to see that after reassigning the node c with the index vector $I(c)$ of length n_c with m_c set bits to the parent of its parent v with index vector $I(v)$ of length n_v with m_v set bits, the node c updates and replaces its index vector $I(c)$ with a vector $I'(c)$ of length n_v with m_c set bits. This provides us with the following simplified formula for estimating the cost of removing a node from the Multi-BRWT tree

$$\mathrm{CostRem}(v) = \sum_{c \in \mathrm{Children}(v)} Size(n_v, m_c)$$

$$- \left[Size(n_v, m_v) + \sum_{c \in \mathrm{Children}(v)} Size(n_c, m_c)\right]. \quad (5)$$

Formula (5) can be efficiently computed without rebuilding the current structure of the Multi-BRWT. As a result, a decision about removing node v from the

Multi-BRWT is made if the cost $CostRem(v)$ is negative, leading thereby to a decrease of the Multi-BRWT in size. In our practical implementation we use the following formula for approximating the size required for storing an RRR bit vector [16] with block size t: $Size(n, m) = \lceil \log_2 \binom{n}{m} \rceil + n \lceil \log_2(t+1)/t \rceil$.

2.4 Implementation Details

We implement the underlying de Bruijn graph as a hash table storing k-mers packed into 64bit integers with 64bit indexes assigned to the k-mers, or as a complete de Bruijn graph represented by a mapping of k-mers to 4^k row indexes of the binary relation matrix.

In the column-major representation, the columns of the binary relation matrix are stored using SD vectors implemented in sdsl-lite [11]. The same data structure is used for storing the single long vector in the row flat representation.

BinRel-WT (sdsl) compressor uses the implementation of wavelet tree from the sdsl-lite library, using an RRR vector to store its underlying bit vector. The delimiter vector uses the RRR vector implementation from sdsl-lite.

The BinRel-WT compressor uses the binary relation implementation from https://github.com/dieram3/binrel_wt. This implementation stores the underlying bit vector of the wavelet tree in uncompressed form.

Our BRWT is implemented as a tree in memory, compressing the index vectors as RRR vectors. To avoid multiple passes through the matrix rows, we construct the BRWT using a bottom-up approach. Given a fixed clustering of the matrix columns, the leaves of the BRWT are constructed first, followed by their parents constructed for the index vectors propagated from the children nodes. To speed up the greedy matching algorithm, we sample randomly 10^6 rows in each experiment and use those to approximate the number of bits shared in the input columns and the index vectors during the Multi-BRWT construction. When optimizing the tree arity (as described in Sect. 2.3), we use the formula $Size(n, m) = \lceil \log_2 \binom{n}{m} \rceil + n \lceil \log_2(t+1)/t \rceil$ as an estimate for the size of bit vector I of length n with m set bits, which is provided by the authors of sdsl-lite for the implementation of RRR vectors [11]. We use a block size of $t = 63$.

All SD vectors are constructed with default template parameters, while all RRR vectors are constructed with a block size of 63.

Code Availability. All methods implemented and evaluated in this paper are available at https://github.com/ratschlab/genome_graph_annotation.

2.5 Data

Simulated Data. To profile our compressors, we generated several different series of synthetic binary matrices of varying densities (see Supplementary Section 1 for a more detailed description). In total we generated three different kinds of series: (i) random matrices with uniformly distributed set bits, (ii) initially generated random matrix rows duplicated and permuted randomly, (iii) initially generated random matrix columns duplicated and permuted randomly. The motivation behind these series is the following. The best performing state-of-the-art compressors exploit redundancy between rows of the binary relation matrix [20]. However, the usual structure of annotated de Bruijn graphs often implies a correlation structure on the columns not necessarily leading to redundant rows, for instance when the sequences of many similar or closely related samples are inserted. While for a small (and sufficiently highly correlated) number of columns this correlation translates into rows and increases the number of redundant ones, for larger label sets this is usually not the case. Thus, approaches exploiting correlation structure on the columns might fare better. To test this hypothesis, we generated three different kinds of synthetic data, reflecting uncorrelated rows/columns, redundant rows, and redundant columns for series (i), (ii), and (iii), respectively. Please note, that approach (ii) is the most favorable for the state-of-the-art, as row redundancy rather then high correlation is simulated.

Real-World Data. For evaluating all approaches in a real-world setting, we have chosen two data sets well-known in the community and representative of typical applications.

Kingsford Human RNA-Seq. This dataset consists of 2,652 Human RNA-Seq experiments originally drawn from [22] and subsequently used in [20] for comparison.

NCBI RefSeq. This dataset consists of all 79,448 reference sequences from Release 88 of the NCBI RefSeq database [19]. Each sequence has been annotated with its associated `family` rank taxonomic ID from the NCBI Taxonomy [2]. This results in a total of 3,173 unique labels for the sequences.

3 Results and Discussion

3.1 Experiments on Artificial Data

Based on the artificial dataset described in Sect. 2.5, we evaluated how the compression performance changes depending on the characteristics of the input binary relation matrix A of a simple structure.

Dependency of Compression Ratio on Matrix Structure. One of the key characteristics of the binary relation matrix A is its density, the number of set bits divided by the total number of entries in A. For reference, the labels for a sequencing-based de Bruijn graphs typically exhibit very low densities, commonly <0.5%. Especially in this low-density region, we find that the properties of the binary relation matrix have a strong effect on the compression ratio of individual methods. A second determinant of performance is whether any assumptions are made on the properties of the data.

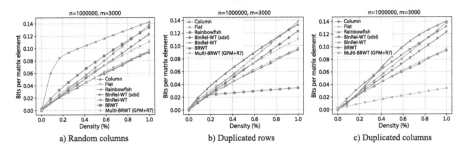

Fig. 3. Size of the representation of $A \in \{0,1\}^{10^6 \times 3 \cdot 10^3}$ with densities $d < 0.01$ using different approaches: (a) uniformly random bits; (b) uniformly random rows with multiplicity 5; (c) uniformly random columns with multiplicity 5. We expect approach (c) to be best reflecting the real-world data of a de Bruijn graph built on related sequences.

On sparse, fully random data, the baseline compressors fare very well (Fig. 3(a)), as no assumptions can be made about relationships. Notably, Rainbowfish, which exploits redundancy among the rows, generates considerable overhead for very low densities. In the field of BRWT methods, the Multi-BRWT is closest to the best performing choices.

In the setting of redundant rows (data set ii; Fig. 3(b)), as expected, Rainbowfish shows the strongest performance, clearly exploiting the row redundancy. Again, among the BRWT methods the Multi-BRWT performs best.

Finally, in the setting that comes closest to a typical task of labeling de Bruijn graphs derived from sequencing data (Fig. 3(c)), the Multi-BRWT approach shows superior performance. Exploiting the correlated columns of the matrix, Multi-BRWT achieves a 5-fold improvement in compression ratio compared to Rainbowfish and more than 2-fold compared to the closest competitor. Notably, the baseline binary BRWT has no advantage over the other baseline methods. Further, we observe that this performance gain increases with the total number of columns in the matrix (Supplemental Figures 1 and 2).

3.2   Experiments on Real-World Data

To compare the compression performance of the considered methods under a variety of conditions, we have constructed two test datasets that exhibit different matrix sparsity characteristics.

Kingsford Human RNAseq (2,652 Read Sets). We filtered the 2,652 raw sequencing read sets with the KMC [13] tool to extract frequent unique *canonical* k-mers (defined as the lexicographical minimum of the k-mer and its reverse complement) from each ($k = 20$). We used the same k value and thresholds for the k-mer frequency level as [20]. Using the k-mers extracted, we constructed a de Bruijn graph with 3,693,178,415 nodes and annotated these with their source read sets, which resulted in 2,586 labels (66 filtered read sets were empty) and a binary relation (annotation) matrix of density $\sim$0.19%. As a baseline for comparison, we used the straightforward column-compressed annotation, which required a total of 36.56 Gigabytes (Gb) of space. We used this as a starting point to convert the annotation into the other formats.

Table 1. The measured size of the compressed binary relation matrix for different representations, in Gigabytes (Gb).

Methods	Kingsford	RefSeq
Column	36.56	80.18
Flat	41.21	121.60
Rainbowfish	23.16	136.65
BinRel-WT	49.57	N/A
BinRel-WT (sdsl)	31.44	150.59
BRWT	**14.05**	**57.24**
Multi-BRWT (Split 3)	13.20	53.95
Multi-BRWT (Split 5)	**13.01**	**53.09**
Multi-BRWT (Split 7)	13.27	53.54
Multi-BRWT (Split 10)	13.54	54.77
Multi-BRWT (Split 13)	14.10	56.25
Multi-BRWT (GPM)	10.60	50.13
Multi-BRWT (GPM + Relax 3)	10.16	47.20
Multi-BRWT (GPM + Relax 5)	**9.94**	44.22
Multi-BRWT (GPM + Relax 7)	**9.94**	44.03
Multi-BRWT (GPM + Relax 10)	9.95	43.73
Multi-BRWT (GPM + Relax 20)	9.95	**43.62**

The results are summarized in Table 1. As expected, the simple row-based and BinRel-WT representations require more than 30 Gb in total. The current state-of-the-art method, Rainbowfish, reduces this by 23% to 23.16 Gb, exploiting the redundancy of rows in the input matrix. The basic BRWT benefits from the column correlation and drastically improves on Rainbowfish, showing a 39% lower size. We further reduce this size through our generalized approach using Multi-BRWT. While some increase in arity reduces size compared to the binary case, a higher arity does not necessarily translate into lower space, as certain submatrices do not benefit from being grouped. The smallest fixed-arity representation is Multi-BRWT (Split 5), requiring 13 Gb of storage space and 222 min of compute time to construct with four threads.

We improved the compression performance of a binary BRWT through the greedy pairwise matching (GPM) procedure described in Sect. 2.3. This strategy further decreases the size by another 18% to 10.6 Gb. Finally, optimizing the tree topology using the GPM procedure and selectively removing internal nodes (reassigning children to their grandparents) while maintaining a constraint on each node's maximum number of children, leads to the smallest space achieved in our experiments. By applying this technique, we decrease the required space to 9.94 Gb (Multi-BRWT (GPM + Relax 5), with at most 5 children for each node). This is a 29% improvement over the basic BRWT representation and a 57% improvement over Rainbowfish. The Multi-BRWT (GPM + Relax 5) representation took 187 and 141 min of the compute time with 30 threads for the first and the second stages of the construction algorithm, respectively.

RefSeq Reference Genomes. Compression of the complete RefSeq genome annotation (release 88) resulted in a de Bruijn graph of dimension $k = 15$ containing $n = 1,073,741,824$ nodes, leading to a binary relation matrix of n rows and $m = 3,173$ columns with density $\sim 3.8\%$, which is relatively high for a genome graph annotation and can be explained by the small k-mer size used.

This is a substantially larger dataset with less dependency between labels (columns). With the Multi-BRWT (GPM + Relax 20) representation, we were able to achieve a compressed storage size of only 43.6 Gb (Table 1). Conversion from the column-compressed representation to Multi-BRWT (GPM + Relax 20) took 625 and 733 min of the compute time with 30 threads for the first and the second Multi-BRWT construction stages, respectively, which is quite reasonable for a real-world setting.

Also here, the basic BRWT method improves drastically over the column compressed baseline (29%), and the Multi-BRWT approach considerably surpasses the basic BRWT method (24% redunction in size). One can see that the state-of-the-art method Rainbowfish performs very poorly on the RefSeq dataset, which can be explained by the high density of the annotation matrix.

The construction of the BinRel-WT representation exceeded our available RAM (2Tb).

All experiments were performed on a Intel(R) Xeon(R) CPU E7-8867 v3 (2.50 GHz) processor from ETH's shared high-performance compute systems.

Supplementary Results. Compression ratios for methods using RRR vectors of block size 127 can be found in the Supplementary Materials.

4   Conclusion

We have presented a series of compressed representation methods for binary relations, building upon and improving on the existing literature. By generalizing BRWTs to multiary trees with improved partitioning schemes and adaptive arity to reduce data representation overhead, we have improved on state-of-the-art compression techniques for both simulated and real-world biological datasets.

We have shown that the structure of the input data has a strong influence on the compression performance and methods such as Rainbowfish benefit from presence of redundancy in rows or their correlations (when multiple objects carry a similar set of labels). It is noteworthy that in a real-world setting, where more and more labels are added to the set, the number of redundant rows decreases (ultimately leading to a set of mostly independent rows) and these methods work less well. Interestingly, it is especially this setting that regularly occurs in the labeling of genome graphs, where an underlying set of (related) sequences is assigned a growing set of different labels.

We have presented a method that copes very well with an increasing number of related columns as well as with the increasing density of the compressed binary matrix, and we showed that this results in considerable performance gains on both synthetic and typical real-world data. Our method, Multi-BRWT, led to a 24–29% reduction in size compared to the basic BRWT scheme on real-world data, and to a 57–68% reduction compared to the closest state-of-the-art method for compressing graph annotations, Rainbowfish.

A natural extension of this work will involve the utilization of dynamic vectors in the underlying storage of BRWTs to allow for their use in dynamic database contexts. Of particular interest are the ability to rearrange columns and use of dynamic compressed structures to avoid expensive decompression and recompression steps when performing updates.

Another interesting direction is the development of hybrid BRWT schemes that take the shape of Multi-BRWT but assign multiple columns to the leaves of the tree, using arbitrary schemes for compressing these. This would take advantage of both column and row structure in the binary relation matrix. These approaches are also beneficial for tackling the problem of achieving similar time complexities for both object and label queries on the compressed representation of the binary relations.

Overall, we conclude that, despite the advancements in compression over the recent years, there is still much room and many degrees of freedom in compressor design for further improvement.

Supplementary Materials. Supplementary materials may be accessed via the bioRxiv pre-print located at https://doi.org/10.1101/468512.

Acknowledgements. We would like to thank the members of the Biomedical Informatics group for fruitful discussions and critical questions, and Torsten Hoefler and Mario Stanke for constructive feedback on the graph setup. Harun Mustafa and Mikhail Karasikov are funded by the Swiss National Science Foundation grant #407540_167331 "Scalable Genome Graph Data Structures for Metagenomics and Genome Annotation" as part of Swiss National Research Programme (NRP) 75 "Big Data".

References

1. UK10K Project (2015). Accessed 2 Nov 2018. https://www.uk10k.org
2. Agarwala, R., et al.: Database resources of the national center for biotechnology information. Nucleic Acids Res. (2017). https://doi.org/10.1093/nar/gkw1071

3. Alipanahi, B., Muggli, M.D., Jundi, M., Noyes, N., Boucher, C.: Resistome SNP Calling via Read Colored de Bruijn Graphs. bioRxiv (2018). https://doi.org/10. 1101/156174

4. Almodaresi, F., Pandey, P., Patro, R.: Rainbowfish: a succinct colored de Bruijn graph representation. In: Schwartz, R., Reinert, K. (eds.) 17th International Workshop on Algorithms in Bioinformatics (WABI 2017). Leibniz International Proceedings in Informatics (LIPIcs), vol. 88, pp. 18:1–18:15. Schloss Dagstuhl–Leibniz-Zentrum fuer Informatik, Dagstuhl (2017). https://doi.org/10.4230/LIPIcs.WABI. 2017.18

5. Álvarez-García, S., Brisaboa, N.: Compressed k2-triples for full-in-memory RDF engines. arXiv preprint arXiv:1105.4004 (2011)

6. Baraniuk, R., Davenport, M., DeVore, R., Wakin, M.: A simple proof of the restricted isometry property for random matrices. Constr. Approximation (2008). https://doi.org/10.1007/s00365-007-9003-x

7. Barbay, J., Claude, F., Navarro, G.: Compact binary relation representations with rich functionality. Inf. Comput. (2013). https://doi.org/10.1016/j.ic.2013.10.003

8. Brisaboa, N.R., Ladra, S., Navarro, G.: k^2-trees for compact web graph representation. In: Karlgren, J., Tarhio, J., Hyyrö, H. (eds.) SPIRE 2009. LNCS, vol. 5721, pp. 18–30. Springer, Heidelberg (2009). https://doi.org/10.1007/978-3-642-03784-9_3

9. Church, D.M., et al.: Modernizing reference genome assemblies. PLoS Biol. **9**(7), e1001,091 (2011). https://doi.org/10.1371/journal.pbio.1001091

10. Ghosh, M., Gupta, I., Gupta, S., Kumar, N.: Fast compaction algorithms for NoSQL databases. In: Proceedings of International Conference on Distributed Computing Systems (2015). https://doi.org/10.1109/ICDCS.2015.53

11. Gog, S., Beller, T., Moffat, A., Petri, M.: From theory to practice: plug and play with succinct data structures. In: Gudmundsson, J., Katajainen, J. (eds.) SEA 2014. LNCS, vol. 8504, pp. 326–337. Springer, Cham (2014). https://doi.org/10. 1007/978-3-319-07959-2_28

12. Iqbal, Z., Caccamo, M., Turner, I., Flicek, P., McVean, G.: De novo assembly and genotyping of variants using colored de Bruijn graphs. Nat. Genet. (2012). https:// doi.org/10.1038/ng.1028

13. Kokot, M., Długosz, M., Deorowicz, S.: KMC 3: counting and manipulating k-mer statistics. Bioinformatics **33**(17), 2759–2761 (2017). https://doi.org/10.1093/ bioinformatics/btx304

14. Muggli, M.D., et al.: Succinct colored de Bruijn graphs. Bioinformatics (2017). https://doi.org/10.1093/bioinformatics/btx067

15. Mustafa, H., et al.: Dynamic compression schemes for graph coloring. Bioinformatics (2018). https://doi.org/10.1093/bioinformatics/bty632

16. Navarro, G., Providel, E.: Fast, small, simple rank/select on bitmaps. In: Klasing, R. (ed.) SEA 2012. LNCS, vol. 7276, pp. 295–306. Springer, Heidelberg (2012). https://doi.org/10.1007/978-3-642-30850-5_26

17. Novak, A., Paten, B.: A graph extension of the positional Burrows Wheeler transform and its applications. Algorithms Mol. Biol. (2016). https://doi.org/10.1101/ 051409

18. Okanohara, D., Sadakane, K.: Practical entropy-compressed rank/select dictionary. In: Proceedings of the Meeting on Algorithm Engineering & Experimiments, pp. 60–70. Society for Industrial and Applied Mathematics (2007). https://dl.acm.org/ citation.cfm?id=2791194

19. O'Leary, N.A., et al.: Reference sequence (RefSeq) database at NCBI: current status, taxonomic expansion, and functional annotation. Nucleic Acids Res. (2016). https://doi.org/10.1093/nar/gkv1189

20. Pandey, P., Almodaresi, F., Bender, M.A., Ferdman, M., Johnson, R., Patro, R.: Mantis: a fast, small, and exact large-scale sequence-search index. Cell Syst. (2018). https://doi.org/10.1016/j.cels.2018.05.021

21. Raman, R., Raman, V., Satti, S.R.: Succinct indexable dictionaries with applications to encoding k-ary trees, prefix sums and multisets. ACM Trans. Algorithms **3**(4) (2007). https://doi.org/10.1145/1290672.1290680

22. Solomon, B., Kingsford, C.: Improved search of large transcriptomic sequencing databases using split sequence bloom trees. J. Comput. Biol. **25**(7), 755–765 (2018). https://doi.org/10.1089/cmb.2017.0265

23. Stephens, Z.D., et al.: Big data: astronomical or genomical? PLoS Biol. (2015). https://doi.org/10.1371/journal.pbio.1002195

24. Weigel, D., Mott, R.: The 1001 genomes project for Arabidopsis thaliana. Genome Biol. **10**(5), 107 (2009). https://doi.org/10.1186/gb-2009-10-5-107

25. Zhang, G.: Bird sequencing project takes off. Nature **522**(7554), 34–34 (2015). https://doi.org/10.1038/522034d, http://www.nature.com/articles/522034d

How Many Subpopulations Is Too Many? Exponential Lower Bounds for Inferring Population Histories

Younhun Kim, Frederic Koehler[✉], Ankur Moitra, Elchanan Mossel, and Govind Ramnarayan

Massachusetts Institute of Technology,
77 Massachusetts Avenue, 02139 Cambridge, USA
{younhun,fkoehler,moitra,elmos,govind}@mit.edu

Abstract. Reconstruction of population histories is a central problem in population genetics. Existing coalescent-based methods, like the seminal work of Li and Durbin (*Nature*, 2011), attempt to solve this problem using sequence data but have no rigorous guarantees. Determining the amount of data needed to *correctly reconstruct* population histories is a major challenge. Using a variety of tools from information theory, the theory of extremal polynomials, and approximation theory, we prove new sharp information-theoretic lower bounds on the problem of reconstructing *population structure*—the history of multiple subpopulations that merge, split and change sizes over time. Our lower bounds are *exponential* in the number of subpopulations, even when reconstructing recent histories. We demonstrate the sharpness of our lower bounds by providing algorithms for distinguishing and learning population histories with matching dependence on the number of subpopulations.

Keywords: Population size histories · Mixtures of exponentials · Sample complexity

1 Introduction

1.1 Background: Inference of Population Size History

A central task in population genetics is to reconstruct a species' *effective population size* over time. Coalescent theory [20] provides a mathematical framework for understanding the relationship between effective population size and genetic variability. In this framework, observations of present-day genetic variability—captured by DNA sequences of individuals—can be used to make inferences about changes in population size over time.

There are many existing methods for estimating the size history of a *single population* from *sequence data*. Some rely on Maximum Likelihood methods [14, 21–23] and others utilize Bayesian inference [5, 8, 19] along with a variety of simplifying assumptions. A well-known work of Li and Durbin [14] is based on

© Springer Nature Switzerland AG 2019
L. J. Cowen (Ed.): RECOMB 2019, LNBI 11467, pp. 136–157, 2019.
https://doi.org/10.1007/978-3-030-17083-7_9

using sequence data from just a single human (a single pair of haplotypes) and revolves around the assumption that coalescent trees of alleles along the genome satisfy a certain conditional independence property [15]. By and large, methods such as these do not have any associated provable guarantees. For example, Expectation-Maximization (EM) is a popular heuristic for maximizing the likelihood but can get stuck in a local maximum. Similarly, Markov Chain Monte Carlo (MCMC) methods are able to sample from complex posterior distributions if they are run for a long enough time, but it is rare to have reasonable bounds on the mixing time. In the absence of provable guarantees, simulations are often used to give some sort of evidence of correctness.

Under what sorts of conditions is it possible to infer a single population history? Kim, Mossel, Rácz and Ross [11] gave a strong lower bound on the number of samples needed even when one is given exact coalesence data. In particular, they showed that the number of samples must be at least exponential in the number of generations. Thus there are serious limitations to what kind of information we can hope to glean from (say) sequence data from a single human individual. In a sense, their work provides a quantitative answer to the question: *How far back into the past can we hope to reliably infer population size, using the data we currently have?* We emphasize that although they work in a highly idealized setting, this only makes their problem easier (e.g. assuming independent inheritance of loci along the genome and assuming that there are no phasing errors) and thus their lower bounds more worrisome.

1.2   Our Setting: Inference of Multiple Subpopulation Histories

A more interesting and challenging task is the reconstruction of *population structure*, which refers to the sub-division of a single population into several subpopulations that merge, split, and change sizes over time. There are two well-known works that attack this problem using coalescent-based approaches. Both use sequence data to infer population histories where present-day subpopulations were formed via divergence events of a single, ancestral population in the distant past. The first is Schiffels and Durbin [21], who used their method to infer the population structure of nine human subpopulations up to about 200,000 years into the past. More recently, Terhorst, Kamm and Song [23] inferred population structures of up to three human subpopulations. Just as in the single population case, these methods do not come with provable guarantees of correctness due to the simplifying assumptions they invoke and the heuristics they employ.

As for theoretical work, the lower bounds proven for single population trivially carry over to the setting of inferring population structure. However, the lower bound in [11] only applies when we are trying to reconstruct events in the distant past, leading us to a natural question: can we infer *recent* population structure, but, when there are multiple subpopulations?

In this paper, we establish strong limitations to inferring the population sizes of multiple subpopulation histories using pairwise coalescent trees. We prove sample complexity lower bounds that are *exponential in the number of subpopulations*, even for reconstructing recent histories. Our results provide a quantitative answer to the question, *Up to what granularity of dividing a population into multiple subpopulations, can we hope to reliably infer population structure?*

Our methods incorporate tools from information theory, approximation theory, and analysis (from [25]). To complement our lower bounds, we also give an algorithm for hypothesis testing based on the celebrated Nazarov-Turán lemma [18]. Our upper and lower bounds match up to constant factors and establish sharp bounds for the number of samples needed to distinguish between two known population structures as a function of the number of subpopulations. Finally, for the more general problem of learning the population structure (as opposed to testing which of two given population structures is more accurate) we give an algorithm with provable guarantees based on the Matrix Pencil Method [9] from signal processing. We elaborate on our results in Sect. 1.4.

1.3   Modeling Assumptions

Our results will apply under the following assumptions: (1) individuals are haploids[1], (2) the genome can be divided into *known* allelic blocks that are inherited independently and (3) for each pair of blocks, we are given the exact coalescence time. Indeed, in practice, one must start with sequenced genomes—and in the context of recovering events in human history, (potentially unphased) genotypes of diploid individuals. The problem of recovering coalescence times from sequences provides a major challenge and often requires one to either know the population history beforehand, or leverage simultaneous recovery of history and coalescence times using various joint models that enable probabilistic inference.

But since the main message of our paper is a *lower bound* on the number of exact pairwise coalescent samples needed to recover population history, in practice it would only be harder. Even in our idealized setting, handling 7 or 8 subpopulations already requires more data than one could reasonably be assumed to possess. Thus, our work provides a rather direct challenge to empirical work in the area: Either results with 7 or 8 subpopulations are not to be trusted or there must be some biological reason why the types of population histories that arise in our lower bounds, that are information-theoretically impossible to distinguish from each other using too few samples, can be ruled out.

The Multiple-Subpopulation Coalescent Model. Consider a panmictic haploid[2] population, such that each subpopulation evolves according to the standard Wright-Fisher dynamics[3]—we direct the reader to [3] for an overview. For simplicity, we assume no admixture between distinct subpopulations as long as

---

[1] Alternatively, diploids whose phasing is provided.

[2] In a diploid population, the exponents are scaled by a constant factor 2. This can be handled easily via scaling and therefore makes little difference in the analysis.

[3] The distinction between the Wright-Fisher and Moran models is of no consequence in this work, as the latter also yields an exponential model in the limit as population size increases [3].

they are separated in the model (i.e. they have not merged into a single population in the time period under consideration).

As a reminder, if one assumes that a single population has size N which is large and constant throughout time, then the time to the most recent common ancestor (TMRCA) of two randomly sampled individuals closely follows the Kingman coalescent [3] with exponential rate N:

$$\Pr(T > t) = \exp(-t/N). \tag{1}$$

where T, the coalescence time for two randomly chosen individuals, is measured in **generations**. Henceforth, we will assume that this is the distribution of T in the single-component case.

If instead we have a population which is partitioned into a collection of distinct subpopulations with non-constant sizes, let $\mathbf{N}(t)$ be the function that describes the sub-population sizes over time. As in [11], we will assume that the function $\mathbf{N}(t)$ is piecewise constant with respect to some unknown collection of intervals $I_1, I_2, \ldots$ partitioning the real line. In particular, for each $t \in I_k$, there is an associated vector of effective subpopulation sizes $\mathbf{N}(t) = (N_1^{(k)}, \ldots, N_{D_k}^{(k)})$, indexed by the D_k subpopulations present at time t. The indexing need not be consistent across different intervals, as their semantic meaning will change as subpopulations merge and split. For example, $N_1^{(k)}$ and $N_1^{(k+1)}$ need not always represent the sizes of the same subpopulation.

Consider the case where $\mathbf{N}(t)$ is constant for all $t \in I = [a, b]$, where $0 < a < b$, with no admixture and no migration in-between subpopulations in the time interval I. In this case, the coalescence time follows the law of a convex combination of exponential functions:

$$\Pr(T > a + t \mid T > a) = \sum_{\ell=0}^{D} p_\ell e^{-\lambda_\ell t} \tag{2}$$

where $p_0 + p_1 + \cdots + p_D = 1$, $\lambda_0 = 0$ and the other λ_i are $\frac{1}{N_i}$ (refer to Appendix A for a more careful treatment).

The population structure is assumed to undergo changes over time, where the positive direction points towards the past. The three possible changes are:

1. (Split) One subpopulation at time t^- becomes two subpopulations at time t (i.e. $D_k = D_{k-1} + 1$).
2. (Merge) Two subpopulations at time t^- join to form one subpopulation at time t (i.e. $D_k = D_{k-1} - 1$).
3. (Change Size) An arbitrary number of subpopulations change size at time t.

Figure 1 provides an illustrative example of a population history with all of these events.

If a lineage at time t^- is from a subpopulation of size M which splits into two subpopulations of sizes M_1, M_2 at time t, then its ancestral subpopulation is random: for $i \in \{1, 2\}$, subpopulation i is chosen with probability M_i/M. In our model, we only allow at most one of these events at any particular time point. For us, a "split" looking backward in time refers to a *convergence* event of two subpopulations going *forward* in time, while a "merge" refers to a divergence event. This convention is chosen because we think of reconstruction as proceeding backwards in time from the present.

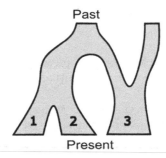

Fig. 1. An example of population structure history, illustrating merges and splits starting with three present-day subpopulations.

Since we are only considering piecewise constant population histories, knowing the exponential mixture for each interval $I_1, I_2, \ldots$ provides a complete description. Namely, if t lies in the interval $I_j = [a, b]$, one can compute the unconditioned probability as $\Pr(T > t) = \Pr(T > t | T > a)\Pr(T > a)$ and iterating on $\Pr(T > a)$ with respect to the interval I_{j-1}.

1.4   Our Results

The main theoretical contribution of this work is an essentially tight bound on the sample complexity of learning population history in the multiple-subpopulation model. In particular, we show sample complexity lower bounds which are *exponential in the number of subpopulations k*. Here is an organized summary of our results:

- First, we show a two-way relationship between the problem of learning a population history (in our simplified model) and the problem of learning a mixture of exponentials. Recall that when the effective subpopulation sizes are all constant, the distribution of coalescence times follows Eq. (2) and thus is equivalent to learning the parameters p_t and λ_t in a mixture of exponentials. Conversely, we show how to use an algorithm for learning mixtures of exponentials to reconstruct the entire population history by locating the intervals where there are no genetic events and then learning the associated parameters in each, separately. (Sect. 2.1 with some details in Appendix B and full details in Appendices B and C of [12]).
- *(Main Result)* Using this equivalence, we show an information-theoretic lower bound on the sample complexity that applies regardless of what algorithm is being used. In particular, we construct a pair of population histories that

have different parameters but which require $\Omega((1/\Delta)^{4k})$ samples in order to tell apart. This lower bound is *exponential in the number of subpopulations k*. Here, $\Delta \le 1/k$ is the smallest gap between any pair of the λ_t's. The proof of this result combines tools spanning information theory, extremal polynomials, and approximation theory. (Sect. 2.3 with details in Appendix D of [12]).

- In the *hypothesis testing* setting where we are given a pair of population histories that we would like to use coalescence statistics to distinguish between, we give an algorithm that succeeds with only $O((1/\Delta)^{4k})$ samples. The key to this result is a powerful tool from analysis, the Nazarov-Turán Lemma [18] which lower bounds the maximum absolute value of a sum of exponentials on a given interval in terms of various parameters. This result matches our lower bounds, thus resolving the sample complexity of hypothesis testing up to constant factors. (Sect. 2.4 with details in Appendix E of [12]).

- In the *parameter learning* setting when we want to directly estimate population history from coalescence times, we give an efficient algorithm which provably learns the parameters of a (possibly truncated) mixture of exponentials given only $O((1/\Delta)^{6k})$ samples. We accomplish this by analyzing the *Matrix Pencil Method* [9], a classical tool from signal processing, in the real-exponent setting. (Sect. 2.2 with details in Appendix F of [12]).

- Finally, we demonstrate using simulated data that our sample complexity lower bounds really do place serious limitations on what can be done in practice. From our plots we see that the sample complexity grows exponentially in the number of subpopulations even in the optimistic case where we have separation $\Delta = 1/k$ which minimizes our lower bounds. In particular, the number of samples we need very quickly exceeds the number of functionally relevant genes (on the order of 10^4) and even the number of SNPs available in the human genome (on the order of 10^7). In fact, via direct numerical analysis of our chosen instances, we can give even stronger sample-complexity lower bounds: for 5 populations, we construct population histories that require over 10 trillion samples to distinguish (Sect. 3, with details in Appendix C.2).

Discussion of Results: In summary, this work highlights some of the fundamental difficulties of reconstructing population histories from pairwise coalescence data. Even for recent histories, the lower bounds grow exponentially in the number of subpopulations. Empirically, and in the absence of provable guarantees, and even with much noisier data than we are assuming, many works suggest that it is possible to reconstruct population histories with as many as nine subpopulations. While testing out heuristics on real data and assessing the biological plausibility of what they find is important, so too is delineating sharp theoretical limitations. Thus we believe that our work is an important contribution to the discussion on reconstructing population histories. It points to the need for the methods that are applied in practice to be able to justify why their findings ought to be believed. Moreover they need to somehow preclude the types of population histories that arise in our lower bounds and are genuinely impossible to distinguish between given the finite amount of data we have access to.

1.5   Related Works

As mentioned in Sect. 1.1, existing methods that attempt to empirically estimate the population history of a *single population* from *sequence data* generally fall into one of two categories: Many are based on (approximately) maximizing the likelihood [14, 21–23] and others perform Bayesian inference [5, 8, 19]. Generally, they are designed to recover a *piecewise constant* function $N(t)$ that describes the size of a population, with the goal of accurately summarizing divergence events, bottleneck events and growth rates throughout time.

Many notable methods that fall into the first category rely on Hidden Markov Models (HMMs), which implicitly make a Markovian assumption on the coalescent trees of alleles across the genome. One notable work is Li and Durbin [14], which gave an HMM-based method (PSMC) that reconstructs the population history of a single population using the genome of a single diploid individual. Later related works gave alternative HMMs that incorporate more than two haplotypes (diCal [22] and MSMC [21]) and improve robustness under phasing errors (SMC++ [23]).

Methods in the second category operate under an assumption about the probability distribution of coalescence events and the observed data. For instance, Drummond [5] prescribes a prior for the distribution of coalescence trees and population sizes, under which MCMC techniques are used to compute both an output and a corresponding 95% credibility interval. However, given the highly idealized nature of their models and the limitations of their methodology (for example, there is no guarantee their MCMC method has actually mixed), it is unclear whether the ground truth actually lies in those credibility intervals.

In the *multiple subpopulations* case, there are two major coalescent-based methods. The first is Schiffels and Durbin [21], which introduced the MSMC model as an improvement over PSMC. These authors used their method to infer the population history of *nine human subpopulations up to about 200,000 years into the past*. Terhorst, Kamm and Song [23] introduced a variant (SMC++) that was directly designed to work on genotypes with missing phase information. In particular, they demonstrate the potential dangers of relying on phase information, by showing that MSMC is sensitive to such errors. In an experiment, SMC++ was used to perform inference of population histories of various combinations of *up to three human subpopulations*. In these experiments, individuals are purposefully chosen from specific subpopulations. We emphasize that in our model, due to the presence of population merges *and* splits, one does not always know what subpopulation an ancestral individual is from.

As a side remark, there are approaches that attempt to infer a (single-component) population history using different types of information. We briefly touch upon some of these known works. One alternative strategy is to use the site frequency spectrum (SFS), e.g. [2, 6]. The earliest theoretical result regarding SFS-based reconstruction is due to Myers, Fefferman and Patterson [17], who proved that *generic* 1-component population histories suffer from unidentifiability issues. Their lower bound constructions have a caveat: They are pathological examples of oscillating functions which are unlikely to be observed in a biological context. Later works [1, 24] prove both identifiability and lower bounds for reconstructing *piecewise constant* population histories using information from the SFS. (In contrast, as our algorithms show, reconstruction from coalescence data does not suffer from the same lack of identifiability issues.)

Most recently, Joseph and Pe'er [10] developed a Bayesian time-series model that incorporates data from *ancient DNA* to recover the history for multiple subpopulations only under size changes, without considering merges or splits. While our analysis does not directly account for such data, the necessity of considering such models is consistent with our assertion: extra information about the ground truth, such as directly observable information about the past (e.g. ancestral DNA), is probably *required* in order for the problem to even be information-theoretically feasible. In addition, [10] does not solve for subpopulation *sizes*, but rather subpopulation *proportions*, which contains less information than what we are after.

Remarks and Comparisons. In the aforementioned works that use genetic sequence data to perform inference, it becomes necessary to consider mutation rates and the genetic identity of ancestral alleles. However, in our version of the problem, each sample is a hypothetical perfect measurement of a pairwise coalescence time. For this setting, alternative models such as Kimura-Crow's infinite alleles model [13] which first condition on coalescent trees, do not provide extra power for inferring population structure.

Looking ahead, we emphasize that our results about hardness rely on a reduction to a statistical problem of learning a certain type of mixture distribution. Informally, such a distribution arises from uncertainty regarding the following query: *Given an individual locus and its lineage, which subpopulation is it contained in at any given moment in time?* This uncertainty is attributed to the following: (1) in the coalescent model, the assumption that a lineage randomly chooses an ancestral subpopulation in the case of a convergence event (split), and (2) the lack of explicit knowledge of which present-day subpopulation the locus belongs to. Statistical approaches to resolving these are confounded by admixture and migration, and ought to be accounted for in any model that attempts to solve this problem. Our model is a natural extension of Kingman's coalescent which inherently incorporates both uncertainties, assuming no admixture only for the sake of theoretical analysis. Previous works such as [10, 21] attempt to correct for admixture during inference and solely consider divergence events.

If the only possible population structure histories are treelike due to the absence of splits—e.g. histories that resemble Fig. 4d of Schiffels and Durbin [21]—and one explicitly knows what subpopulation each lineage belongs to at every point in time, then the problem is significantly easier. However, such reconstructions likely do not provide the complete picture. It is natural to consider more complex histories, and ask how fine-grained of a structure one can obtain from data.

2   Theoretical Discussion

2.1   Reductions Between Mixtures of Exponentials and Population History

In the rest of our theoretical analysis, we will focus on the mixture of exponentials viewpoint of population history. To justify this, note that if we can learn truncated mixtures of exponentials, then we can easily learn population history. Details are given in the full version (Appendix F of [12]), including a concrete algorithm based on our analysis of the Matrix Pencil Method. Conversely, we observe that an arbitrary mixture of exponentials can be embedded as a sub-mixture of a simple population history with two time periods, so that recovering the population history requires in particular learning the mixture of exponentials. The following theorem makes this precise; its proof is delegated to Appendix B.

Theorem 1. *Let* P *with* $P(T > t) = \sum_{i=1}^{k} p_i e^{-\lambda_i t}$ *be the distribution of an arbitrary mixture of* k *exponentials (over random variable* T *) with all* $\lambda_i > 0$ *and* $\sum_i p_i = 1$. *Then for any* $t_0 > 0$, *there exists a two-period population history with* k *populations which induces a distribution* Q *on coalescence times such that*

$$Q(T > t + t_0 | T \neq \infty, T > t_0) = P(T > t).$$

Remark 1. By choosing a small value for t_0, we ensure that very few coalescence times occur in the more recent period, so that the reconstruction algorithm must rely on the information from the second (less recent) period with our planted mixture of exponentials.

Additionally, we provide a more sophisticated version of this reduction which maps two mixtures of exponentials to different population histories simultaneously, while preserving statistical indistinguishability.

Theorem 2. *Let* P *with* $P(T > t) = \sum_{i=1}^{k} p_i e^{-\lambda_i t}$ *and* Q *with* $Q(T > t) = \sum_{j=1}^{\ell} q_j e^{-\mu_j t}$ *be arbitrary mixtures of exponentials with all* $\lambda_i, \mu_j > 0$. *Then for all sufficiently small* $t_0 > 0$, *there exist two distinct 2-period population histories* R *with* $k + 2$ *subpopulations and* S *with* $\ell + 2$ *subpopulations such that:*

1. For any $t > 0$:

$$R(T > t + t_0 \mid T \neq \infty, T > t_0) = P(T > t)$$

and

$$S(T > t + t_0 \mid T \neq \infty, T > t_0) = Q(T > t).$$

2. $R[T = t_0] = S[T = t_0]$ *and* $R[T = \infty] = S[T = \infty]$.

Again, if we take t_0 small enough, we ensure that any distinguishing algorithm must rely on information from the second (less recent) period, and hence because the probability of all other events match, must distinguish between the mixtures of exponentials Q and R. The main idea is to use the construction from Theorem 1 on both mixtures of exponentials, and ensure that the two population histories cannot be distinguished via coalescence statistics by adding extra subpopulations as necessary. We give some high-level details about Theorem 2 and its proof in Appendix B.2. A proof of Theorem 2 can be found in the full version of our paper [12].

2.2 Guaranteed Recovery of Exponential Mixtures via the Matrix Pencil Method

Given samples from a probability distribution

$$\Pr(T > t) = \sum_{i=1}^{k} p_i e^{-\lambda_i t}, \tag{3}$$

can we learn the parameters $p_1, \ldots, p_k, \lambda_1, \ldots, \lambda_k$? In Sect. 2.1, we established the equivalence between solving this problem and learning population history. Suppose for now we we are given access to the exact values of probabilities $v_t := \Pr(T \geq t)$ for $t \in \mathbb{R}_{\geq 0}$, i.e. $v_t = \sum_{j=1}^{k} p_i \alpha_i^t$ where $\alpha_i = e^{-\lambda_i}$. The *Matrix Pencil Method* is the following linear-algebraic method, originating in the signal processing literature [9], which solves for the parameters $\{p_i, \lambda_i\}_{i=1}^{k}$:

1. Let A, B be $k \times k$ matrices where $A_{ij} = v_{i+j-1}$ and $B_{ij} = v_{i+j-2}$.
2. Solve the generalized eigenvalue equation $\det(A - \gamma B) = 0$ for the pair (A, B). The γ which solve $\det(A - \gamma B) = 0$ are the α's.
3. Finish by solving for the p's in a linear system of equations $v = Vp$, where $v = (v_0, \ldots, v_{k-1})$, V is the $k \times k$ Vandermonde matrix generated by $\alpha_1, \ldots, \alpha_k$ and p is the vector of unknowns $(p_1, \ldots, p_k)$.

To understand why the algorithm works in the noiseless setting, consider the decomposition $A = V D_p D_\alpha V^T$ and $B = V D_p V^T$ where $V = V_k(\alpha_1, \ldots, \alpha_k)$ is the $k \times k$ Vandermonde matrix whose (i, j) entry is α_j^{i-1}, $D_\alpha = \mathrm{diag}(\alpha_1, \ldots, \alpha_k)$ and $D_p = \mathrm{diag}(p_1, \ldots, p_k)$. Then it's clear that the α_i are indeed the generalized eigenvalues of the pair (A, B). However, in our setting, we do not have access to the exact measurements v_t, but instead have *noisy* empirical measurements $\tilde{v}_t$; in practice, the output of the MPM can be very sensitive to noise.

Analysis of MPM Under Noise. We now describe our analysis of the MPM in the more realistic setup where the CDF is estimated from sample data. First note that the model (Eq. 3) is statistically unidentifiable if there exist two identical λ's. Indeed, the mixture $\frac{1}{2}e^{-\lambda t} + \frac{1}{2}e^{-\lambda t}$ is exactly same as the single-component model $e^{-\lambda t}$, as is any other re-weighting of the coefficients into arbitrarily many components with exponent λ. Therefore it is natural to introduce a *gap parameter* $\Delta := \min_{i \neq j} |\lambda_i - \lambda_j|$ which is required to be nonzero, as in the work on super-resolution (e.g. [4, 16]).

Without loss of generality, we also assume that: (1) the components are sorted in decreasing order of exponents, so that $\lambda_1 > \cdots > \lambda_k > 0$, and (2) time has been re-scaled[4] by a constant factor, so that $\lambda_i \in (0, 1)$ for each i. Now we can state our guarantee for the MPM under noise:

Theorem 3. *Let $\Delta = \min_{i \neq j} |\lambda_i - \lambda_j|$ and let $p_{\min} = \min_i p_i$. For all $\delta > 0$, there exists $N_0 = O\left(\frac{k^{10}}{p_{\min}^4}\left(\frac{2e}{\Delta}\right)^{6k} \log \frac{1}{\delta}\right)$ such that, with probability $1 - \delta$, using empirical estimates $\widetilde{v}_0, \ldots, \widetilde{v}_{2k-1}$ from $N \geq N_0$ samples, the matrix pencil method outputs $\{(\widetilde{\lambda}_j, \widetilde{p}_j)\}_{j=1}^k$ satisfying*

$$|\widetilde{\lambda}_j - \lambda_j| = O\left(\frac{k^{3.5}}{p_{\min}^2}\left(\frac{2e}{\Delta}\right)^{2k}\sqrt{\frac{1}{N}\log\frac{1}{\delta}}\right) \text{ and } |\widetilde{p}_j - p_j| = O\left(\frac{k^5}{p_{\min}^2}\left(\frac{2e}{\Delta}\right)^{3k}\sqrt{\frac{1}{N}\log\frac{1}{\delta}}\right)$$

for all j.

Remark 2. Letting α_i denote $e^{-\lambda_i}$, we note that we can equivalently focus on learning the α_i's, and that guarantees for recovering λ_i and α_i are equivalent up to constants: $e^{-1}|\alpha_i - \widetilde{\alpha}_i| \leq |\lambda_i - \widetilde{\lambda}_i| \leq |\alpha_i - \widetilde{\alpha}_i|$. since e^{-x} is monotone decreasing on $[0, 1]$ with derivative lying in $[-1, -1/e]$.

The full proof of Theorem 3 is given in the full version [12]. As in previous work analyzing the MPM in the *super-resolution* setting with imaginary exponents [16], we see that the stability of MPM ultimately comes down to analyzing the condition number of the corresponding Vandermonde matrix, which in our case is very well-understood [7].

2.3   Strong Information-Theoretic Lower Bounds

In this section we describe our main results, strong information theoretic lower bounds establishing the difficulty of learning mixtures of exponentials (and hence, by our reductions, population histories). The full proofs of all results found in this section are given in Appendix D of [12]. First, we state a lower bound on learning the exponents λ_j.

Theorem 4. *For any $k > 1$, there exists an infinite family of parameters $a_1, \ldots, a_k, \lambda_1, \ldots, \lambda_k$ and $b_1, \ldots, b_k, \mu_1, \ldots, \mu_k$ parametrized by integers $m > 2(k-1)$ and $\alpha \in (0, \frac{1}{2})$ such that:*

1. *Each λ_i and μ_j is in $(0, 1]$, $\lambda_1 = \mu_1$, and the elements of $\{\lambda_i\}_{i=2}^k \cup \{\mu_i\}_{i=2}^k$ are all distinct and separated by at least $\Delta = 1/(m + 2k)$. Furthermore $\lambda_2, \mu_2 > \alpha/k$.*

---

[4] In practice, even if this scaling is unknown, this is easily handled by e.g. trying powers of 2 and picking the best result in CDF distance, for instance $\|F - G\|_\infty = \sup_t |F(t) - G(t)|$.

2. *Let H_1 and H_2 be hypotheses, under which the random variable T respectively follows the distributions*

$$\Pr_{H_1}[T \geq t] = \sum_{i=1}^{k} a_i e^{-\lambda_i t} \quad and \quad \Pr_{H_2}[T \geq t] = \sum_{i=1}^{k} b_i e^{-\mu_i t}.$$

If N samples are observed from either H_1 or H_2, each with prior probability $1/2$, then the Bayes error rate for any classifier that distinguishes H_1 from H_2 is at least $\frac{1-\delta}{2}$, where

$$\delta = \frac{\alpha\sqrt{2N}}{2k-3}[\Delta(2k-3)]^{2k-4}. \tag{4}$$

Remark 3. From the square-root dependence of N in Theorem 3, the required number of samples N_0 has rate $4k$ in the exponent of $\frac{2e}{\Delta}$ if one just wants to learn the λ's, and Theorem 4 confirms that the exponent $4k$ is tight for learning the λ's.

Table 1. Sample lower bound from Theorem 4 illustrated for $k = 5, 7$, and 9. We instantiate Theorem 4 with $\alpha = 1/k$, $m = 2k$, $\Delta = 1/(4k)$, and $\delta = 1/2$, and solve for N in Eq. 4 to get the required number of samples N_0.

k	Sample lower bound (from Theorem 4)
5	4.531×10^7
7	9.665×10^{10}
9	1.008×10^{14}

In Table 1, we give a table illustrating the theoretical sample lower bound from Theorem 4 for a few sample values of k, the number of distinct exponentials in the mixture. It can be easily verified that the number of samples required to hypothesis test two mixtures of exponentials blows up exponentially as the number of exponentials in the mixture increases, quickly getting completely infeasible at $k = 9$. The parameters in Table 1 are selected to approximately match the parameters of our simulated findings in Sect. 3. In fact, while the sample complexity in Theorem 4 has a tight exponent, for $k = 5$ we manage to get stronger lower bounds via a direct analysis of simple mixtures found via simulation (in Sect. 3). Indeed, we find a pair of mixtures for $k = 5$ that requires at least $\approx 1.4 \times 10^{13}$ samples to distinguish, which is much greater than the value in Table 1. We refer the reader to Sect. 3 and Appendix C.2 for details.

Next we state an additional information-theoretic lower bound showing that the information-theoretic (minimax) rate is necessarily of the form $\frac{1}{\sqrt{N}}\Delta^{-O(k)}$ up to lower order terms, *even if* all of the λ_i are already known and we are only asked to reconstruct the mixing weights p_j.

Theorem 5. *Let m, k be positive integers such that $m > k > 3$ and let $\Delta = 2/(m + k)$. There exists a fixed choice of $\lambda_1, \ldots, \lambda_k$ which are Δ-separated such that*

$$\inf_{\hat{p}} \max_{p} \mathbb{E}_p \|p - \hat{p}\|_1 \geq \frac{1}{4} \min \left(1, \frac{k-3}{\sqrt{2N}} \left(\frac{2}{\Delta(k-3)} \right)^{k-4} \right) \qquad (5)$$

where the max is taken over feasible choices of p, and the infimum is taken over possible estimators $\hat{p}$ from N samples of the mixture of exponentials with CDF $F(t) = 1 - \sum_j p_j e^{-\lambda_j t}$.

Remark 4. Recall that in Theorem 3, the number of samples needed was exponential in $4k$ when learning just the λ's and in $6k$ for learning both the λ's and the p's. The exponent of $2k$ in Theorem 5 suggests that the discrepancy of $2k$ for MPM in Theorem 3 is tight.

As expected, our lower bounds show that the learning problem becomes harder as Δ approaches 0. The "easiest" case, then, ought to be when Δ is as large as possible, so that the λ_i are equally spaced apart in the unit interval. This raises the following question: as Δ grows, does the sample complexity remains exponential in k, or is there a phase transition (as in super-resolution [16]) where the problem becomes easier? In the full version of our paper [12], we completely resolve this question: the sample complexity still grows exponentially in $4k$ when Δ is maximally large.

2.4 A Tight Upper Bound: Nazarov-Turán-Based Hypothesis Testing

As an alternative to the learning problem that the Matrix Pencil Method solves, we also consider the hypothesis testing scenario in which we want to test if the sampled data matches a hypothesized mixture distribution. In this case, we can give guarantees from weaker assumptions and requiring smaller numbers of samples. To state our guarantee, we need the following additional notation: for P a mixture of exponentials, let $p_\lambda(P)$ denote the coefficient of $e^{-\lambda t}$, which is 0 if this component is not present in the mixture. We study the following simple-versus-composite hypothesis testing problem using N samples:

Problem 1. Fix $k_0, k_1, \delta, \Delta > 0$ and let P be a known mixture of k_0 exponentials.

- H_0: The sampled data is drawn from P.
- H_1: The sampled data is drawn from a different, unknown mixture of at most k_1 exponentials Q. Let $\nu_1 := \max\{\lambda : p_\lambda(P) > p_\lambda(Q)\}$ and $\nu_2 := \max\{\lambda : p_\lambda(Q) > p_\lambda(P)\}$. We assume that $\min\{|p_{\nu_1}(P) - p_{\nu_1}(Q)|, |p_{\nu_2}(P) - p_{\nu_2}(Q)|\} \geq \delta$ and $|\nu_1 - \nu_2| \geq \Delta$.

Henceforth, we will refer to H_0 as the *null hypothesis* and H_1 as the *alternative hypothesis* (note that H_1 is a composite hypothesis). To solve this hypothesis testing problem, we propose a finite-sample variant of the Kolmogorov-Smirnov test:

---

1. Let $\alpha > 0$ be the significance level.
2. Let F_N be the empirical CDF and let F be the CDF under the null hypothesis H_0.
3. Reject H_0 if $\sup_t |F_n(t) - F(t)| > \sqrt{\log(2/\alpha)/2N}$.

We show that this test comes with a provable finite-sample guarantee.

Theorem 6. *Consider the problem setup as in Problem 1 and fix a significance level $\alpha > 0$. Let $k := \frac{k_0 + k_1}{2}$ and $c_\Delta = 8e^2/\min(1/\Delta, 2k - 1)$. Then:*

1. *(Type I Error) Under the null hypothesis, the above test rejects H_0 with probability at most α.*
2. *(Type II Error) There exists $N_0(\alpha) = O((c_\Delta/\Delta)^{4k-2}\log(2/\alpha)/\delta^2)$ such that if $N \geq N_0$, then the power of the test at significance level α is at least:*

$$\Pr_{Q \in H_1}[Reject\ H_0] \geq 1 - 2\exp\left(-N\delta^2(\Delta/c_\Delta)^{4k-2}/8\right). \tag{6}$$

The full proof of Theorem 6 is given in the full version [12]. The key step in the proof is a careful application of the celebrated Nazarov-Turán Lemma [18].

Remark 5. This improves upon the Matrix Pencil Method upper bound (Theorem 3), in terms of the exponent found above Δ (Δ^{-6k} versus Δ^{-4k}) and above the mixing weights ($p_{\min}^4$ versus δ^2). Even when the alternative Q is fixed and known, we see from Theorem 4 that $\Omega((1/\delta^2)(1/\Delta)^{4k})$ many samples are information-theoretically required, which matches Theorem 6.

3   Simulations and Indistinguishability in Simple Examples

Our theoretical analysis rigorously establishes the worst-case dependence on the number of samples needed in order to learn the parameters of a single period of population history under our model – recall the construction of Theorem 4 of two hard-to-distinguish mixtures of exponentials and the result Theorem Theorem 2 converting these to population histories.

In our simulations, we will analyze both the performance and information-theoretic difficulty of learning not a specially constructed worst-case instance, but instead an *extremely simple population history* with k populations. More precisely we consider the following instance:

Simulation Instance(k):

Population history description: We consider reconstructing a single period model with k populations in which the ratio of the population sizes is $1 : 2 : \cdots : k$ and the relative probability of tracing back to each of these populations (i.e. $\Pr(\mathcal{E}_{i,i}|T > t_0)$ from Appendix A) are all equal to $1/k^2$. This

can easily be realized as a one period of a 2-period population history model, in which in the second (more recent) era all populations are the same size[5]
Mixture of exponentials description: We consider the following mixture of exponentials:

$$\Pr(T > t) = (1 - 1/k) + \sum_{i=1}^{k}(1/k^2)e^{-t/k}.$$

The constant term represents atomic mass at ∞ and corresponds to no coalescence. When $k = 1$ this is a standard exponential distribution, otherwise it is a mixture of $k + 1$ exponentials, counting the degenerate constant term.

We do not believe that this is an unusually difficult instance of a mixture of exponentials on k components. If anything, the situation is likely the opposite: our worst-case analysis (Theorems 5, 3) suggests that this is comparatively *easy* as the gap parameter Δ is maximally large.

In order to evaluate the error in parameter space from the result of the learning algorithm, we adopted a natural metric, the well-known *Earthmover's distance*. Informally, this measures the minimum distance (weighted by p_i and recovered $\tilde{p}_i$) that the recovered exponents must be moved to agree with the ground truth; we give the precise definition in Appendix C.2.

For a point of comparison to MPM, we also tested a natural convex programming formulation which essentially minimizes $\| \int e^{-\lambda t}d\mu(\lambda) - (1 - \tilde{F}(t))\|_\infty$ over probability measures μ on $\mathbb{R}_{\geq 0}$, where $\tilde{F}$ is the empirical CDF – refer to Appendix C.1 for details.

The results of running both the convex program and the MPM are shown in Fig. 2 (blue and green lines) plotted on a log (base 10) scale; details of the setup are provided in Appendix C.2. As expected based on our theoretical analysis, the number of samples needed scaled exponentially in k, the number of populations in our instance. Details of the setup are provided in Appendix C.2; due to limitations of machine precision, the convex program could not reliably reconstruct at 5 components with any noise level and so this point is omitted.

Besides showing the performance of the algorithms, we were able to deduce *rigorous, unconditional* lower bounds on the information-theoretic difficulty of these particular instances. Each point on the red line corresponds to the existence of a different mixture of exponentials (found by examining the output of the convex program), with a comparable number of mixture components[6], which is far in parameter space[7] from the ground truth and yet the distribution of N

---

[5] As in Remark 1, we can optionally make the more recent era short so that almost all samples will be from the earlier period.

[6] The alternative hypothesis had no more than a few additional mixture components. A byproduct of this analysis is that even estimating the number of populations is in these examples requires a very large number of samples.

[7] More precisely, with Earthmover's distance in parameter space greater than 0.01. For comparison, an estimator which only gets the (easy) constant component correct already has Earthmover distance at most $1/k$.

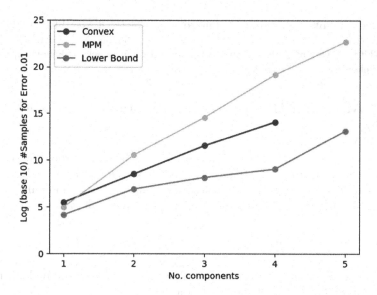

Fig. 2. Plot of #components versus log (base 10) number of samples needed for accurate reconstruction (parameters within Earthmover's distance 0.01). Below the red line, it is mathematically impossible for *any method* to distinguish with greater than 75% success between the ground truth and a fixed alternative instance which has significantly different parameters. (Color figure online)

samples from this model (where $N = 10^y$ and y is the y-coordinate in the plot) has total-variation (TV) distance at most 0.5 from the distribution of N samples from the true distribution. By the Neyman-Pearson Lemma, this implies that if the prior distribution is $\left(\frac{1}{2}, \frac{1}{2}\right)$ between these two distributions, then we cannot successfully distinguish them with greater than 75% probability. We describe the mathematical derivation of the TV bound in Appendix C.2, and illustrate such a hard-to-distinguish pair in Example 1. Recall that by Theorem 2, such a hard to distinguish pair of mixtures can automatically be converted into a pair of hard-to-distinguish population histories.

Notably, the lower bound shows that reliably learning the underlying parameters in this simple model with 5 components necessarily requires at least 10 trillion samples from the true coalescence distribution. In reality, since we do not truly have access to clean i.i.d. samples from the distribution, this is likely a significant underestimate.

Example 1. Consider the mixtures of exponentials with CDFs $F(t)$ and $G(t)$, where $1 - F(t) = 0.5 + 0.25e^{-0.5t} + 0.25e^{-t}$ and

$$1 - G(t) = 0.49975946 + 0.15359557e^{-0.45t} + 0.30642727e^{-0.81t} + 0.0402177e^{-1.55t}.$$

Despite being very different in parameter space, their H^2 distance is $7.9727 \cdot 10^{-6}$ so any learning algorithm requires at least 15660 samples to distinguish them with better than 75% success rate.

Acknowledgements. This work was funded in part by ONR N00014-16-1-2227, NSF CCF1665252, NSF DMS-1737944, NSF Large CCF-1565235, NSF CAREER Award CCF-1453261, as well as Ankur Moitra's David and Lucile Packard Fellowship, Alfred P. Sloan Fellowship, and ONR Young Investigator Award.

A   Derivation of the Multiple-Subpopulation Coalescent Model

For $b > a > 0$, let $I = [a, b]$ be an interval such that the population structure $\mathbf{N}$ is constant over I. Then Eq. (1), together with the Markov property of Kingman's coalescent model tells us that the coalescence time T of two randomly sampled individuals in the ith sub-population is given, for any $t \in [0, b - a]$, by

$$\Pr\left(T > a + t \mid \mathcal{E}_{i,i} \wedge \{T > a\}\right) = \exp\left(-\frac{1}{N_i}t\right). \tag{7}$$

Here, $\mathcal{E}_{i,j}$ represents the event where the ancestry of one of the individuals traces back to subpopulation i, and the other traces back to j.

Let D be the number of subpopulations restricted to the interval I. By the law of total probability, the random variable T satisfies, again for any $t \in [0, b - a]$,

$$\Pr(T > a + t \mid T > a) = \sum_{i < j} \Pr(\mathcal{E}_{i,j} \mid T > a) + \sum_{i=1}^{D} \Pr(\mathcal{E}_{i,i} \mid T > a) \Pr(T > a + t \mid \mathcal{E}_{i,i} \wedge \{T > a\})$$

The first summation over $i < j$ uses the fact that $\Pr(T > a + t \mid \mathcal{E}_{i,j} \wedge \{T > a\}) = 1$, via the "no admixture" assumption; whenever the two individuals' lineages at time a lie in distinct subpopulations, they do not coalesce anywhere in I. Via Eq. (7), the right hand side can be re-written as seen in Eq. (2), i.e.

$$\Pr(T > a + t \mid T > a) = \sum_{\ell=0}^{D} p_\ell e^{-\lambda_\ell t}.$$

B   Reduction from Mixtures of Exponentials to Population History

B.1   Proof of Theorem 1

Proof. We consider the following population history:

- In the (more recent) period $[0, t_0]$ there are k populations and population i has size $\sqrt{q_i}$, where q_i is the (unique) nonnegative solution to

$$p_i = q_i e^{-t_0/\sqrt{q_i}}.$$

To see that the solution exists and is unique, observe that the rhs of this equation is a strictly increasing function in q_i which maps $(0, \infty)$ to $(0, \infty)$.

– In the (less recent) period $[t_0, \infty)$ each of the k populations changes to size $1/\lambda_i$.

By construction, the probability of two independently sampled individuals being in the same population is proportional to $\sqrt{q_i} \cdot \sqrt{q_i} = q_i$, and conditioned on no coalescence before time t_0 this probability is proportional to p_i (since the event of no coalescence before time t_0 given that the two individuals come from population i is $e^{-t_0/\sqrt{q_i}}$). Therefore the distribution Q of coalescence times satisfies

$$Q(T > t + t_0 | T > t_0) = q_0 + \frac{1}{1 - q_0} \sum_{i=1}^{k} p_i e^{-\lambda_i t} \qquad (8)$$

where q_0 is the probability that the two individuals sampled were in different populations.

B.2   Hardness for Distinguishing Population Histories

Theorem 1 shows that any arbitrary mixture of exponentials can be embedded as a sub-mixture of a simple population history with two time periods. We can leverage this equivalence to reduce distinguishing two mixtures of exponentials to distinguishing two population histories, and hence conclude from Theorem 4 that distinguishing two population histories is exponentially hard in the number of subpopulations.

The high-level idea is to take the reduction from Theorem 1 and apply it to two arbitrary mixtures of exponentials, and argue that the problem of distinguishing the resulting two population histories is at least as hard as distinguishing the two mixtures they came from. This suffices to achieve Condition 1 of Theorem 1.

The problem that arises is that other statistics about coalescence can still distinguish the two population histories; at a high level, Condition 2 of Theorem 1 is meant to rule this out. In particular, note that the constant term (q_0 in Eq. 8) is exactly the probability of no coalescence, which is fixed by the subpopulation sizes, and therefore fixed by the desired mixture. Therefore, if we are not careful, the probability of no coalescence will be significantly different between our two population histories, making them easily distinguishable.

To remedy this, we add extra "dummy" subpopulations to the two population histories, with sizes that we set in order to equalize the probabilities of no coalescence (i.e. make $R[T = \infty] = S[T = \infty]$). In order to prevent these dummy subpopulations from affecting condition 1 (i.e. that $R[T = t|t > t_0] = S[T = t|t > t_0]$), we ensure that coalescence in these subpopulations *cannot* occur at a time greater than t_0 by shrinking these subpopulations to size zero at time t_0 (and therefore forcing coalescence). However, this means that we will additionally have a lot of coalescence occurring at *exactly* t_0 from these dummy subpopulations, and so this introduces yet another constraint: we have to ensure that $R[T = t_0] = S[T = t_0]$. Once again, we need to tune the sizes of the dummy subpopulations correctly so that this happens.

The proof of Theorem 2 can be found in the full version of our paper [12].

C   Simulation Methods

C.1   A Convex Programming Approach to Learning

In addition to using the Matrix Pencil Method to learn mixtures of exponentials in each interval, we also implemented a convex program. Here, the goal is to learn a mixture of exponentials, whose support is perhaps restricted to an interval $I = [a, b]$. The idea is as follows: assume that we know the interval $\Lambda = [0, c]$ for which we can assume $\lambda_1, \ldots, \lambda_n \in \Lambda$. We first discretize the space of possible exponents by choosing n equally spaced points $\lambda_1, \ldots, \lambda_n$ inside Λ. Solve the convex program

$$\underset{p}{\text{minimize}} \quad \sup_{t \in I} \left| \sum_{i=1}^{n} p_i e^{\lambda_i t} - v_t \right|$$

$$\text{subject to} \quad \sum_i p_i = 1$$

$$p_i \geq 0, i = 1, \ldots, n.$$

In practice, we replace $\sup_{t \in I}$ with the discretization $\max_{t \in \mathcal{S}}$, where $\mathcal{S} \subset I$ is a finite mesh of points in I. Since we are learning from samples, we also substitute v_t with $\tilde{v}_t$, the empirical estimate of the tail probability $\Pr[T = t \mid T \geq a]$. Since the ℓ_1-norm of the p_i is fixed to be 1, we do not expect to need additional regularization to get sparse output.

For small instances (see Sect. 3), the convex program is more sample-efficient than the Matrix Pencil Method. In the context of this paper, however, it does not come with robustness guarantees. Results on convex programming approaches for super-resolution are known, due to Candès and Fernandez-Granda [4]; for our (real-exponent) setting, a different analysis will be required and we leave this to future work. If we assume that the program does return a sparse output (which occurs in practice), some guarantees for the accuracy of the output follow automatically from the analysis of Theorems 3 and 6, since for sparse mixtures they (implicitly) bound parameter error in terms of the closeness in CDF-distance.

Implementation in Simulations: In our experiments, we solved the above convex program using the barrier method of CPLEX version 12.8 with numerical emphasis enabled.

C.2   Simulations: Additional Details

Earthmover's Distance Between Parameters: The Earthmover's (or 1-Wasserstein) distance between P and Q measures the minimum transport cost to move the "mass" corresponding to probability distribution P to that of Q. Rigorously, in one dimension it can be defined by

$$EMD(P, Q) := \min_{\pi : \pi|_X = P, \pi|_Y = Q} \mathbb{E}_{(X,Y) \sim \pi}[|X - Y|]$$

where here π ranges over all possible couplings of marginal distributions P and Q. The following definition makes the notion of Earthmover's distance between the parameters of two mixtures of exponentials precise:

Definition 1. *Let P and Q be two mixtures of exponentials $P(T > t) = \sum_i p_i e^{-\lambda_i t}$ and $Q(T > t) = \sum_i q_i e^{-\gamma_i t}$. The* Earthmover's distance in parameter space *between P and Q is the Earthmover's distance between corresponding atomic measures $\mu_P := \sum_i p_i \delta_{\lambda_i}$ and $\mu_Q := \sum_i q_i \delta_{\gamma_i}$ where δ_x represents a Dirac mass at point x.*

Derivation of Per-Instance Information-Theoretic Lower bounds: Given the alternative instance, we derived the bound by computing the H^2 (Hellinger squared) distance between the true distribution and the alternative distribution, and then applying standard tensorization and comparison inequalities to bound the TV.

Upper Bound Simulations: We ran 300 trials for each setting of k and number of samples; in order to run the simulation for very large numbers of samples, we directly generated the corresponding noisy CDF estimates by adding Gaussian noise of order $O(1/\sqrt{N})$ where N is the number of samples. For reasonable size N we also ran the methods using actual sample-estimated CDFs and the results were consistent with the simulated Gaussian-noise CDFs. The lower bound is analytically computed, not simulated, so it is unaffected by this Gaussian-noise approximation.

Plotted Data: Here we provide the data plotted in Fig. 2, that was found via simulation as described above (Table 2).

Table 2. Values plotted on a log (base 10) scale in Fig. 2.

k	CVX	MPM	LB
1	2.98×10^5	9.28×10^4	1.34×10^4
2	3.25×10^8	3.45×10^{10}	8.18×10^6
3	3.55×10^{11}	3.87×10^{14}	1.44×10^8
4	1.21×10^{14}	1.40×10^{19}	1.13×10^9
5	N/A	4.89×10^{22}	1.43×10^{13}

References

1. Bhaskar, A., Song, Y.S.: Descartes' rule of signs and the identifiability of population demographic models from genomic variation data. Ann. Stat. **42**(6), 2469 (2014)
2. Bhaskar, A., Wang, Y.R., Song, Y.S.: Efficient inference of population size histories and locus-specific mutation rates from large-sample genomic variation data. Genome Res. **25**(2), 268–279 (2015). gr-178756

3. Blythe, R.A., McKane, A.J.: Stochastic models of evolution in genetics, ecology and linguistics. J. Stat. Mech.: Theory Exp. **2007**(07), P07018 (2007)
4. Candès, E.J., Fernandez-Granda, C.: Super-resolution from noisy data. J. Fourier Anal. Appl. **19**(6), 1229–1254 (2013)
5. Drummond, A., Rambaut, A., Shapiro, B., Pybus, O.: Bayesian coalescent inference of past population dynamics from molecular sequences. Mol. Biol. Evol. **22**(5), 1185–1192 (2005)
6. Excoffier, L., Dupanloup, I., Huerta-Sánchez, E., Sousa, V.C., Foll, M.: Robust demographic inference from genomic and SNP data. PLoS Genet. **9**(10), e1003905 (2013)
7. Gautschi, W.: On inverses of vandermonde and confluent vandermonde matrices. Numer. Math. **4**(1), 117–123 (1962)
8. Heled, J., Drummond, A.: Bayesian inference of population size history from multiple loci. BMC Evol. Biol. **8**(1), 289 (2008)
9. Hua, Y., Sarkar, T.K.: Matrix pencil method for estimating parameters of exponentially damped/undamped sinusoids in noise. IEEE Trans. Acoust. Speech Signal Process. **38**(5), 814–824 (1990)
10. Joseph, T.A., Pe'er, I.: Inference of population structure from ancient DNA. In: Raphael, B.J. (ed.) RECOMB 2018. LNCS, vol. 10812, pp. 90–104. Springer, Cham (2018). https://doi.org/10.1007/978-3-319-89929-9_6
11. Kim, J., Mossel, E., Rácz, M.Z., Ross, N.: Can one hear the shape of a population history? Theor. Popul. Biol. **100**, 26–38 (2015)
12. Kim, Y., Koehler, F., Moitra, A., Mossel, E., Ramnarayan, G.: How many subpopulations is too many? Exponential lower bounds for inferring population histories. arXiv preprint arXiv:1811.03177 (2018)
13. Kimura, M., Crow, J.F.: The number of alleles that can be maintained in a finite population. Genetics **49**(4), 725 (1964)
14. Li, H., Durbin, R.: Inference of human population history from individual whole-genome sequences. Nature **475**(7357), 493 (2011)
15. McVean, G.A., Cardin, N.J.: Approximating the coalescent with recombination. Philos. Trans. Roy. Soc. London B: Biol. Sci. **360**(1459), 1387–1393 (2005)
16. Moitra, A.: Super-resolution, extremal functions and the condition number of vandermonde matrices. In: Proceedings of the Forty-seventh Annual ACM Symposium on Theory of Computing, STOC 2015, pp. 821–830. ACM, New York (2015). https://doi.org/10.1145/2746539.2746561
17. Myers, S., Fefferman, C., Patterson, N.: Can one learn history from the allelic spectrum? Theor. Popul. Biol. **73**(3), 342–348 (2008)
18. Nazarov, F.L.: Local estimates for exponential polynomials and their applications to inequalities of the uncertainty principle type. Algebra i analiz **5**(4), 3–66 (1993)
19. Nielsen, R.: Estimation of population parameters and recombination rates from single nucleotide polymorphisms. Genetics **154**(2), 931–942 (2000)
20. Nordborg, M.: Coalescent theory. Handb. Stat. Genet. **2**, 843–877 (2001)
21. Schiffels, S., Durbin, R.: Inferring human population size and separation history from multiple genome sequences. Nat. Genet. **46**(8), 919 (2014)
22. Sheehan, S., Harris, K., Song, Y.S.: Estimating variable effective population sizes from multiple genomes: a sequentially markov conditional sampling distribution approach. Genetics **194**, 647–662 (2013)
23. Terhorst, J., Kamm, J.A., Song, Y.S.: Robust and scalable inference of population history from hundreds of unphased whole genomes. Nat. Genet. **49**(2), 303 (2017)

24. Terhorst, J., Song, Y.S.: Fundamental limits on the accuracy of demographic inference based on the sample frequency spectrum. Proc. Nat. Acad. Sci. **112**(25), 7677–7682 (2015)
25. Turán, P.: On a New Method of Analysis and Its Applications. Wiley, New York (1984)

Efficient Construction of a Complete Index for Pan-Genomics Read Alignment

Alan Kuhnle[1]([✉]), Taher Mun[2], Christina Boucher[1], Travis Gagie[3], Ben Langmead[2], and Giovanni Manzini[4]

[1] Department of Computer and Information Science and Engineering, University of Florida, Gainesville, FL, USA
akuhnle418@gmail.com
[2] Department of Computer Science, John Hopkins University, Baltimore, MD, USA
[3] School of Computer Science and Telecommunications, Universidad Diego Portales and CeBiB, Santiago, Chile
[4] Department of Science and Technological Innovation, University of Eastern Piedmont, Alessandria, Italy

Abstract. While short read aligners, which predominantly use the FM-index, are able to easily index one or a few human genomes, they do not scale well to indexing databases containing thousands of genomes. To understand why, it helps to examine the main components of the FM-index in more detail, which is a rank data structure over the Burrows-Wheeler Transform (BWT) of the string that will allow us to find the interval in the string's suffix array (SA) containing pointers to starting positions of occurrences of a given pattern; second, a sample of the SA that—when used with the rank data structure—allows us access to the SA. The rank data structure can be kept small even for large genomic databases, by run-length compressing the BWT, but until recently there was no means known to keep the SA sample small without greatly slowing down access to the SA. Now that Gagie et al. (SODA 2018) have defined an SA sample that takes about the same space as the run-length compressed BWT—we have the design for efficient FM-indexes of genomic databases but are faced with the problem of building them. In 2018 we showed how to build the BWT of large genomic databases efficiently (WABI 2018) but the problem of building Gagie et al.'s SA sample efficiently was left open. We compare our approach to state-of-the-art methods for constructing the SA sample, and demonstrate that it is the fastest and most space-efficient method on highly repetitive genomic databases. Lastly, we apply our method for indexing partial and whole human genomes and show that it improves over Bowtie with respect to both memory and time.

A. Kuhnle and T. Mun—Equal contribution, ordered alphabetically.
C. Boucher, T. Gagie, B. Langmead and G. Manzini—Equal contribution, ordered alphabetically.

L. J. Cowen (Ed.): RECOMB 2019, LNBI 11467, pp. 158–173, 2019.
https://doi.org/10.1007/978-3-030-17083-7_10

1   Introduction

The FM-index, which is a compressed subsequence index based on Burrows Wheeler transform (BWT), is the primary data structure of the majority of short read aligners—including Bowtie [19], BWA [13] and SOAP2 [21]. These aligners build a FM-index based data structure of sequences from a given genomic database and then use the index to perform queries that find approximate matches of sequences to the database. While these methods can easily index one or a few human genomes, they do not scale well to indexing the databases of thousands of genomes. This is problematic in analysis of the data produced by consortium projects, which routinely have several thousand genomes.

In this paper, we address this need by introducing and implementing an algorithm for efficiently constructing the FM-index, which allows for the FM-index construction to scale to larger sets of genomes. To understand the challenge and solution behind our method, consider the two principal components of the FM-index: first, a rank data structure over the BWT of the string that enables us to find the interval in the string's suffix array (SA) containing pointers to starting positions of occurrences of a given pattern (and to compute how many such occurrences there are); second, a sample of the SA that, when used with the rank data structure, allows us access the SA (so we can list those starting positions). Searching with an FM-index can be summarized as follows: starting with the empty suffix, for each proper suffix of the given pattern we use rank queries at the ends of the BWT interval containing the characters immediately preceding occurrences of that suffix in the string, to compute the interval containing the characters immediately preceding occurrences of the suffix of length 1 greater; when we have the interval containing the characters immediately preceding occurrences of the whole pattern, we use a SA sample to list the contexts of the corresponding interval in the SA, which are the locations of those occurrences.

Although it is possible to use a compressed implementation of the rank data structure that does not become much slower or larger even for thousands of genomes, the same cannot be said for the SA sample. The product of the size and the access time must be at least linear in the length of the string for the standard SA sample. This implies that the FM-index will become much slower and/or much larger as the number of genomes in the databases grows significantly. This bottleneck has forced researchers to consider variations of FM-indexes adapted for massive genomic datasets, such as Valenzuela et al.'s pan-genomic index [32] or Garrison et al.'s variation graphs [7]. Some of these proposals use elements of the FM-index, but all deviate in substantial ways from the description above. Not only does this mean they lack the FM-index's long and successful track record, it also means they usually do not give us the BWT intervals for all the suffixes as we search (whose lengths are the suffixes' frequencies, and thus a tightening sequence of upper bounds on the whole pattern's frequency), nor even the final interval in the suffix array (which is an important input in other string processing tasks).

Recently, Gagie, Navarro and Prezza [11] proposed a different approach to SA sampling, that takes space proportional to that of the compressed rank data structure while still allowing reasonable access times. While their result yields a potentially practical FM-index on massive databases, it does not directly lead to a solution since the problem of how to efficiently construct the BWT and SA sample remained open. In a direction toward to fully realizing the theoretical result of Gagie et al. [11], Boucher et al. [2] showed how to build the BWT of large genomic databases efficiently. We refer to this construction as *prefix-free parsing*. It takes as input string S and in one-pass generates a dictionary and a parse of S with the property that the BWT can be constructed from dictionary and parse using workspace proportional to their total size and $O(|S|)$ time. Yet, the resulting index of Boucher et al. [2] has no SA sample, and therefore, it only supports counting and not locating. This makes this index not directly applicable to many bioinformatic applications, such as sequence alignment.

Our Contributions. In this paper, we present a solution for building the FM-index[1] for very large datasets by showing that we can build the BWT and Gagie et al.'s SA sample together in roughly the same time and memory needed to construct the BWT alone. We note that this algorithm is also based on prefix-free parsing. Thus, we begin by describing how to construct the BWT from the prefix-free parse, and then we show that it can be modified to build the SA sample in addition to the BWT in roughly the same time and space. We implement this approach, and we refer to the resulting implementation as `bigbwt`. We compare it to state-of-the-art methods for constructing the SA sample and demonstrate that `bigbwt` is currently the fastest and most space-efficient method for constructing the SA sample on large genomic databases.

Next, we demonstrate the applicability of our method to short read alignment. In particular, we compare the memory and time needed by our method to build an index for collections of chromosome 19 with that of Bowtie. Through these experiments, we show that Bowtie was unable to build indexes for our largest collections (500 or more) because it exhausted memory, whereas our method was able to build indexes up to 1,000 chromosome 19s (and likely beyond). At 250 chromosome 19 sequences, our method required only about 2% of the time and 6% the peak memory of Bowtie's. Lastly, we demonstrate that it is possible to index collections of whole human genome assemblies with sub-linear scaling as the size of the collection grows.

Related Work. The development of methods for building the FM-index on large datasets is closely related to the development short-read aligners for pan-genomics—an area where there is growing interest [5,12,27]. Here, we briefly describe some previous approaches to this problem and detail their connection to the work in this paper. We note that majority of pan-genomic aligners require

---

[1] With the SA sample of Gagie et al. [11], this index is termed the r-index.

building the FM-index for a population of genomes and thus could increase proficiency by using the methods described in this paper.

GenomeMapper [27], the method of Danek et al. [5], and GCSA [29] represent the genomes in a population as a graph and then reduce the alignment problem to finding a path within the graph. Hence, these methods require all possible paths to be identified, which is exponential in the worst case. Some of these methods—such as GCSA—use the FM-index to store and query the graph and could capitalize on our approach by building the index in the manner described here. Another set of approaches [8,12,24,33] considers the reference pan-genome as the concatenation of individual genomes and exploits redundancy by using a compressed index. The hybrid index [8] operates on a Lempel-Ziv compression of the reference pan-genome. An input parameter M sets the maximum length of reads that can be aligned; the parameter M has a large impact on the final size of the index. For this reason, the hybrid index is suitable for short-read alignment only, although there have been recent heuristic modifications to allow longer alignments [9]. In contrast, the r-index, of which we provide an implementation in this work, has no such length limitation. The most recent implementation of the hybrid index is CHIC [32]. Although CHIC has support for counting multiple occurrences of a pattern within a genomic database, it is an expensive operation, namely $O(\ell \log \log n)$, where ℓ is the number of occurrences in the databases and n is the length of the database. However, the r-index is capable of counting all occurrences of a pattern of length m in $O(m)$ time up to polylog factors. There are a number of other approaches building off the hybrid index or similar ideas [5,34]; for an extended discussion, we refer the reader to the survey of Gagie and Puglisi [12].

Finally, a third set of approaches [14,23] attempts to encode variants within a single reference genome. BWBBLE by Huang et al. [14] follows this by supplementing the alphabet to indicate if multiple variants occur at a single location. This approach does not support counting of the number of variants matching a specific alignment; also, it suffers from memory blow-up when larger structural variations occur.

2 Background

2.1 BWT and FM Indexes

Consider a string S of length n from a totally ordered alphabet Σ, such that the last character of S is lexicographically less than any other character in S. Let F be the list of S's characters sorted lexicographically by the suffixes starting at those characters, and let L be the list of S's characters sorted lexicographically by the suffixes starting immediately after those characters. The list L is termed the Burrows-Wheeler Transform [3] of S and denoted BWT. If $S[i]$ is in position p in F then $S[i-1]$ is in position p in L. Moreover, if $S[i] = S[j]$ then $S[i]$ and $S[j]$ have the same relative order in both lists; otherwise, their relative order in F is the same as their lexicographic order. This means that if $S[i]$ is in position p in L then, assuming arrays are indexed from 0 and $\prec$ denotes lexicographic precedence, in F it is in position $j_i = |\{h : S[h] \prec S[i]\}| + |\{h : L[h] = S[i], h \leq p\}| - 1$. The mapping $i \mapsto j_i$ is termed the LF mapping. Finally, notice that the last character in S always appears first in L. By repeated application of the LF mapping, we can invert the BWT, that is, recover S from L. Formally, the *suffix array* SA of the string S is an array such that entry i is the starting position in S of the ith largest suffix in lexicographical order. The above definition of the BWT is equivalent to the following:

$$\mathsf{BWT}[i] = S[(\mathsf{SA}[i] - 1) \mod n]. \tag{1}$$

The BWT was introduced as an aid to data compression: it moves characters followed by similar contexts together and thus makes many strings encountered in practice locally homogeneous and easily compressible. Ferragina and Manzini [10] showed how the BWT may be used for *indexing* a string S: given a pattern P of length $m < n$, find the number and location of all occurrences of P within S. If we know the range $\mathsf{BWT}(S)[i..j]$ occupied by characters immediately preceding occurrences of a pattern Q in S, then we can compute the range $\mathsf{BWT}(S)[i'..j']$ occupied by characters immediately preceding occurrences of cQ in S, for any character $c \in \Sigma$, since

$$i' = |\{h : S[h] \prec c\}| + |\{h : S[h] = c, h < i\}|$$
$$j' = |\{h : S[h] \prec c\}| + |\{h : S[h] = c, h \leq j\}| - 1.$$

Notice $j' - i' + 1$ is the number of occurrences of cQ in S. The essential components of an FM-index for S are, first, an array storing $|\{h : S[h] \prec c\}|$ for each character c and, second, a rank data structure for BWT that quickly tells us how often any given character occurs up to any given position[2]. To be able to locate the occurrences of patterns in S (in addition to just counting them), the FM-index uses a sampled[3] suffix array of S and a bit vector indicating the positions in the BWT of the characters preceding the sampled suffixes.

2.2 Prefix-Free Parsing

Next, we give an overview of prefix-free parsing, which produces a dictionary $\mathcal{D}$ and a parse $\mathcal{P}$ by sliding a window of fixed width through the input string S and dividing it into variable-length overlapping with delimiting prefixes and suffixes. We refer the reader to Boucher et al. [2] for the formal proofs and Sect. 3.1 for the algorithmic details. A rolling hash function identifies when substrings are parsed into elements of a dictionary, which is a set of substrings of S. Intuitively, for a repetitive string, the same dictionary phrases will be encountered frequently.

We now formally define the dictionary $\mathcal{D}$ and parse $\mathcal{P}$. Given a string[4] S of length n, window size $w \in \mathbb{N}$ and modulus $p \in \mathbb{N}$, we construct the dictionary $\mathcal{D}$ of

---

[2] Given a sequence (string) $S[1, n]$ over an alphabet $\Sigma = \{1, \ldots, \sigma\}$, a character $c \in \Sigma$, and an integer i, $\mathsf{rank}_c(S, i)$ is the number of times that c appears in $S[1, i]$.

[3] *Sampled* means that only some fraction of entries of the suffix array are stored.

[4] For technical reasons, the string S must terminate with w copies of lexicographically least \$ symbol.

substrings of S and the parse $\mathcal{P}$ as follows. We let f be a hash function on strings of length w, and let $\mathcal{T}$ be the sequence of substrings $W = S[s, s+w-1]$ such that $f(W) \equiv 0 \pmod{p}$ or $W = S[0, w-1]$ or $W = S[n-w+1, n-1]$, ordered by initial position in S; let $\mathcal{T} = (W_1 = S[s_1, s_1 + w - 1], \ldots, W_k = [s_k, s_k + w - 1])$. By construction, the strings

$$S[s_1, s_2 + w - 1], S[s_2, s_3 + w - 1], \ldots, S[s_{k-1}, s_k + w - 1]$$

form a parsing of S in which each pair of consecutive strings $S[s_i, s_{i+1}+w-1]$ and $S[s_{i+1}, s_{i+2}+w-1]$ overlaps by exactly w characters. We define $\mathcal{D} = \{S[s_i, s_{i+1} + w - 1] : 1 \leq i < k\}$; that is, $\mathcal{D}$ consists of the set of the unique substrings s of S such that $|s| > w$ and the first and last w characters of s form consecutive elements in $\mathcal{T}$. If S has many repetitions we expect that $|\mathcal{D}| \ll k$. With a little abuse of notation we define the parsing $\mathcal{P}$ as the sequence of lexicographic ranks of substrings in $\mathcal{D}$: $\mathcal{P} = (\text{rank}_{\mathcal{D}}(S[s_i, s_{i+1} + w - 1]))_{i=1}^{k-1}$. The parse $\mathcal{P}$ indicates how S may be reconstructed using elements of $\mathcal{D}$. The dictionary $\mathcal{D}$ and parse $\mathcal{P}$ may be constructed in one pass over S in $O(n + |\mathcal{D}| \log |\mathcal{D}|)$ time if the hash function f can be computed in constant time.

2.3 r-Index Locating

Policriti and Prezza [26] showed that if we have stored $\text{SA}[k]$ for each value k such that $\text{BWT}[k]$ is the beginning or end of a run (i.e., a maximal nonempty unary substring) in BWT, and we know both the range $\text{BWT}[i..j]$ occupied by characters immediately preceding occurrences of a pattern Q in S and the starting position of one of those occurrences of Q, then when we compute the range $\text{BWT}[i'..j']$ occupied by characters immediately preceding occurrences of cQ in S, we can also compute the starting position of one of those occurrences of cQ. Bannai et al. [1] then showed that even if we have stored only $\text{SA}[k]$ for each value k such that $\text{BWT}[k]$ is the beginning of a run, then as long as we know $\text{SA}[i]$ we can compute $\text{SA}[i']$.

Gagie, Navarro and Prezza [11] showed that if we have stored in a predecessor data structure $\text{SA}[k]$ for each value k such that $\text{BWT}[k]$ is the beginning of a run in BWT, with $\phi^{-1}(\text{SA}[k]) = \text{SA}[k+1]$ stored as satellite data, then given $\text{SA}[h]$ we can compute $\text{SA}[h+1]$ in $O(\log \log n)$ time as $\text{SA}[h+1] = \phi^{-1}(\text{pred}(\text{SA}[h])) + \text{SA}[h] - \text{pred}(\text{SA}[h])$, where $\text{pred}(\cdot)$ is a query to the predecessor data structure. Combined with Bannai et al.'s result, this means that while finding the range $\text{BWT}[i..j]$ occupied by characters immediately preceding occurrences of a pattern Q, we can also find $\text{SA}[i]$ and then report $\text{SA}[i+1..j]$ in $O((j-i) \log \log n)$-time, that is, $O(\log \log n)$-time per occurrence.

Gagie et al. gave the name r-index to the index resulting from combining a rank data structure over the run-length compressed BWT with their SA sample, and Bannai et al. used the same name for their index. Since our index is an implementation of theirs, we keep this name; on the other hand, we do not apply it to indexes based on run-length compressed BWTs that have standard SA samples or no SA samples at all.

3   Methods

Here, we describe our algorithm for building the SA or the sampled SA from the prefix free parse of a input string S, which is used to build the r-index. We first review the algorithm from [2] for building the BWT of S from the prefix free parse. Next, we show how to modify this construction to compute the SA or the sampled SA along with the BWT.

3.1   Construction of BWT from Prefix-Free Parse

We assume we are given a prefix-free parse of $S[1..n]$ with window size w consisting of a dictionary $\mathcal{D}$ and a parse $\mathcal{P}$. We represent the dictionary as a string $\mathcal{D}[1..\ell] = t_1 \# t_2 \# \cdots t_{d-1} \# t_d \#$ where t_i's are the dictionary phrases in lexicographic order and $\#$ is a unique separator. We assume we have computed the SA of $\mathcal{D}$, denoted by $\mathsf{SA}_\mathcal{D}[1..\ell]$ in the following, and the BWT of $\mathcal{P}$, denoted $\mathsf{BWT}_\mathcal{P}$, and the array $\mathrm{Occ}[1,d]$ such that $\mathrm{Occ}[i]$ stores the number of occurrences of the dictionary phrase t_i in the parse. These preliminary computations take $O(|\mathcal{D}| + |\mathcal{P}|)$ time.

By the properties of the prefix-free parsing, each suffix of S is prefixed by *exactly one* suffix α of a dictionary phrase t_j with $|\alpha| > w$. We call α_i the *representative prefix* of the suffix $S[i..n]$. From the uniqueness of the representative prefix we can partition S's suffix array $\mathsf{SA}[1..n]$ into k ranges

$$[b_1, e_1], \quad [b_2, e_2], \quad [b_3, e_3], \quad \ldots, \quad [b_k, e_k]$$

with $b_1 = 1$, $b_i = e_{i-1} + 1$ for $i = 2, \ldots, k$, and $e_k = n$, such that for $i = 1, \ldots, k$ all suffixes

$$S[\mathsf{SA}[b_i]..n], \quad S[\mathsf{SA}[b_i + 1]..n], \quad \ldots, \quad S[\mathsf{SA}[e_i]..n]$$

have the same representative prefix α_i. By construction $\alpha_1 \prec \alpha_2 \prec \cdots \prec \alpha_k$.

By construction, any suffix $\mathcal{D}[i..\ell]$ of the dictionary $\mathcal{D}$ is also prefixed by the suffix of a dictionary phrase. For $j = 1, \ldots, \ell$, let β_j denote the longest prefix of $\mathcal{D}[\mathsf{SA}_\mathcal{D}[j]..\ell]$ which is the suffix of a phrase (i.e. $\mathcal{D}[\mathsf{SA}_\mathcal{D}[j] + |\beta_j|] = \#$). By construction the strings β_j's are lexicographically sorted $\beta_1 \prec \beta_2 \prec \cdots \prec \beta_\ell$. Clearly, if we compute $\beta_1, \ldots, \beta_\ell$ and discard those such that $|\beta_j| \leq w$, the remaining β_j's will coincide with the representative prefixes α_i's. Since both β_j's and α_i's are lexicographically sorted, this procedure will generate the representative prefixes in the order $\alpha_1, \alpha_2, \ldots, \alpha_k$. We note that more than one β_j can be equal to some α_i since different dictionary phrases can have the same suffix.

We scan $\mathsf{SA}_\mathcal{D}[1..\ell]$, compute $\beta_1, \ldots \beta_\ell$ and use these strings to find the representative prefixes. As soon as we generate an α_i we compute and output the portion $\mathsf{BWT}[b_i, e_i]$ corresponding to the range $[b_i, e_i]$ associated to α_i. To implement the above strategy, assume there are exactly k entries in $\mathsf{SA}_\mathcal{D}[1..\ell]$ prefixed by α_i. This means that there are k distinct dictionary phrases $t_{i_1}, t_{i_2}, \ldots, t_{i_k}$ that end with α_i. Hence, the range $[b_i, e_i]$ contains $z_i = e_i - b_i + 1 = \sum_{h=1}^{k} \mathrm{Occ}[i_h]$

elements. To compute $\mathsf{BWT}[b_i, e_i]$ we need to: (1) find the symbol immediately preceding each occurrence of α_i in S, and (2) find the lexicographic ordering of S's suffixes prefixed by α_i. We consider the latter problem first.

Computing the Lexicographic Ordering of Suffixes. For $j = 1, \ldots, z_i$ consider the j-th occurrence of α_i in S and let i_j denote the position in the parsing of S of the phrase ending with the j-th occurrence of α_i. In other words, $\mathcal{P}[i_j]$ is a dictionary phrase ending with α_i and $i_1 < i_2 < \cdots < i_{z_i}$. By the properties of $\mathsf{BWT}_\mathcal{P}$ the lexicographic ordering of S's suffixes prefixed by α_i coincides with the ordering of the symbols $\mathcal{P}[i_j]$ in $\mathsf{BWT}_\mathcal{P}$. In other words, $\mathcal{P}[i_j]$ precedes $\mathcal{P}[i_h]$ in $\mathsf{BWT}_\mathcal{P}$ if and only if S's suffix prefixed by the j-th occurrence of α_i is lexicographically smaller than S's suffix prefixed by the h-th occurrence of α_i.

We could determine the desired lexicographic ordering by scanning $\mathsf{BWT}_\mathcal{P}$ and noticing which entries coincide with one of the dictionary phrases $t_{i_1}, \ldots, t_{i_k}$ that end with α_i but this would clearly be inefficient. Instead, for each dictionary phrase t_i we maintain an array IL_i of length $\mathrm{Occ}[i]$ containing the indexes j such that $\mathsf{BWT}_\mathcal{P}[j] = i$. These sorts of "inverted lists" are computed at the beginning of the algorithm and replace the $\mathsf{BWT}_\mathcal{P}$ which can be discarded.

Finding the Symbol Preceding α_i. Given a representative prefix α_i from $\mathrm{SA}_\mathcal{D}$ we retrieve the indexes $i_1, \ldots, i_k$ of the dictionary phrases $t_{i_1}, \ldots, t_{i_k}$ that end with α_i. Then, we retrieve the inverted lists $\mathrm{IL}_{i_1}, \ldots \mathrm{IL}_{i_k}$ and we merge them obtaining the list of the z_i positions $y_1 < y_2 < \cdots < y_{z_i}$ such that $\mathsf{BWT}_P[y_j]$ is a dictionary phrase ending with α_i. Such list implicitly provides the lexicographic order of S's suffixes starting with α_i.

To compute the BWT we need to retrieve the symbols preceding such occurrences of α_i. If α_i *is not* a dictionary phrase, then α_i is a proper suffix of the phrases $t_{i_1}, \ldots, t_{i_k}$ and the symbols preceding α_i in S are those preceding α_i in $t_{i_1}, \ldots, t_{i_k}$ that we can retrieve from $\mathcal{D}[1..\ell]$ and $\mathrm{SA}_\mathcal{D}[1..\ell]$. If α_i *coincides* with a dictionary phrase t_j, then it cannot be a suffix of another phrase. Hence, the symbols preceding α_i in S are those preceding t_j in S that we store at the beginning of the algorithm in an auxiliary array PR_j along with the inverted list IL_j.

3.2   Construction of **SA** and **SA** Sample Along with the **BWT**

We now show how to modify the above algorithm so that, along with BWT, it computes the full SA of S or the sampled SA consisting of the values $\mathsf{SA}[s_1], \ldots, \mathsf{SA}[s_r]$ and $\mathsf{SA}[e_1], \ldots, \mathsf{SA}[e_r]$, where r is the number of maximal non-empty runs in BWT and s_i and e_i are the starting and ending positions in BWT of the i-th such run, respectively. Note that if we compute the sampled SA the actual output will consist of r start-run pairs $\langle s_i, \mathsf{SA}[s_i] \rangle$ and r end-run pairs $\langle e_i, \mathsf{SA}[e_i] \rangle$ since the SA values alone are not enough for the construction of the r-index.

We solve both problems using the following strategy. Simultaneously to each entry $\mathsf{BWT}[j]$, we compute the corresponding entry $\mathsf{SA}[j]$. Then, if we need the sampled SA, we compare $\mathsf{BWT}[j-1]$ and $\mathsf{BWT}[j]$ and if they differ, we output the pair $\langle j-1, \mathsf{SA}[j-1]\rangle$ among the end-runs and the pair $\langle j, \mathsf{SA}[j]\rangle$ among the start-runs. To compute the SA entries, we only need d additional arrays $\mathrm{EP}_1, \ldots \mathrm{EP}_d$ (one for each dictionary phrase), where $|\mathrm{EP}_i| = |\mathrm{IL}_i| = \mathrm{Occ}[i]$, and $\mathrm{EP}_i[j]$ contains the ending position in S of the dictionary phrase which is in position $\mathrm{IL}_i[j]$ of $\mathsf{BWT}_\mathcal{P}$.

Recall that in the above algorithm for each occurrence of a representative prefix α_i, we compute the indexes $i_1, \ldots, i_k$ of the dictionary phrases $t_{i_1}, \ldots, t_{i_k}$ that end with α_i. Then, we use the lists $\mathrm{IL}_{i_1}, \ldots, \mathrm{IL}_{i_k}$ to retrieve the positions of all the occurrences of $t_{i_1}, \ldots, t_{i_k}$ in $\mathsf{BWT}_\mathcal{P}$, thus establishing the relative lexicographic order of the occurrences of the dictionary phrases ending with α_i. To compute the corresponding SA entries, we need the starting position in S of each occurrence of α_i. Since the ending position in S of the phrase with relative lexicographic rank $\mathrm{IL}_{i_h}[j]$ is $\mathrm{EP}_{i_h}[j]$, the corresponding SA entry is $\mathrm{EP}_{i_h}[j] - |\alpha_i| + 1$. Hence, along with each BWT entry we obtain the corresponding SA entry which is saved to the output file if the full SA is needed, or further processed as described above if we need the sampled SA.

4   Time and Memory Usage for SA and SA Sample Construction

We compare the running time and memory usage of bigbwt with the following methods, which represent the current state-of-the-art.

bwt2sa Once the BWT has been computed, the SA or SA sample may be computed by applying the LF mapping to invert the BWT and the application of Eq. 1. Therefore, as a baseline, we use bigbwt to construct the BWT only, as in Boucher et al. [2]; we use bigbwt since it seems best suited to the inputs we consider. Next, we load the BWT as a Huffman-compressed string with access, rank, and select support to compute the LF mapping. We step backwards through the BWT and compute the entries of the SA in non-consecutive order. Finally, these entries are sorted in external memory to produce the SA or SA sample. This method may be parallelized when the input consists of multiple strings by stepping backwards from the end of each string in parallel.

pSAscan A second baseline is to compute the SA directly from the input; for this computation, we use the external-memory algorithm pSAscan [17], with available memory set to the memory required by bigbwt on the specific input; with the ratio of memory to input size obtained from bigbwt, pSAscan is the current state-of-the-art method to compute the SA. Once pSAscan has computed the full SA, the SA sample may be constructed by loading the input text T into memory, streaming the SA from the disk, and the application of Eq. 1 to detect run boundaries. We denote this method of computing the SA sample by pSAscan+.

We compared the performance of all the methods on two datasets: (1) Salmonella genomes obtained from GenomeTrakr [31]; and (2) chromosome 19 haplotypes derived from the 1000 Genomes Project phase 3 data

[4]. The Salmonella strains were downloaded from NCBI (NCBI BioProject PRJNA183844) and preprocessed by assembling each individual sample with IDBA-UD [25] and counting k-mers ($k = 32$) using KMC [6]. We modified IDBA by setting kMaxShortSequence to 1024 per public advice from the author to accommodate the longer paired end reads that modern sequencers produce. We sorted the full set of samples by the size of their k-mer counts and selected 1,000 samples about the median. This avoids exceptionally short assemblies, which may be due to low read coverage, and exceptionally long assemblies which may be due to contamination.

Next, we downloaded and preprocessed a collection of chromosome 19 haplotypes from 1000 Genomes Project. Chromosome 19 is 58 million base pairs in length and makes up around 1.9% of the total human genome sequence. Each sequence was derived by using the `bcftools consensus` tool to combine the haplotype-specific (maternal or paternal) variant calls for an individual in the 1KG project with the chr19 sequence in the GRCH37 human reference, producing a FASTA record per sequence. All DNA characters besides A, C, G, T and N were removed from the sequences before construction.

We performed all experiments in this section on a machine with Intel(R) Xeon(R) CPU E5-2680 v2 @ 2.80 GHz and 324 GB RAM. We measured running time and peak memory footprint using `/usr/bin/time -v`, with peak memory footprint captured by the `Maximum resident set size (kbytes)` field and running time by the `User Time` and `System Time` field.

We witnessed that the running time of each method to construct the full SA is shown in Figs. 1(a)–(c). On both the Salmonella and chr19 datasets, `bigbwt` ran the fastest, often by more than an order of magnitude. In Fig. 1(d), we show the peak memory usage of `bigbwt` as a function of input size. Empirically, the peak memory usage was sublinear in input size, especially on the chr19 data, which exhibited a high degree of repetition. Despite the higher diversity of the Salmonella genomes, `bigbwt` remained space-efficient and the fastest method for construction of the full SA. Furthermore, we found qualitatively similar results for construction of the SA sample, shown in Fig. 2. Similar to the results on full SA construction, `bigbwt` outperformed both baseline methods and exhibited sublinear memory scaling on both types of databases.

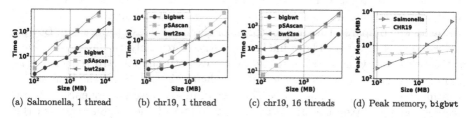

(a) Salmonella, 1 thread   (b) chr19, 1 thread   (c) chr19, 16 threads   (d) Peak memory, `bigbwt`

Fig. 1. Runtime and peak memory usage for construction of full SA.

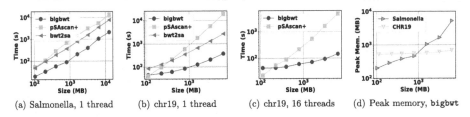

(a) Salmonella, 1 thread   (b) chr19, 1 thread   (c) chr19, 16 threads   (d) Peak memory, bigbwt

Fig. 2. Runtime and peak memory usage for construction of SA sample.

5   Application to Many Human Genome Sequences

We studied how the r-index scales to repetitive texts consisting of many similar genomic sequences. Since an ultimate goal is to improve read alignment, we benchmark against Bowtie (version 1.2.2) [19]. We ran Bowtie with the -v 0 and --norc options; -v 0 disables approximate matching, while --norc causes Bowtie (like r-index) to perform the locate query with respect to the query sequence only and not its reverse complement.

5.1   Indexing Chromosome 19s

We performed our experiments on collections of one or more versions of chromosome 19. These versions were obtained from 1000 Genomes Project haplotypes in the manner described in the previous section. We used 10 collections of chromosome 19 haplotypes, containing 1, 2, 10, 30, 50, 100, 250, 500, and 1000 sequences, respectively. Each collection is a superset of the previous. Again, all DNA characters besides A, C, G, T and N were removed from the sequences before construction. All experiments in this section were ran on a Intel(R) Xeon(R) CPU E5-2680 v3 @ 2.50 GHz machine with 512 GB memory. We measured running time and peak memory footprint as described in the previous section.

First we constructed r-index and Bowtie indexes on successively larger chromosome 19 collections (Fig. 3(a) and (b)). The r-index's peak memory is substantially smaller than Bowtie's for larger collections, and the gap grows with the collection size. At 250 chr19s, the r-index procedure takes about 2% of the time and 6% the peak memory of Bowtie's procedure. Bowtie fails to construct collections of more than 250 sequences due to memory exhaustion.

Next, we compared the disk footprint of the index files produced by Bowtie and r-index (Fig. 3(c)). The r-index currently stores only the forward strand of the sequence, while the Bowtie index stores both the forward sequence and its reverse as needed by its double-indexing heuristic [19]. Since the heuristic is relevant only for approximate matching, we omit the reverse sequence in these size comparisons. We also omit the 2-bit encoding of the original text (in the *.3.ebwt and *.4.ebwt files) as these too are used only for approximate matching. Specifically, the Bowtie index size was calculated by adding the sizes of the forward *.1.ebwt and *.2.ebwt files, which contain the BWT, SA sample, and auxiliary data structures for the forward sequence. The size of the r-index

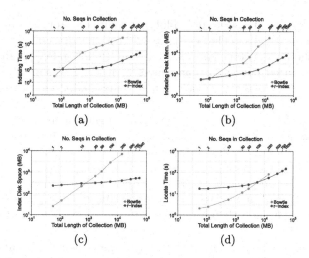

Fig. 3. Scalability of r-index and bowtie indexes against chr19 haplotype collection size and total sequence length (megabases) with respect to index construction time (seconds) (a), index construction peak memory (megabytes) (b), index disk space (megabytes) (c), and locate time (seconds) of 100,000 100bp queries (d).

increased more slowly than Bowtie's, though the r-index was larger for the smallest collections. This is because, unlike Bowtie which samples a constant fraction of the SA elements (every 32nd by default), the density of the r-index SA sample depends on the ratio n/r. When the collection is small, n/r is small and more SA samples must be stored per base. At 250 sequences, the r-index index takes 6% the space of the Bowtie index.

We then compared the speed of the locate query for r-index and Bowtie. We extracted 100,000 100-character substrings from the chr19 collection of size 1, which is also contained in all larger collections. We queried these against both the Bowtie and r-indexes. We used the --max-hits option for r-index and the -k option for Bowtie to set the maximum number of hits reported to be equal to the collection size. The actual number of hits reported will often equal this number, but could be smaller (if the substring differs between individuals due to genetic variation) or larger (if the substring is from a repetitive portion of the genome). Since the source of the substrings is present in all the collections, every query is guaranteed to match at least once. As seen in Fig. 3(d), the r-index locate query was faster for the collection of 250 chr19s. No comparison was possible for larger collections because Bowtie could not build the indexes.

5.2   Indexing Whole Human Genomes

Lastly, we used r-index to index many human genomes at once. We repeated our measurements for successively larger collections of (concatenated) genomes.

Thus, we first evaluated a series of haplotypes extracted from the 1000 Genomes Project [4] phase 3 callset (1KG). These collections ranged from 1 up to 10 genomes. As the first genome, we selected the GRCh37 reference itself. For the remaining 9, we used `bcftools consensus` to insert SNVs and other variants called by the 1000 Genomes Project for a single haplotype into the GRCh37 reference.

Second, we evaluated a series of whole-human genome assemblies from 6 different long-read assembly projects ("LRA"). We selected GRCh37 reference as the first genome, so that the first data point would coincide with that of the previous series. We then added long-read assemblies from a Chinese genome assembly project [28], a Korean genome assembly project [16] a project to assemble the well-studied NA12878 individual [15], a hydatidiform mole (known as CHM1) assembly project [30] and the Celera human genome project [20]. Compared to the series with only 1000 Genomes Project individuals, this series allowed us to measure scaling while capturing a wider range of genetic variation between humans. This is important since *de novo* human assembly projects regularly produce assemblies that differ from the human genome reference by megabases of sequence (12 megabases in the case of the Chinese assembly [28]), likely due to prevalent but hard-to-profile large-scale structural variation. Such variation was not comprehensively profiled in the 1000 Genomes Project, which relied on short reads.

The 1KG and LRA series were evaluated twice, once on the forward genome sequences and once on both the forward and reverse-complement sequences. This accounts for the fact that different *de novo* assemblies make different decisions about how to orient contigs. The r-index method achieves compression only with respect to the forward-oriented versions of the sequences indexed. That is, if two contigs are reverse complements of each other but otherwise identical, r-index achieves less compression than if their orientations matched. A more practical approach would be to index both forward and reverse-complement sequences, as Bowtie 2 [18] and BWA [22] do.

We measured the peak memory footprint when indexing these collections (Fig. 4). We ran these experiments on an Intel(R) Xeon(R) CPU E5-2650 v4 @ 2.20 GHz system with 256 GB memory. Memory footprints for LRA grew more quickly than those for 1KG. This was expected due to the greater genetic diversity captured in the assemblies. This may also be due in part to the presence of sequencing errors in the long-read assembles; long-read technologies are more prone to indel errors than short-read technologies, for examples, and some may survive in the assemblies. Also as expected, memory footprints for the LRA series that included both forward and reverse complement sequences grew more slowly than when just the forward sequence was included. This is due to sequences that differ only (or primarily) in their orientation between assemblies. All series exhibit sublinear trends, highlighting the efficacy of r-index compression even when indexing genetically diverse whole-genome assemblies. Indexing the forward and reverse complement strands of 10 1KG individuals took about 6 h and 20 min and the final index size was 36 GB.

Table 1. Sequence length and n/r statistic with respect to number of whole genomes for the first 6 collections in the 1000 Genomes (1KG) and long-read assembly (LRA) series.

# Genomes	Sequence			
	Length (MB)		n/r	
	1KG	LRA	1KG	LRA
1	6,072	6,072	1.86	1.86
2	12,144	12,484	3.70	3.58
3	18,217	17,006	5.38	4.83
4	24,408	22,739	7.13	6.25
5	30,480	28,732	8.87	7.80
6	36,671	34,420	10.63	9.28

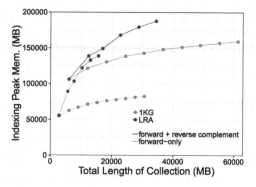

Fig. 4. Peak index-building memory for r-index when indexing successively large collections of 1000-Genomes individuals (1KG) and long-read whole-genome assemblies (LRA).

We also measured lengths and n/r ratios for each collection of whole genomes (Table 1). Consistent with the memory-scaling results, we see that the n/r ratios are somewhat lower for the LRA series than for the 1KG series, likely due to greater genetic diversity in the assemblies.

6   Conclusions and Future Work

We give an algorithm for building the SA and SA sample from the prefix-free parse of an input string S, which fully completes the practical challenge of building the index proposed by Gagie et al. [11]. This leads to a mechanism for building a complete index of large databases—which is the linchpin in developing practical means for pan-genomics short read alignment. In fact, we apply our method for indexing partial and whole human genomes, and show that it scales better than Bowtie with respect to both memory and time. This allows for an index to be constructed for large collections of chromosome 19s (500 or more); a task that is out of reach of Bowtie—as exceeded our limit of 512 GB of memory.

Even though this work opens up doors to indexing large collections of genomes, it also highlights problems that warrant further investigation. For example, there still remains a significant amount of work in adapting the index to work well on large sets of sequence reads. This problem not only requires the construction of the r-index but also an efficient means to update the index as new datasets become available. Moreover, there is interest in supporting more sophisticated queries than just pattern matching, which would allow for more complex searches of large databases.

Acknowledgements. AK and CB were supported by the National Institute of Allergy and Infectious Diseases of the National Institutes of Health (1R01AI141810-01) and NSF-IIS (1618814). TM and BL were supported by the National Institutes of Health

(R01GM118568) and NSF-IIS (1349906). TG was supported by FONDECYT grant 1171058 *Compression-aware algorithmics*. GM was partially supported by PRIN grant 201534HNXC and by INdAM-GNCS Project 2019 *Innovative methods for the solution of medical and biological big data.*

References

1. Bannai, H., Gagie, T., I, T.: Online LZ77 parsing and matching statistics with RLBWTs. In: Proceedings of the 29th Annual Symposium on Combinatorial Pattern Matching, (CPM), vol. 105, pp. 7:1–7:12 (2018)
2. Boucher, C., Gagie, T., Kuhnle, A., Manzini, G.: Prefix-free parsing for building big BWTs. In: Proceedings of 18th International Workshop on Algorithms in Bioinformatics, WABI, vol. 113, pp. 2:1–2:16 (2018)
3. Burrows, M., Wheeler, D.J.: A block sorting lossless data compression algorithm. Technical report 124. Digital Equipment Corporation (1994)
4. The 1000 Genomes Project Consortium: A global reference for human genetic variation. Nature **526**(7571), 68–74 (2015)
5. Danek, A., Deorowicz, S., Grabowski, S.: Indexes of large genome collections on a PC. PLoS ONE **9**(10), e109384 (2014)
6. Deorowicz, S., Kokot, M., Grabowski, S., Debudaj-Grabysz, A.: KMC 2: fast and resource-frugal k-mer counting. Bioinformatics **31**(10), 1569–1576 (2015)
7. Garrison, E., et al.: Variation graph toolkit improves read mapping by representing genetic variation in the reference. Nat. Biotechnol. **36**(9), 875–879 (2018)
8. Ferrada, H., Gagie, T., Hirvola, T., Puglisi, S.J.: Hybrid indexes for repetitive datasets. Philos. Trans. Roy. Soc. A: Math. Phys. Eng. Sci. **372**(2016), 1–9 (2014)
9. Ferrada, H., Kempa, D., Puglisi, S.J.: Hybrid indexing revisited. In: Proceedings of the 21st Algorithm Engineering and Experiments, ALENEX, pp. 1–8 (2018)
10. Ferragina, P., Manzini, G.: Opportunistic data structures with applications. In: Proceedings of the 41st Annual Symposium on Foundations of Computer Science, FOCS, pp. 390–398 (2000)
11. Gagie, T., Navarro, G., Prezza, N.: Optimal-time text indexing in BWT-runs bounded space. In: Proceedings of the 29th Annual Symposium on Discrete Algorithms, SODA, pp. 1459–1477 (2018)
12. Gagie, T., Puglisi, S.J.: Searching and indexing genomic databases via kernelization. Front. Bioeng. Biotechnol. **3**, 10–13 (2015)
13. Li, H., Durbin, R.: Fast and accurate short read alignment with Burrows-Wheeler Transform. Bioinformatics **25**, 1754–1760 (2009)
14. Huang, L., Popic, V., Batzoglou, S.: Short read alignment with populations of genomes. Bioinformatics **29**(13), i361–i370 (2013)
15. Jain, M., et al.: Nanopore sequencing and assembly of a human genome with ultra-long reads. Nat. Biotechnol. **36**(4), 338–345 (2018)
16. Jeong-Sun, S., et al.: De novo assembly and phasing of a Korean human genome. Nature **538**(7624), 243–247 (2016)
17. Kärkkäinen, J., Kempa, D., Puglisi, S.J.: Parallel external memory suffix sorting. In: Cicalese, F., Porat, E., Vaccaro, U. (eds.) CPM 2015. LNCS, vol. 9133, pp. 329–342. Springer, Cham (2015). https://doi.org/10.1007/978-3-319-19929-0_28
18. Langmead, B., Salzberg, S.L.: Fast gapped-read alignment with Bowtie 2. Nat. Methods **9**(4), 357 (2012)

19. Langmead, B., Trapnell, C., Pop, M., Salzberg, S.L.: Ultrafast and memory-efficient alignment of short DNA sequences to the human genome. Genome Biol. **10**, R25 (2008)
20. Levy, S., et al.: The diploid genome sequence of an individual human. PLoS Biol. **5**(10), e254 (2007)
21. Li, R., et al.: SOAP2: an improved tool for short read alignment. Bioinformatics **25**(15), 1966–1967 (2009)
22. Li, H.: Aligning sequence reads, clone sequences and assembly contigs with BWA-MEM. arXiv:1303.3997 [q-bio], March 2013
23. Maciuca, S., del Ojo Elias, C., McVean, G., Iqbal, Z.: A natural encoding of genetic variation in a Burrows-Wheeler Transform to enable mapping and genome inference. In: Frith, M., Storm Pedersen, C.N. (eds.) WABI 2016. LNCS, vol. 9838, pp. 222–233. Springer, Cham (2016). https://doi.org/10.1007/978-3-319-43681-4_18
24. Mäkinen, V., Navarro, G., Sirén, J., Välimäki, N.: Storage and retrieval of highly repetitive sequence collections. J. Comput. Biol. **17**(3), 281–308 (2010)
25. Peng, Y., Leung, H.C.M., Yiu, S.M., Chin, F.Y.L.: IDBA-UD: a de novo assembler for single-cell and metagenomic sequencing data with highly uneven depth. Bioinformatics **28**(11), 1420–1428 (2012)
26. Policriti, A., Prezza, N.: LZ77 computation based on the run-length encoded BWT. Algorithmica **80**(7), 1986–2011 (2018)
27. Schneeberger, K., et al.: Simultaneous alignment of short reads against multiple genomes. Genome Biol. **10**(9), R98 (2009)
28. Shi, L., et al.: Long-read sequencing and *de novo* assembly of a Chinese genome. Nat. Commun. **7**, 12065 (2016)
29. Sirén, J., Välimäki, N., Mäkinen, V.: Indexing graphs for path queries with applications in genome research. IEEE/ACM Trans. Comput. Biol. Bioinform. **11**(2), 375–388 (2014)
30. Steinberg, K.M., et al.: Single haplotype assembly of the human genome from a hydatidiform mole. Genome Res. **24**, 2066–2076 (2014). p. gr.180893.114
31. Stevens, E.L., et al.: The public health impact of a publically available, environmental database of microbial genomes. Front. Microbiol. **8**, 808 (2017)
32. Valenzuela, D., Norri, T., Välimäki, N., Pitkänen, E., Mäkinen, V.: Towards pan-genome read alignment to improve variation calling. BMC Genomics **19**(2), 87 (2018)
33. Valenzuela, D., Mäkinen, V.: CHIC: a short read aligner for pan-genomic references. Technical report, biorxiv.org (2017)
34. Wandelt, S., Starlinger, J., Bux, M., Leser, U.: RCSI: scalable similarity search in thousand(s) of genomes. Proc. VLDB Endow. **6**(13), 1534–1545 (2013)

Tumor Copy Number Deconvolution Integrating Bulk and Single-Cell Sequencing Data

Haoyun Lei[1], Bochuan Lyu[2], E. Michael Gertz[3,4], Alejandro A. Schäffer[3,4], Xulian Shi[5], Kui Wu[5], Guibo Li[5], Liqin Xu[5], Yong Hou[5], Michael Dean[6], and Russell Schwartz[1,7(✉)]

[1] Computational Biology Department, Carnegie Mellon University, Pittsburgh, PA, USA

[2] Department of Mathematics, Rose-Hulman Institute of Technology, Terre Haute, IN, USA

[3] National Center for Biotechnology Information, U.S. National Institutes of Health, Bethesda, MD, USA

[4] Cancer Data Science Laboratory, National Cancer Institute, U.S. National Institutes of Health, Bethesda, MD, USA

[5] BGI-Shenzhen, Shenzhen, China

[6] Laboratory of Translational Genomics, Division of Cancer Epidemiology and Genetics, National Cancer Institute, U.S. National Institutes of Health, Gaithersburg, MD, USA

[7] Department of Biological Sciences, Carnegie Mellon University, Pittsburgh, PA, USA
russells@andrew.cmu.edu

Abstract. Characterizing intratumor heterogeneity (ITH) is crucial to understanding cancer development, but it is hampered by limits of available data sources. Bulk DNA sequencing is the most common technology to assess ITH, but mixes many genetically distinct cells in each sample, which must then be computationally deconvolved. Single-cell sequencing (SCS) is a promising alternative, but its limitations—e.g., high noise, difficulty scaling to large populations, technical artifacts, and large data sets—have so far made it impractical for studying cohorts of sufficient size to identify statistically robust features of tumor evolution. We have developed strategies for deconvolution and tumor phylogenetics combining limited amounts of bulk and single-cell data to gain some advantages of single-cell resolution with much lower cost, with specific focus on deconvolving genomic copy number data. We developed a mixed membership model for clonal deconvolution via non-negative matrix factorization (NMF) balancing deconvolution quality with similarity to single-cell samples via an associated efficient coordinate descent algorithm. We then improve on that algorithm by integrating deconvolution with clonal phylogeny inference, using a mixed integer linear programming (MILP) model to incorporate a minimum evolution phylogenetic tree cost in the

Supplementary material is available at bioRxiv (https://doi.org/10.1101/519892/).

© Springer Nature Switzerland AG 2019
L. J. Cowen (Ed.): RECOMB 2019, LNBI 11467, pp. 174–189, 2019.
https://doi.org/10.1007/978-3-030-17083-7_11

problem objective. We demonstrate the effectiveness of these methods on semi-simulated data of known ground truth, showing improved deconvolution accuracy relative to bulk data alone.

Keywords: Cancer · Heterogeneity · Genomic deconvolution · Copy number alteration (CNA) · Non-negative matrix factorization (NMF)

1 Introduction

Cancer is one of the most lethal terminal diseases in the world, resulting for example in approximately 600,000 deaths in the U.S.A. in the past year [34]. Nevertheless, the age-adjusted rate of cancer deaths in the U.S.A. has been declining, partly due to the invention of new cancer treatments. Recent work in developing cancer therapeutics is based on the notion of personalized or precision medicine [6] to target driver alterations in specific cancer genes. Such targeted treatments have shown success in prolonging life but rarely lead to durable cures [12], largely because tumors are not normally static or homogeneous entities [8]. Most cancers exhibit phenotypes of hypermutability [20] that result in a process of continuing evolution of clonal populations of tumor cells [27], creating the opportunity for continuing acquisition of adaptive mutations as well as putatively selectively neutral genetic variants [41]. As a consequence, different cells in the same tumor may acquire distinct sets of somatic alterations, including single nucleotide variants (SNVs), copy number alterations (CNAs), and structural variations (SVs) such as gene fusions or chromosomal rearrangements. This phenomenon, called intratumor heterogeneity (ITH) [24], allows tumors to develop resistance to targeted treatments, as treatment-resistant subclones emerge within the tumor [12,27] or expand from initially rare subpopulations within the tumor's clonal diversity. Considerable recent research into the molecular mechanisms of cancer has concentrated on characterizing ITH and reconstructing the processes of clonal evolution by which it develops across tumor progression (see, for example, [30]).

Currently, the most common technology to profile ITH is bulk DNA sequencing, which allows one to observe aggregate genetic variation in tumors and possibly matched normal tissue from the same patients. Bulk DNA sequencing allows one to identify reasonably common genetic lesions and estimate their variant allele fractions (VAFs). Resolving these VAFs into models of clonal heterogeneity, however, requires solving a challenging computational inference problem, known as *genomic deconvolution*, which strives to explain VAFs as mixtures of unobserved clonal sequences occurring at varying frequencies within the tumor. These methods have limited accuracy and resolution, particularly with respect to rare clonal subpopulations [1], and reveal far less clonal heterogeneity than is evident from direct single-cell analysis (e.g., [13,26]). Genomic deconvolution is particularly challenging in cancers exhibiting CNAs [38], a significant limitation given that CNAs are the primary mechanism of functional adaptation in at least some cancer types [21,44] and that CNAs at specific loci can have important consequences for treatment outcome (e.g., [25]).

Single cell sequencing (SCS) has emerged as an alternative allowing for the direct inference of clonal genotypes [26]. SCS itself is limited by difficult technical artifacts, however, such as the phenomenon of allelic dropout [14] and distortion of copy numbers due to the amplification steps used in most SCS methods to date [19]. Moreover, SCS is relatively costly in comparison to bulk sequencing. As a result, SCS studies to date have involved only small cohorts [28].

The tradeoffs between bulk sequencing and SCS have recently led to the idea that we might combine them to reconstruct ITH with both accuracy and scale [22,23], yielding improved performance in bulk data deconvolution and relative to using SCS data alone. To date, though, such work has focused on SNVs specifically. There is substantial value in developing comparable methods for CNAs given their biological importance, the greater difficulty of CNA deconvolution, and their suitability for phylogenetics from low-coverage SCS [26].

In this work, we develop methods for combining bulk and single-cell data to characterize ITH by CNAs specifically, both as a stand-alone inference and joint with phylogenetic inference on clonal subpopulations. We pose the problem of inferring the tumor subpopulations and their representation across genomic samples using a variant of non-negative matrix factorization (NMF). We seek solutions that deconvolve bulk data while achieving consistency between inferred single cells and limited SCS data. We consider two problem variants, one minimizing genomic distance between SCS-observed single-cells and inferred clones and the other explicitly incorporating a tumor phylogeny model to favor solutions that yield parsimonious evolution models relating observed cells and inferred clones. We characterize performance of the methods on semi-simulated data generated from low-coverage SCS. We show that both methods are effective at improving clonal deconvolution of CNAs with limited amounts of SCS data, with increasing accuracy as the number of genomic samples grows. We further show that explicitly modeling clonal evolution notably improves accuracy, suggesting the value of accounting for the process of tumor evolution in characterizing clonal structure.

2   Methods

2.1   Non-negative Matrix Factorization (NMF) Deconvolution Model

As in previous work [31], we formalize the generic problem of genomic deconvolution in terms of a mixed membership model but here relating bulk and SCS data. We focus here specifically on deconvolution of copy number data, as in [38,43], which we assume is profiled on a set of m genomic regions. In the pure deconvolution problem, we assume a set of n bulk samples, which might correspond to measurements from distinct tumor sites or regions in one patient. These bulk samples are collectively encoded in an $m \times n$ matrix B, where element b_{ij} corresponds to the mean copy number of locus i in sample j. Our goal is to identify an

$m \times k$ matrix of mixture components C, representing copy numbers of inferred common clones, and a $k \times n$ matrix of mixture fractions, F, describing the degree to which each column of C is represented in each column of B. B is presumed to be approximated by the product of C and F. We seek to minimize the deviation between B and $C \times F$ by some measure, such as the Frobenius norm. With the additional constraints that B, C, and F are non-negative, the problem is known as non-negative matrix factorization (NMF) [40]. More formally, we seek:

$$\min_{C,F} \quad ||B - CF||_{\mathrm{Fr}}^2 \tag{1}$$

where $|| \cdot ||_{\mathrm{Fr}}$ is the Frobenius norm of the matrix,

$||B - CF||_{\mathrm{Fr}} = \sqrt{\sum_{i=1}^{m} \sum_{j=1}^{n} \left(b_{ij} - \sum_{\ell=1}^{k} c_{i\ell} \cdot f_{\ell j} \right)^2}$, subject to the constraints

$f_{\ell j} \geq 0, \forall \ell \in \{1, ..., k\}, j \in \{1, ..., n\}; \sum_{\ell=1}^{k} f_{\ell j} = 1, \forall j \in \{1, ..., n\}; c_{i\ell} \in \mathbb{N}_0, \forall i \in \{1, ..., m\}, \ell \in \{1, ..., k\}$.

This optimization problem is non-convex, but prior work showed that the Euclidean distance between B and CF is non-increasing under the following multiplicative update rules [18]:

$$f_{\ell j} \leftarrow f_{\ell j} \frac{(C^T B)_{\ell j}}{(C^T CF)_{\ell j}}, \qquad c_{i\ell} \leftarrow c_{i\ell} \frac{(BF^T)_{i\ell}}{(CFF^T)_{i\ell}},$$

providing formulas for iterative local optimization by fixing C or F on alternate steps. In practice, we modify this process heuristically to renormalize columns of F after each iteration to ensure they add to 1. Since this heuristic might undermine the guarantee of monotonicity, we manually verify that $||B - CF||_{\mathrm{Fr}}$ decreases on each iteration, terminating the optimization if it fails to yield continuing improvements. More details are provided in the Supplementary Methods.

Figure 1 provides an illustrative example of the deconvolution model. Suppose we have a possible B, C, and F. The two data points B_1 and B_2 represent bulk tumor samples combining three mixture components C_1, C_2 and C_3. For ease of illustration, we assume data are assayed on the copy numbers of just two genomic loci, G_1 and G_2. The matrix B represents the average copy numbers of G_1 and G_2 in the bulk tumor samples B_1 and B_2. In each component of C, the copy numbers should be integers, but since the bulk tumors are weighted mixtures of components, the values in B need not be integers. The matrix F represents the fractional weights used to generate B_1 and B_2 from the pure components in C. For example, the first column of F indicates that B_1 is a mixture of equal parts of C_1 and C_2. This relationship can be expressed via matrix multiplication, $B = CF$, as shown in the right part of Fig. 1.

2.2 Extending NMF with Single Cell Sequence (SCS) Data

The multiplicative update algorithm is a standard method for the pure NMF optimization problem, provided the number of samples n is large compared to the intrinsic dimension k of the mixture. We would, however, expect it to perform

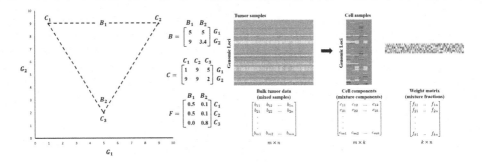

Fig. 1. Illustration of the mathematical formulation for the mixed membership modeling problem. The model implies that each entry of B, C and F is non-negative, each entry of C is integer, and each column of F must sum up to 1. (Color figure online)

poorly for our problem, in part because real tumor data generally include few samples per patient and in part because deconvolution of copy numbers is an underdetermined problem. We sought to improve the optimization by biasing the objective function to favor inferred clones similar to the observed SCS data via an auxiliary penalty in the objective function, similar to the approach of [22,23]. Intuitively, we assume that the inferred clones (C) should be closely related to one or more of the observed single cells, which we call observed cell components ($C^{(observed)}$). While any given single cell may not exactly match a consensus clone, we propose that the method will be able to approximately infer mixture components reflecting dominant clones by balancing quality of deconvolution against similarity to observed single cells. We quantify this intuition using the Euclidean distance between the inferred clones and observed cells, introducing a regularization parameter α to balance the weight of this penalty relative to the prior cost based on deconvolution quality. The resulting combined objective appears as (Eq. (2)):

$$\min_{C,F} \quad ||B - CF||^2_{Fr} + \frac{1}{2}\alpha||C - C^{(observed)}||^2_{Fr} \tag{2}$$

which we optimize subject to the constraints $f_{\ell j} \geq 0, \forall \ell \in \{1, ..., k\}, j \in \{1, ..., n\}; \sum_{\ell=1}^{k} f_{\ell j} = 1, \forall j \in \{1, ..., n\}; c_{i\ell} \in \mathbb{N}_0, \forall i \in \{1, ..., m\}, \ell \in \{1, ..., k\}$.

We solve for the revised model through an extension of the iterative update algorithm [3,18]:

$$f_{\ell j} \leftarrow f_{\ell j} \frac{(C^T B)_{\ell j}}{(C^T C F)_{\ell j}}, \qquad c_{i\ell} \leftarrow c_{i\ell} \frac{(BF^T)_{i\ell}}{(CFF^T + \alpha(C - C^{(observed)}))_{i\ell}},$$

adding the constraints on C and F to the update rules [37]. We further heuristically improve on the standard practice of random initialization by initializing the cell component matrix C with true SCS data. Pseudocode for the complete algorithm is provided in Supplementary Methods as Algorithm 1. Collectively, these additions to the pure NMF iterative update algorithm constitute our first approach to integrating SCS data for improved deconvolution of CNAs from bulk DNA-seq, which we dub our phylogeny-free method.

2.3 Extending the NMF Model with a Single-Cell Phylogeny Objective

We next developed an alternative phylogeny-based approach, seeking to deconvolve the bulk data into clonal subpopulations while simultaneously inferring a phylogeny on those deconvolved clones, similar to the SNV PHiSCS method of Malikic et al. [22]. Intuitively, evolutionary distance provides a more biologically motivated measure of what we mean in asserting that inferred single cells should be similar to observed single cells. As with the phylogeny-free method, we would expect that any small sample of single cells will not exactly reflect the spectrum of dominant clones, but that the method will be able to approximately infer dominant clones by balancing deconvolution quality against evolutionary distance of mixture components to observed single cells. This approach trades off a more principled measure of solution quality for a harder optimization problem.

We quantify phylogenetic distance as the minimum over evolutionary trees incorporating both observed single cells and inferred clones of the L_1 distance between copy number vectors describing each tree edge. Let $C^* = [C, C^{(observed)}]$ be a $m \times k^*$ matrix consisting of columns representing inferred clonal copy numbers followed by columns representing the copy numbers of the observed cells. Let c_u^* denote column u of C^*. We introduce a $k^* \times k^*$ matrix of binary variables S. A value of $s_{uv} = 1$ indicates the existence of a directed edge from node u to node v, and a value of $s_{uv} = 0$ indicates the absence of such a edge; we set $s_{uu} = 0$ to avoid self loops. In other words, S is an adjacency matrix for a directed graph; in the full formulation (Supplementary Methods) we introduce constraints that ensure the graph is a tree. We define our measure of tree cost to be

$$J(S, C, C^{(observed)}) = \sum_{u=1}^{k^*} \sum_{v=1}^{k^*} s_{uv} \cdot \|c_u^* - c_v^*\|_1. \tag{3}$$

Intuitively, $J(S, C, C^{(observed)})$ is a form of minimum evolution model on a phylogeny defined by S. While there are more sophisticated and realistic models for CNA distance (e.g., [4,5,10]), we favored L_1 distance here as a tractable approximation easily incorporated into the overall ILP framework. Similarly, while there are now a number of sophisticated methods available specifically for phylogenetics of single-cell sequences (c.f., [17]) these are largely focused on SNV rather than CNA phylogenetics (e.g.,[15,29,45]) with limited exceptions [36,39].

More specifically, we modify the NMF objective function as follows:

$$\min_{C,F,S} \left(\|B - CF\|_1 + \beta \cdot J(S, C, C^{(observed)}) \right), \tag{4}$$

where $\|B - CF\|_1 = \sum_{i=1}^{m} \sum_{j=1}^{n} \left| b_{ij} - \sum_{\ell=1}^{k} c_{i\ell} \cdot f_{\ell j} \right|$ and β is a regularization parameter to balance deconvolution quality against parsimony of the evolutionary model. The norm $||\cdot||_1$ is the element-wise L_1 matrix norm, i.e., the sum of the absolute values of matrix elements, rather than the induced L_1 matrix norm for which the same notation is sometimes used. These are optimized subject to the same constraints as in the previous formulations: $f_{\ell j} \geq 0, \forall \ell \in \{1, ..., k\}, j \in \{1, ..., n\}; \sum_{\ell=1}^{k} f_{\ell j} = 1, \forall j \in \{1, ..., n\}; c_{i\ell} \in \mathbb{N}, \forall i \in \{1, ..., m\}, \ell \in \{1, ..., k\}$.

The discrete tree optimization term lacks an analytic expression and hence does not lend itself to the prior iterative update strategy. We therefore employ a different computational strategy based on integer linear programming (ILP) to replace the linear algebra steps of the Lee and Seung method [18], similar to other recent work in joint deconvolution and phylogenetics [9,43].

For this optimization problem, we use an iterative coordinate descent approach. There are three sets of variables over which to optimize: the weight matrix $\boldsymbol{F}$, the tree structure $\boldsymbol{S}$, and the inferred copy numbers $\boldsymbol{C}$. We solve for variables $\boldsymbol{F}$, $\boldsymbol{S}$, and $\boldsymbol{C}$ alternately, in this order, while holding all other variables as constant. The iterative coordinate descent continues until the decrease between successive values of $\boldsymbol{C}$ falls below some threshold. To initialize $\boldsymbol{C}$, we used observed single cell data. Whenever two of the three sets of variables is held constant, the resulting optimization problems can each be expressed as either a linear program (LP) or an integer linear program (ILP).

When certain subsets of the variables are fixed, the resulting LP or ILP may be simplified. When solving for $\boldsymbol{F}$ with fixed values of $\boldsymbol{S}$ and $\boldsymbol{C}$, the term $J(\boldsymbol{S}, \boldsymbol{C}, \boldsymbol{C}^{(observed)})$ is constant and the value of $\boldsymbol{S}$ is irrelevant. Similarly, when solving for $\boldsymbol{S}$ for fixed values of $\boldsymbol{F}$ and $\boldsymbol{C}$, the term $||\boldsymbol{B} - \boldsymbol{CF}||_1$ is constant and therefore $\boldsymbol{F}$ is irrelevant. The optimal value of $\boldsymbol{C}$ for fixed values of $\boldsymbol{F}$ and $\boldsymbol{S}$, however, depends both on $\boldsymbol{F}$ and $\boldsymbol{S}$. We note that in the limit of using no single-cell data, our problem statement and method is similar to that of Zaccaria et al. [43] for incorporating tree mixtures into purely-bulk CNA deconvolution.

We solve for $\boldsymbol{S}$ via an ILP that uses a flow model to constrain solutions to a minimum evolution tree, adapting a similar ILP method originally developed for finding maximum parsimony character-based phylogenies [35]. Intuitively, the model forces a tree structure by setting up a flow from an arbitrary root to each other clone in the tree and minimizing the cost of edges needed to accommodate all such flows. The full ILP is described in the Supplementary Methods.

2.4   Validation via Observed Single-Cell Data

To validate the method, we require bulk data for which clone copy number vectors and frequencies are known. As this is unavailable for any real dataset, we use semi-simulated data generated from CNV calls [2] from real SCS data from two human glioblastoma cases [42]. The full single cell data set consists of low-depth SCS DNA-seq used to establish mean copy numbers at 9934 genomic positions throughout the genome, at intervals of approximately 40 kbp. Each tumor was subdivided into three regions (i.e., samples), with each single cell labeled by its region (1, 2, or 3) of origin. We used these true SCS CNA data to generate a series of synthetic bulk data sets, simulating either one, two, or

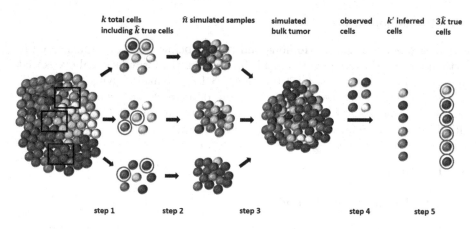

Fig. 2. Work-flow for the simulation and validation. We separate the whole process into 5 main steps: in step 1, we randomly chose k total single cells from each region (indicated by the black frames), where we can pick $\hat{k}$ dominant clones (indicated by red circles, also called true cells); in step 2, we simulated $\hat{n}$ tumor samples from each region using the k cells; in step 3, we combined the $\hat{n}$ tumor samples to get a simulated bulk tumor; in step 4, we deconvolved the bulk tumor integrating observed cells to get $k' = 3\hat{k}$ inferred clones; and in step 5, we assessed the performance using the k' inferred clones and $3\hat{k}$ true cells. (Color figure online)

three bulk samples from each region for a total of three, six, or nine bulk samples per trial. Each simulated sample is generated by sampling two dominant cells from a region to represent major clones, twenty three other cells from the same region to represent minor clones, and 50 cells from the other regions to represent contamination, which are mixed with Dirichlet-sampled proportions with weight parameters for dominant, minor, and contaminant clones in the ratio 10 to 0.1 to 0.01. We then assessed our ability to deconvolve the bulk data across a range of regularization parameter values and random replicates of the chosen single cells. We assessed accuracy by the fraction of genomic positions assigned correct copy number and by the root mean square deviation (RMSD) between true and inferred cell components and mixture fractions. Figure 2 summarizes the overall experimental design, which is described in more detail in the Supplementary Methods in Sect. A.4. This design treats observed SCS as the ground truth, allowing us to ignore the problem of doublet cells that typically must be addressed with SCS data. We would normally require that likely doublets be removed from SCS data in preprocessing before applying our method. This design also does not explicitly include calling CNA markers on bulk data, itself a hard problem that would need to be performed in preprocessing before applying our method.

We were unable to identify any competitive tool for bulk deconvolution of purely CNA data applicable to small numbers of bulk samples and for which software is publicly available. We therefore compare our methods to standard NMF, as implemented by our code with zero regularization parameters.

2.5   Implementation

The methods described in Methods and refined below were all implemented in Python3, using Gurobi. One practical change from the formulation above is that we replaced the theoretical $f_{\ell j} \geq 0$ with $f_{\ell j} > 10^{-4}$ to avoid having the f values trapped at 0. The observed human subjects data cannot be redistributed, but code for the methods is available along with artificial data on Github (https://github.com/CMUSchwartzLab/SCS_deconvolution).

3   Results

3.1   Phylogeny-Free Method

We first assessed the accuracy of the phylogeny-free method relative to pure NMF and simple heuristic improvements. Figure 3 provides a summary of accuracy and RMSD for inference of true SCS components via the method of Sect. 2.2 for variations in the number of tumor samples (3, 6, 9) and regularization parameter α (0–1) over 40 replicates per condition. To provide a baseline for comparison, each plot provides equivalent accuracy measures for NMF [18] (i.e., Algorithm 1 with $\alpha = 0$) with random initial integer valued C (red dashed line in Fig. 3) and with the proposed solution that all copy numbers have the normal value of 2, which we call the "all-diploid baseline" (black dashed line in Fig. 3 and Fig. S3). In each case, the bulk data is simulated from $k' = 6$ fundamental cell components (2 out of a random 25 cells selected in each region).

Pure NMF with random initialization performed poorly, which is unsurprising since NMF on CNA data is an underdetermined problem, although the simple heuristic of biasing the search toward biologically plausible solutions by initializing with real SCS data improves accuracy. Bringing true SCS data into the objective function yielded modest improvements in accuracy over using SCS data solely for initialization for at least some values of the regularization parameter. The phylogeny-free method with $\alpha = 0$ corresponds to pure NMF initialized with true SCS data, and this performed slightly worse than the all-diploid baseline solution. Modestly increasing α led to some improvement in accuracy, but above some value, α put too much weight on similarity to observed SCS data and too little weight on quality of the deconvolution, giving worse overall results. The best value of α depended on sample size, which we attribute again to NMF being underdetermined if the number of desired components is larger than the number of samples. The plots suggest that the method is fairly robust to α if the number of samples exceeds the intrinsic dimension of the data (six), but that SCS data can overcome that limit for small numbers of samples with a well-tuned regularization term. Additional Supplementary Results show minimal additional improvement even with unrealistically large sample sizes (Fig. S4), and also show the performance is consistent across individual inferred clones (Fig. S5).

Figure 4 provides an illustrative example of performance for a single selected clone inferred from three, six, or nine samples, intended to demonstrate kinds of errors the method tends to produce. We chose the one cell component with

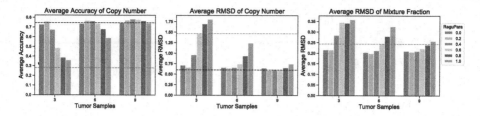

Fig. 3. Accuracy and RMSD of the phylogeny-free method as functions of tumor samples and regularization parameter. The red dashed line shows average overall accuracy (left panel) or RMSD (center and right panels) of NMF with random initialization. The black dashed line shows the performance of the all-diploid baseline solution. Since we cannot resolve mixture fractions for an all-diploid solution, we omit it from the mixture fraction results. Different bars show performance as a function of regularization parameter α of Eq. 2 from 0.0 to 1.0 in increments of 0.2. The X-axis shows the number of tumor samples and the Y-axis the average accuracy or RMSD. (Color figure online)

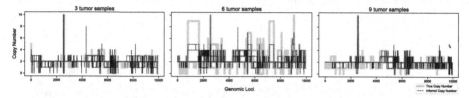

Fig. 4. Visualization of copy number as a function of genomic locus for single examples of inferred and true clones for the phylogeny-free method for three, six, and nine samples. The figure uses the minimum-RMSD pair for each case. The black dashed line shows the copy number inferred by modified NMF and the orange bar shows the true copy number in that position. (Color figure online)

smallest RMSD for each sample size to simplify visual inspection. We see that at least in these high-quality cases, the distributions of copy numbers are similar for the inferred and true cells. For loci at or just above diploid, the modified NMF can usually infer the exact copy number. Where errors occur, they tend to be in loci with large (5–10) or smaller copy numbers (0–1).

3.2 Phylogeny-Based Method

We next examined results of the phylogeny-based method of Sect. 2.3 under the same conditions used to assess the phylogeny-free method. Figure 5 summarizes average accuracy and RMSD as a function of regularization parameter β. The figure compares the results of pure NMF with the all-diploid baseline. Setting $\beta = 0$ provides poor performance, substantially below the all-diploid baseline solution. Making $\beta = 0$ for Fig. 5 represents the same optimization problem as $\alpha = 0$ for Fig. 3, but solved by the coordinate descent method we developed to accommodate the ILP phylogeny objective rather than by the modified iterative update algorithm with the simpler L_2 objective. Figure 5 thus suggests

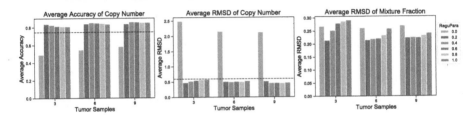

Fig. 5. Average accuracy and RMSD for the phylogeny-based method as functions of tumor samples and regularization parameter. The left panel shows the average accuracy of inferred copy numbers, the center panel average RMSD between inferred and true copy numbers, and the right panel average RMSD between the inferred and true mixture fractions. The black dashed line shows the performance of the all-diploid baseline solution. Since we cannot resolve mixture fractions for an all-diploid baseline, we omit it from the mixture fraction results. Bar plots show performance with different regularization parameters β of Eq. 3 from 0.0 to 1.0 with increment of 0.2. The X-axis shows the number of tumor samples and the Y-axis the average accuracy or RMSD. (Color figure online)

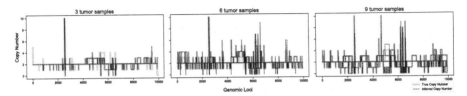

Fig. 6. Visualization of copy number as a function of genomic locus for single examples of inferred and true clones for the phylogeny-based method for three, six, and nine samples. The figure uses the minimum-RMSD pair for each case. The black dashed line shows the inferred copy number and the orange bars show the true copy number in each position. (Color figure online)

that the new coordinate descent method is less effective at pure NMF than is the prior iterative update algorithm. Despite that observation, the results on $\beta \geq 0.2$ show substantially better accuracy than was achieved by pure NMF or the phylogeny-free algorithm. Further, the results appear robust to variation in β across the range examined. Supplementary Results distinguishing accuracy across cells (Fig. S6) support the robustness of the phylogeny-based method to a range of β values in cell-to-cell inferences.

Figure 6 shows copy numbers for a single minimum-RMSD pair for inferred and true clones for each number of samples, again to visualize the nature of inference errors. The results again show exact fitting for most loci, as well as better fitting for both large (5–10) and small (0–1) copy numbers than the phylogeny-free method of Fig. 4. There is no evident pattern to the smaller number of errors that do occur for the phylogeny-based versus phylogeny-free method, which are observed for a range of low and high copy number values.

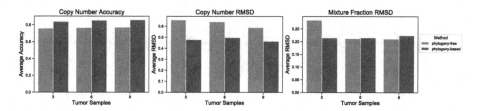

Fig. 7. Comparison between phylogeny-free and phylogeny-based methods. Bar graphs show average accuracy and RMSD over all cell components and replicates using the optimal regularization parameter for the given method, measure, and number of samples. The left panel shows accuracy in copy numbers for $\alpha = 0.2$, 0.2, 0.4 for the phylogeny-free method and $\beta = 0.2$, 0.4, 0.4 for the phylogeny-based method for 3, 6 and 9 tumor samples, respectively. The center panel shows RMSD of copy numbers for $\alpha = 0.2$, 0.2, 0.2 for the phylogeny-free method and $\beta = 0.2$, 0.4, 0.4 for the phylogeny-based method for 3, 6, and 9 tumor samples, respectively. The right panel shows RMSD of mixture fractions for $\alpha = 0.2$, 0.2, 0.2 for the phylogeny-free method and $\beta = 0.2$, 0.2, 0.2 for the phylogeny-based method for 3, 6, and 9 tumor samples, respectively. The X-axis shows the number of tumor samples and the Y-axis the average accuracy or RMSD. (Color figure online)

Figure 7 compares the two methods at their optimal regularization parameters for three, six, and nine tumor samples. The phylogeny-based method outperforms the phylogeny-free method in accuracy and copy number RMSD in all cases. It is slightly better in mixture fraction RMSD for three samples, but worse for six and nine samples. Figure S7 in the Supplementary Results shows comparative performance of the two methods in individual cell components. Given the poorer performance at pure NMF of the phylogeny-based method's algorithm versus the phylogeny-free method's algorithm, we tentatively attribute the phylogeny-based method's better overall performance to better evolutionary distance estimates and not to a better optimization algorithm.

The phylogeny-based method also provides as output the phylogeny. While we cannot exhaustively show trees across all replicates, we provide three representative examples in Fig. 8. Because we use true SCS data to generate our synthetic mixtures, we do not know the full ground truth trees for the data and do not attach any biological meaning to the inferred trees. We can partially validate correctness of the trees using the fact that the cells were gathered from distinct tumor regions, and while we would not expect clonal ancestry to segregate perfectly by region we should see a trend towards closer evolutionary relationship among cells in spatial proximity. We tested whether pairs of cells from distinct regions cluster together in disjoint subtrees (a kind of partial-information quartet distance); we found that a significant majority of pairs-of-pairs do (79% for 3-sample data, 74% each for 6- or 9-sample data) providing some support for the biological relevance of the trees.

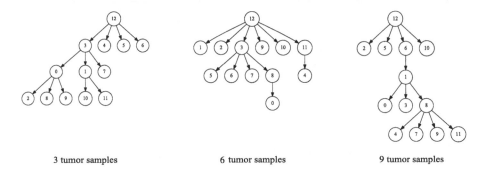

3 tumor samples   6 tumor samples   9 tumor samples

Fig. 8. Tree structure inferred via the phylogeny-based method ILP method for three problem instances. The examples come from the same instances used to pick the representative copy number profiles in Fig. 6. In each tree, nodes 0–5 are inferred cells, nodes 6–11 are observed cells, and node 12 is the diploid root.

4   Conclusions and Discussion

We presented two novel methods for deconvolving clonal copy number variation from bulk tumor genomic data assisted by small amounts of SCS data. The work is intended to provide a practical strategy for producing high-quality clonal CNA deconvolution scalable to large tumor cohorts in the face of still high costs of single-cell DNA sequencing. Validation on semi-simulated data shows that limited amounts of SCS copy number data can be productively used to improve upon pure bulk deconvolution, as assessed by accuracy in inferring clonal copy number profiles and their proportions in single- or multi-sample tumor genomic data. We showed substantial improvement by explicitly constructing clonal phylogenies jointly with deconvolution, suggesting the value of a principled evolutionary model in inferring accurate clonal structure.

While this work provides a proof-of-principle demonstration for combining bulk and SCS data for CNA deconvolution, it also suggests a need for future work. Data of the kind needed by this study remain rare, largely because current SCS studies have not been designed for such a hybrid approach. Most studies to date have profiled many single cells from few patients rather than few cells from larger cohorts, as the current work proposes. We hope that demonstrating the effectiveness of the strategy will promote its use in future study designs, and stimulate new thinking on how most effectively to use single-cell sequencing technologies to solve the underlying data science problems, in turn creating more data on which similar algorithms can be improved. The framework might also be improved in a variety of ways, including more realistic tree models and consideration of other constraints one can extract from SCS data. For example, we considered only penalty terms on C but might also use SCS to improve estimates of the clonal frequency matrix F [33]. The method might also be improved by replacing L1 distance with measures reflecting more sophisticated models of CNA-driven evolution [4,5,10,32]. It could be useful to identify minor clones that have likely loss of heterozygosity events, since these may influence clinical outcomes, and to automate inference of the number of dominant clones. It would also be useful to combine the CNAs of this work with the SNVs of Malikic et al. [22,23], as is commonly done now for bulk deconvolution (e.g., [7,11,16]), and to leverage more effectively data from new low-coverage SCS DNA-seq methods [46] or long-read sequencing. In addition, our algorithms for solving for these models are heuristic and we might productively consider alternative methods to approach true global optima or to improve scalability to larger datasets.

Acknowledgements. This research was supported in part by the Intramural Research Program of the National Institutes of Health, National Library of Medicine and both Center for Cancer Research and Division of Cancer Epidemiology and Genetics within the National Cancer Institute. This research was supported in part by the Exploration Program of the Shenzhen Science and Technology Innovation Committee [JCYJ20170303151334808]. Portions of this work have been funded by U.S. N.I.H. award R21CA216452 and Pennsylvania Dept. of Health award 4100070287. The Pennsylvania Department of Health specifically disclaims responsibility for any analyses, interpretations or conclusions.

References

1. Barber, L.J., Davies, M.N., Gerlinger, M.: Dissecting cancer evolution at the macro-heterogeneity and micro-heterogeneity scale. Curr. Opin. Genet. Dev. **30**, 1–6 (2015)
2. Baslan, T., et al.: Genome-wide copy number analysis of single cells. Nat. Protoc. **7**(6), 1024 (2012)
3. Berry, M.W., Browne, M., Langville, A.N., Pauca, V.P., Plemmons, R.: Algorithms and applications for approximate nonnegative matrix factorization. Comput. Stat. Data Anal. **52**(1), 155–173 (2007)
4. Chowdhury, S.A., et al.: Inferring models of multiscale copy number evolution for single-tumor phylogenetics. Bioinformatics **31**(12), i258–i267 (2015)
5. Chowdhury, S., Shackney, S., Heselmeyer-Haddad, K., Ried, T., Schäffer, A., Schwartz, R.: Algorithms to model single gene, single chromosome, and whole genome copy number changes jointly in tumor phylogenetics. PLoS Comput. Biol. **10**(7), e1003740 (2014)
6. Coyne, G.O., Takebe, N., Chen, A.P.: Defining precision: the precision medicine initiative trials NCI-IMPACT and NCI-MATCH. Curr. Probl. Cancer **41**, 182–193 (2017)
7. Deshwar, A.G., Vembu, S., Yung, C.K., Yang, G.H., Stein, L., Morris, Q.: PhyloWGS: reconstructing subclonal composition and evolution from whole-genome sequencing of tumors. Genome Biol. **16**, 35 (2015)
8. Dexter, D.L., Leith, J.T.: Tumor heterogeneity and drug resistance. J. Clin. Oncol. **4**(2), 244–257 (1986)
9. Eaton, J., Wang, J., Schwartz, R.: Deconvolution and phylogeny inference of structural variations in tumor genomic samples. Bioinformatics **34**, i357–i365 (2018)
10. El-Kebir, M., et al.: Complexity and algorithms for copy-number evolution problems. Algorithms Mol. Biol. **12**(1), 13 (2017)
11. El-Kebir, M., Satas, G., Oesper, L., Raphael, B.J.: Inferring the mutational history of a tumor using multi-state perfect phylogeny mixtures. Cell Syst. **3**(1), 43–53 (2016)

12. Fisher, R., Pusztai, L., Swanton, C.: Cancer heterogeneity: implications for targeted therapeutics. Br. J. Cancer **108**(3), 479–485 (2013)
13. Heselmeyer-Haddad, K., et al.: Single-cell genetic analysis of ductal carcinoma in situ and invasive breast cancer reveals enormous tumor heterogeneity yet conserved genomic imbalances and gain of MYC during progression. Am. J. Pathol. **181**(5), 1807–1822 (2012)
14. Hou, Y., et al.: Single-cell exome sequencing and monoclonal evolution of a JAK-2 negative myeloproliferative neoplasm. Cell **148**(5), 873–885 (2012)
15. Jahn, K., Kuipers, J., Beerenwinkel, N.: Tree inference for single-cell data. Genome Biol. **17**(1), 86 (2016)
16. Jiang, Y., Qiu, Y., Minn, A.J., Zhang, N.R.: Assessing intratumor heterogeneity and tracking longitudinal and spatial clonal evolutionary history by next-generation sequencing. Proc. Natl. Acad. Sci. **113**(37), E5528–E5537 (2016)
17. Kuipers, J., Jahn, K., Beerenwinkel, N.: Advances in understanding tumour evolution through single-cell sequencing. Biochimica et Biophysica Acta (BBA)-Rev. Cancer **1867**(2), 127–138 (2017)
18. Lee, D.D., Seung, H.S.: Algorithms for non-negative matrix factorization. In: Advances in Neural Information Processing Systems, pp. 556–562 (2001)
19. Lei, H., Ma, F., Chapman, A., Lu, S., Xie, X.S.: Single-cell whole-genome amplification and sequencing: methodology and applications. Ann. Rev. Genomics Hum. Genet. **16**, 79–102 (2015)
20. Loeb, L.A.: A mutator phenotype in cancer. Cancer Res. **61**(8), 3230–3239 (2001)
21. Macintyre, G., et al.: Copy number signatures and mutational processes in ovarian carcinoma. Nat. Genet. **50**(9), 1262–1270 (2018)
22. Malikic, S., et al.: PhISCS-a combinatorial approach for sub-perfect tumor phylogeny reconstruction via integrative use of single cell and bulk sequencing data. bioRxiv p. 376996 (2018)
23. Malikic, S., Jahn, K., Kuipers, J., Sahinalp, C., Beerenwinkel, N.: Integrative inference of subclonal tumour evolution from single-cell and bulk sequencing data. bioRxiv p. 234914 (2017)
24. Marusyk, A., Polyak, K.: Tumor heterogeneity: causes and consequences. Biochimica et Biophysica Acta (BBA)-Rev. Cancer **1805**(1), 105–117 (2010)
25. McGranahan, N., et al.: Allele-specific HLA loss and immune escape in lung cancer evolution. Cell **171**(6), 1259–1271 (2017)
26. Navin, N., et al.: Tumour evolution inferred by single-cell sequencing. Nature **472**(7341), 90–94 (2011)
27. Nowell, P.C.: The clonal evolution of tumor cell populations. Science **194**(4260), 23–28 (1976)
28. Ortega, M.A., et al.: Using single-cell multiple omics approaches to resolve tumor heterogeneity. Clin. Transl. Med. **6**, 46 (2017)
29. Ross, E.M., Markowetz, F.: OncoNEM: inferring tumor evolution from single-cell sequencing data. Genome Biol. **17**(1), 69 (2016)
30. Schwartz, R., Schäffer, A.A.: The evolution of tumour phylogenetics: principles and practice. Nat. Rev. Genet. **18**(4), 213–229 (2017)
31. Schwartz, R., Shackney, S.E.: Applying unmixing to gene expression data for tumor phylogeny inference. BMC Bioinform. **11**(1), 42 (2010)
32. Schwarz, R.F., et al.: Spatial and temporal heterogeneity in high-grade serous ovarian cancer: a phylogenetic analysis. PLoS Med. **12**(2), e1001789 (2015)
33. Shackleton, M., Quintana, E., Fearon, E.R., Morrison, S.J.: Heterogeneity in cancer: cancer stem cells versus clonal evolution. Cell **138**(5), 822–829 (2009)

34. Siegel, R.L., et al.: Colorectal cancer statistics, 2017. CA: Cancer J. Clin. **67**(3), 177–193 (2017)

35. Sridhar, S., Lam, F., Blelloch, G.E., Ravi, R., Schwartz, R.: Efficiently finding the most parsimonious phylogenetic tree via linear programming. In: Măndoiu, I., Zelikovsky, A. (eds.) ISBRA 2007. LNCS, vol. 4463, pp. 37–48. Springer, Heidelberg (2007). https://doi.org/10.1007/978-3-540-72031-7_4

36. Subramanian, A., Schwartz, R.: Reference-free inference of tumor phylogenies from single-cell sequencing data. BMC Genomics **16**(11), S7 (2015)

37. Thurau, C., Kersting, K., Bauckhage, C.: Convex non-negative matrix factorization in the wild. In: 2009 Ninth IEEE International Conference on Data Mining, pp. 523–532, December 2009. https://doi.org/10.1109/ICDM.2009.55

38. Tolliver, D., Tsourakakis, C., Subramanian, A., Shackney, S., Schwartz, R.: Robust unmixing of tumor states in array comparative genomic hybridization data. Bioinformatics **26**(12), i106–i114 (2010)

39. Wang, Y., et al.: Clonal evolution in breast cancer revealed by single nucleus genome sequencing. Nature **512**(7513), 155–160 (2014)

40. Wang, Y., Zhang, Y.: Nonnegative matrix factorization: a comprehensive review. IEEE Trans. Knowl. Data Eng. **25**(6), 1336–1353 (2013)

41. Williams, M.J., Werner, B., Barnes, C.P., Graham, T.A., Sottoriva, A.: Identification of neutral tumor evolution across cancer types. Nat. Genet. **48**(3), 238–244 (2016)

42. Wu, K., et al.: Diverse evolutionary dynamics in glioblastoma inference by multi-region and single-cell sequencing. J. Clin. Oncol. **34**(15_suppl), 11580 (2016)

43. Zaccaria, S., El-Kebir, M., Klau, G.W., Raphael, B.J.: Phylogenetic copy-number factorization of multiple tumor samples. J. Comput. Biol. **25**(7), 689–708 (2018)

44. Zack, T.I., et al.: Pan-cancer patterns of somatic copy number alteration. Nat. Genet. **45**(10), 1134–1140 (2013)

45. Zafar, H., Tzen, A., Navin, N., Chen, K., Nakhleh, L.: SiFit: inferring tumor trees from single-cell sequencing data under finite-sites models. Genome Biol. **18**(1), 178 (2017)

46. Zahn, H., et al.: Scalable whole-genome single-cell library preparation without preamplification. Nat. Methods **14**(2), 167 (2017)

OMGS: Optical Map-Based Genome Scaffolding

Weihua Pan, Tao Jiang, and Stefano Lonardi$^{(\boxtimes)}$

Department of Computer Science and Engineering, University of California,
Riverside, CA 92521, USA
`stelo@cs.ucr.edu`

Abstract. Due to the current limitations of sequencing technologies, *de novo* genome assembly is typically carried out in two stages, namely contig (sequence) assembly and scaffolding. While scaffolding is computationally easier than sequence assembly, the scaffolding problem can be challenging due to the high repetitive content of eukaryotic genomes, possible mis-joins in assembled contigs and inaccuracies in the linkage information. Genome scaffolding tools either use paired-end/mate-pair/linked/Hi-C reads or genome-wide maps (optical, physical or genetic) as linkage information. Optical maps (in particular Bionano Genomics maps) have been extensively used in many recent large-scale genome assembly projects (e.g., goat, apple, barley, maize, quinoa, sea bass, among others). However, the most commonly used scaffolding tools have a serious limitation: they can only deal with one optical map at a time, forcing users to alternate or iterate over multiple maps. In this paper, we introduce a novel scaffolding algorithm called OMGS that for the first time can take advantages of multiple optical maps. OMGS solves several optimization problems to generate scaffolds with optimal contiguity and correctness. Extensive experimental results demonstrate that our tool outperforms existing methods when multiple optical maps are available, and produces comparable scaffolds using a single optical map. OMGS can be obtained from https://github.com/ucrbioinfo/OMGS.

Keywords: *De novo* genome assembly · Scaffolding · Optical maps · Combinatorial optimization

1 Introduction

Genome assembly is a fundamental problem in genomics and computational biology. Due to the current limitations of sequencing technologies, the assembly is typically carried out in two stages, namely contig (sequence) assembly and scaffolding. Scaffolds are arrangements of oriented contigs with gaps representing the estimated distance separating them. The scaffolding process can vastly improve the assembly contiguity and can produce chromosome-level assemblies. Despite significant algorithmic progress, the scaffolding problem can be challenging due to the high repetitive content of eukaryotic genomes, possible mis-joins in assembled contigs and the inaccuracies of the linkage information.

© Springer Nature Switzerland AG 2019
L. J. Cowen (Ed.): RECOMB 2019, LNBI 11467, pp. 190–207, 2019.
https://doi.org/10.1007/978-3-030-17083-7_12

Genome scaffolding tools either use paired-end/mate-pair/linked/Hi-C reads or genome-wide maps. The first group includes scaffolding tools for second generation sequencing data, such as Bambus [17,29], GRASS [13], MIP [31], Opera [12], SCARPA [11], SOPRA [8] and SSPACE [5] and the scaffolding modules from assemblers ABySS [35], SGA [34] and SOAPdenovo2 [22]. Since the relative orientation and approximate distance between paired-end/mate-pair/linked/Hi-C reads are known, the consistent alignment of a sufficient number of reads to two contigs can indicate their relative order, their orientation and the distance between them. An extensive comparison of scaffolding methods in this first group of tools can be found in [14].

The second group uses genome-wide maps such as genetic maps [37], physical maps, or optical maps. According to the markers provided by these maps, contigs can be anchored to specific positions so that their order and orientations can be determined. The distance between contigs can also be estimated with varying degree of accuracy depending on the density of the map.

The optical mapping technologies currently on the market (e.g., BioNano Genomics Irys systems, OpGen Argus) allow computational biologists to produce genome-wide maps by fingerprinting long DNA molecules (up to 1 Mb), via nicking restriction enzymes [32]. Linear DNA fragments are stretched on a glass surface or in a nano-channel array, then the locations of restriction sites are identified with the help of dyes or fluorescent labels. The results are imaged and aligned to each other to map the locations of the restriction sites relative to each other. While the assembly process for optical molecules is highly reliable, there is clear evidence that a small fraction of the optical molecules is chimeric [15].

A few scaffolding algorithms that use optical maps are available. SOMA appears to be the first published tool that can take advantage of optical maps but it can only deal with a non-fragmented optical map [25]. The scaffolding tool proposed in [30] was used for two bacterial genomes *Yersinia pestis* and *Yersinia enterocolitica*, but the software is no longer publicly available. In the last few years, Bionano optical maps have become very popular, and have been used to improve the assembly contiguity in many large-scale *de novo* genome assembly projects (e.g., goat, apple, barley, maize, quinoa, sea bass [4,7,23,28]). To the best of our knowledge, the main tools used to generate scaffolds using Bionano optical maps are SEWINGMACHINE from KSU [33] and HYBRIDSCAFFOLD from Bionano Genomics (unpublished, 2016). SEWINGMACHINE seems to be favored by practitioners over HYBRIDSCAFFOLD.

Both HYBRIDSCAFFOLD and SEWINGMACHINE have, however, a serious limitation: they can only deal with one optical map at a time, forcing users to alternate or iterate over optical maps when multiple maps are available. In this paper, we introduce a novel scaffolding algorithm called OMGS that for the first time can take advantage of any number of optical maps. OMGS solves several optimization problems to generate scaffolds with optimal contiguity and correctness.

2   Problem Definition

The input to the problem is the genome assembly to be scaffolded (represented by a set of assembled contigs), and one or more optical maps (represented by a set of sets of genomic distances). We use $C = \{c_i | i = 1, \ldots, l\}$ to denote the set of contigs in the genome assembly, where each c_i is a string over the alphabet $\{A, C, G, T\}$. Henceforth, we assume that the contigs in C are chimera-free.

An optical map is composed by a set of optical molecules, each of which is represented by an ordered set of positions for the restriction enzyme sites. As said, optical molecules are obtained by an assembly process similar to sequence assembly, but we will reserve the term *contig* for sequenced contigs. We use $M = \{m_i | i = 1, \ldots, n\}$ to denote the optical map, where each optical molecule m_i is an ordered set of integers, corresponding to the distances in base pairs between two adjacent restriction enzyme sites on molecule m_i. By digesting *in silico* the contigs in C using the same restriction enzyme used to produce the optical map and matching the sequence of adjacent distances between sites, one can align the contigs in C to the optical map M. If one is given multiple optical maps obtained using different restriction enzymes, M will be the union of the molecules from all optical maps. In this case, each genomic location is expected to be covered by multiple molecules in M. As said, high quality alignments allows one to anchor and orient contigs to specific coordinates on the optical map. When multiple contigs align to the same optical map molecule, one can order them and estimate the distance between them. By filling these gaps with a number of N's equal to the estimated distance, longer DNA sequences called *scaffolds* can be obtained.

A series of practical factors make the problem of scaffolding non-trivial. These factors include imprecisions in optical maps (e.g., mis-joins introduced during the assembly of the optical map [15]), unreliable alignments between contigs and optical molecules, and multiple inconsistent anchoring positions for the same contigs. As a consequence, it is appropriate to frame this scaffolding problem as an optimization problem.

We are now ready to define the problem. We are given an assembly represented by a set of contigs C, a set of optical map molecules M and a set of alignments $A = \{a_{1,1}, a_{1,2}, \ldots a_{l,n}\}$ of C to M, where $a_{i,j}$ is the alignment of contig c_i to optical map molecule o_j. The problem is to obtain a set of scaffolds $S = \{s_1, s_2, \ldots s_k\}$ where each s_i is a string over the alphabet $\{A, C, G, T, N\}$, such that (i) each contig c_i is contained/assigned to exactly one scaffold, (ii) the *contiguity* of S is maximized and (iii) the conflicts of S with respect to A are minimized. This optimization problem is not rigorously defined unless one defines precisely the concepts of *contiguity* and *conflict*, but this description captures the spirit of what we want to accomplish. In genome assembly, the assembly contiguity is usually captured by statistical measures like the N50/L50 or the NG50/LG50. The notion of conflict is not easily quantified, and even if it was made precise, this multi-objective optimization problem would be hard to solve. We decompose this problem into two separate steps, namely (a) scaffold detection and (b) gap estimation, as explained below.

3   Method

As said, our proposed method is composed of two phases: scaffold detection and gap estimation. In the first phase, contigs are grouped into scaffolds and the order of contigs in each scaffold is determined. In the second phase, distances between neighboring contigs assigned to scaffolds are estimated. The pipeline of the proposed algorithm is illustrated in Fig. 1.

3.1   Phase 1: Detecting Scaffolds

Phase 1 has three major steps. In Step 1, we align *in silico*-digested chimeric-free contigs to the optical maps (e.g., for a Bionano optical map, we use REFALIGNER), but not all alignments are used in Step 2. We only consider alignments that (i) exceed a minimum confidence level (e.g., confidence 15 in the case of REFALIGNER); (ii) do not overlap each other more than a given genomic distance (e.g., 20 kbp) and (iii) do not create conflict with each other. The method we use here to select conflict-free alignments was introduced in our previous work [27]. In Step 2, we compute candidate scaffolds by building the *order graph* and formulating an optimization problem on it. In Step 3, either the exhaustive algorithm or a log n-approximation algorithm is used to solve the optimization problem (depending on the size of the graph) and produce the final scaffolds.

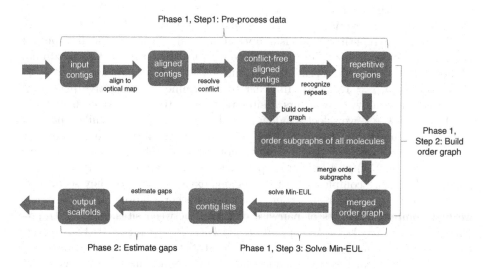

Fig. 1. Pipeline of the proposed algorithm

3.1.1   Building the Order Graph

The order graph O is a directed weighted graph in which each vertex represents a contig. Given two contigs c_i and c_j aligned to an optical molecule o with alignments a_i and a_j, we create a directed edge (c_i, c_j) in O if (i) the starting coordinate of alignment a_i (that we call $a_i.start$ henceforth) is smaller than the starting coordinate of alignment a_j (that we call $a_j.start$ henceforth) and (ii) there is no other alignment a_k such that $a_k.start$ is between $a_i.start$ and $a_j.start$ and (iii) there are no conflict sites between $a_i.end$ and $a_j.start$ on the optical molecule, as defined below. For each alignment a between optical molecule o and contig c, we compute the left overhang l_o and right overhang r_o from o and the left overhang l_c and right overhang r_c from c. The left-end of alignment a is declared a *conflict site* if (i) both l_o and l_c are longer than some minimum length (e.g., 50 kbp) and (ii) at least one restriction enzyme sites appear in both l_o and l_c. A symmetric argument applies to the right-end of the alignment, which determines the values for r_o and r_c.

Directed edge (c_i, c_j) is assigned a weight equal to qual$(o, a_i.end, a_j.start)$ * $(\text{conf}(a_i) + \text{conf}(a_j))$, where (i) qual$(o, a_i.end, a_j.start)$ is the *quality* of the region between $a_i.end$ and $a_j.start$ on molecule o (higher is better, defined next) and (ii) conf(a) is the confidence score provided by REFALIGNER alignment a (higher is better). The quantity qual(o, s, t) is defined based on the length of a repetitive region between coordinates (s, t). Based on our experience, assembly mis-joins on optical molecule almost always happen in repetitive regions [15]. Given the length of repetitive region len_rep(o, s, t) in base pairs (defined below), we define the quality of o in the interval (s, t) as qual$(o, s, t) = e^{-\text{len\_rep}(o,s,t)/100000}$. When a_i and a_j have a small overlap (e.g., shorter than 20 kbp), we set len_rep$(o, s, t) = 0$.

We recognize repetitive regions in optical molecules based on the distribution of restriction enzyme sites. For a molecule o with n sites, let m_i be the coordinate of the i-th site for $i = 1, \ldots, n$. As said, molecule o can be represented as a list of positions $\{m_i | i = 1, \ldots, n\}$. In order to determine the repetitive regions in o, we slide a window that covers k sites (e.g., $k = 10$ sites). At each position $j = 1, \ldots, n - k + 1$, we select window $w_j = \{m_j, \ldots, m_{j+k-1}\}$. While repetitive regions in genome can be highly complex (see, e.g., [40]), we observed only two types of repetitive regions in optical molecules, namely single-site repetitive region (see Fig. 2-A) and two-site repetitive region (see Fig. 2-B). It is entirely possible that more complex repetitive regions exist: if they do, they seem rare. Based on this observation, in order to decide whether window w_j is repetitive, we first compute two lists of pairwise distances between sites, namely $D_{j,1} = \{m_{j+l} - m_{j+l-1} | l = 1, \ldots, k-1\}$ and $D_{j,2} = \{m_{j+l+1} - m_{j+l-1} | l = 1, \ldots, k-2\}$ that we call *distance lists*, then we apply the statistical test described next.

In our statistical test we assume that the values in the distance lists that belong to repetitive regions are independent and identically distributed as a Gaussian. We further assume that each specific distance list ($D_{j,1}$ or $D_{j,2}$) is associated with a Gaussian with a specific mean $\mu_{j,q}$ ($q \in \{1, 2\}$). Finally, we assume that the variance σ^2 is globally shared by all molecules. An estimator of the mean is $\mu_{j,q}$ is $\hat{\mu}_{j,q} = \sum_{i=1}^{k-q} d_i/(k-q)$, where $d_i \in D_{j,q}$ and k is the window size. To

(A)

(B)

Fig. 2. Examples of single-site repetitive region (A) and two-site repetitive region (B) in optical maps. Observe the small variations in the repetitive patterns in (B)

estimate σ^2, we first get an initial (rough) estimate of the repetitive regions on all molecules. Given a particular $D_{j,q}$, let d_{max} and d_{min} be the maximum and minimum distance in $D_{j,q}$. We declare a distance list $D_{j,q}$ to be *estimated repetitive* if $d_{max} - d_{min}$ is smaller than a given distance (e.g., 1.5 kbp). We collect all estimated repetitive lists in set $R = \{D_p$ is estimated repetitive$|p = 1, \ldots, P\}$ and the estimated mean $\hat{\mu}_p$ for each distance list D_p in the set R, where P is the total number of estimated repetitive lists. Then, we define the log likelihood function L as follows (additional details can be found in Appendix, Sect. B)

$$\log L(\sigma^2) = -\frac{\log \sigma^2}{2} \sum_{p=1}^{P} |D_p| - \frac{1}{2\sigma^2} \sum_{p=1}^{P} \sum_{d_i \in D_p} (d_i - \hat{\mu}_p)^2.$$

By maximizing $\log L(\sigma^2)$, the estimator for the variance becomes

$$\hat{\sigma}^2 = \sum_{p=1}^{P} \sum_{d_i \in D_p} (d_i - \hat{\mu}_p)^2 / \sum_{p=1}^{P} |D_p|.$$

Then, we carry out the test on the statistic $d_{max} - d_{min}$ for each $D_{j,q}$. The joint density function of (d_{max}, d_{min}) is

$$f_{d_{max}, d_{min}}(u, v) = n(n - 1) f_{d_i}(u) f_{d_i}(v) [F_{d_i}(v) - F_{d_i}(u)]^{n-2}$$

for $-\infty < u < v < +\infty$, where F_{d_i} and f_{d_i} are the distribution function and density function of $d_i \sim N(\hat{\mu}_{j,q}, \hat{\sigma}^2)$, respectively. The density function of $d_{max} - d_{min}$ is

$$f_{d_{max} - d_{min}}(x) = \int_{-\infty}^{+\infty} n(n - 1) f_{d_i}(y) f_{d_i}(x + y) [F_{d_i}(x + y) - F_{d_i}(y)]^{n-2} dy,$$

defined when $x \geq 0$ (additional details can be found in Appendix, Sect. C). Let now X be a random variable associated with the distribution $f_{d_{max} - d_{min}}$. If the p-value $P(X > d_{max} - d_{min})$ is greater than a predefined threshold (e.g., 0.001), we accept the null hypothesis and declare that window w_j is repetitive. The repetitive regions for the entire molecule o is the union of all the windows w_j's recognized as repetitive according to the test above.

Once the order graph of each optical molecule is built, we connect all the order graphs which share the same contigs using the association graph introduced in [27]. The association graph is an undirected graph in which each vertex represents an optical molecule and an edge indicates that the two molecules share at least one contig aligned to both of them. We use depth first search (DFS) to first build a spanning forest of the association graph. Then, we traverse each spanning tree and connect the corresponding order subgraph to the final order graph. Every time we add a new graph, new vertices and new edges might be added. If an edge already exist, the weights of the new edges are added to the weights of existing edges.

3.1.2 Generating Scaffolds

Once the order graph O is finalized, we generate the ordered sequence of contigs in each scaffold. In the ideal case, each connected component O_i of O is a directed acyclic graph (DAG) because the genome is one-dimensional and the order of any pair of contigs is unique. In practice however, O_i may contain cycles caused by the inaccuracy of the alignments and mis-joins in optical molecules. To convert each cyclic component O_i into a DAG, we solve the MINIMUM FEEDBACK ARC SET problem on O_i. In this problem, the objective is to find the minimum subset of edges (called *feedback arc set*) containing at least one edge of every cycle in the input graph. Since the minimum feedback edge set problem is APX-hard, we use the greedy local heuristics introduced in [2] to solve it.

We then break each DAG G_i of connected component O_i into subgraphs as follows. In each subgraph, we require that the order of every pair of vertices to be uniquely determined by the directed edges. This allows us to uniquely determine the order of the contigs for each scaffold. The formal definition of this optimization problem is as follows.

Definition 1 (MINIMUM EDGE UNIQUE LINEARIZATION problem). INPUT: A weighted directed acyclic graph $G = (V, E)$. OUTPUT: A subset of edges $E' \subseteq E$ such that (i) in each connected component G'_i of the graph $G' = (V, E - E')$ obtained after removing E', the order of all vertices can be uniquely determined, and (ii) the total weights of the edges in E' is the minimum among all the subset of edges satisfying (i).

In Theorem 1 below, we show that the MINIMUM EDGE UNIQUE LINEARIZATION problem (MIN-EUL) is NP-hard by proving that it is equivalent to the MINIMUM EDGE CLIQUE PARTITION problem (MIN-ECP), which is know to be NP-hard [10]. In MIN-ECP, we are given a general undirected graph, and we need to partition its vertices into disjoint clusters such that each cluster forms a clique and the total weight of the edges between clusters is minimized.

Theorem 1. MIN-EUL *is equivalent to* MIN-ECP.

Proof. First, we show that MIN-EUL polynomially reduces to MIN-ECP. Given an instance $G = (V, E)$ of MIN-EUL, we build an instance $G' = (V', E')$ of MIN-ECP as follows. Let $V' = V$. For each pair of vertices $u, v \in V'$ where v is reachable from u, define an undirected edge between u and v in E'. For each directed edge $(u, v) \in E$, set the weight of the corresponding undirected edge

$(u, v) \in E'$ as 1. Set the weights of the other edges in E' as 0. Then it is easy to see that a MIN-EUL solution to G' is equivalent to a MIN-ECP solution to G and vice versa.

Now we show that MIN-ECP polynomially reduces to MIN-EUL. Given an instance $G' = (V', E')$ (assuming G' is connected) of MIN-ECP, we build an instance $G = (V, E)$ of MIN-EUL as follows. Let $V = V'$. Pick any total linear order O of all vertices in V'. For each undirected edge $(u, v) \in E'$ where rank$(u) <$ rank(v) in O, define a directed edge from u to v in E and set its weight to be the same as its corresponding undirected edge in E'. For any two vertices $u, v \in V$, where rank$(u) <$ rank(v) and $(u, v) \notin E'$, add a new vertex $x_{uv} \in V$ with rank$(x_{uv}) =$ rank(v) and a directed edge u to x_{uv} of weight 1 in E. Now for each pair of vertices $u, v \in V$ where rank$(u) <$ rank(v) and $(u, v) \notin E$, add a directed edge u to v with weight zero in E. Then it is easy to see that a MIN-EUL solution to G corresponds to a MIN-ECP solution to G' and vice versa. □

Given the complexity of MIN-EUL, we propose an exponential time exact algorithm and a polynomial time $\log n$-approximation algorithm for solving it. To describe the exact algorithm, we need to introduce some notations. A *conjunction* vertex in a DAG is a vertex which has more than one incoming edge or outgoing edge. A *candidate* edge is an edge which connects at least one conjunction vertex. In Theorem 2 below, we prove that the optimal solution E' of MIN-EUL must only contain candidate edges. Let E_c be the set of all candidate edges in the DAG G, for each subset E'_j of E_c, we check whether the graph $G' = (V, E - E'_j)$ satisfies requirement (i) in Definition 1 after removing E'_j from G. Among all the feasible E'_j, we produce the set of edges with minimum total weights. To check whether E'_j is feasible, we use a variant of topological sorting which requires one to produce a unique topological ordering. To do so, we require that in every iteration of topological sorting, the candidate node to be added to sorted graph is always unique. Details of this algorithm are shown as Algorithm 1 in Appendix A.

Theorem 2. *The optimal solution E' of* MIN-EUL *only contains candidate edges.*

Proof. For sake of contradiction, we assume that E' contains a non-candidate edges (u, v). Since E' is optimal, $G' = (V, E - E')$ satisfies condition (i) in Definition 1. Since both u and v are conjunction vertices, u has only one incoming edge and v has only one outgoing edge. Therefore, by adding (u, v) to $G' = (V, E - E')$, we still satisfy condition (i) in Definition 1. Since the weight of (u, v) is positive, the total weight of $E - E' + \{(u, v)\}$ is larger than $E - E'$. Therefore $E' - \{(u, v)\}$ is optimal, contradicting the optimality of E'. □

As said, MIN-EUL is equivalent to MIN-ECP (Theorem 1). In addition, the authors of [10] showed that for any instance of MIN-ECP one can find an equivalent instance of the MINIMUM DISAGREEMENT CORRELATION CLUSTERING problem. As a consequence, any algorithm for the MINIMUM DISAGREEMENT CORRELATION CLUSTERING problem could be used to solve MIN-EUL. In our tool OMGS, we implemented a $O(\log n)$-approximation algorithm based on linear programming, originally proposed in [9]. Standard linear programming packages (e.g., GLPK or CPLEX) are used to solve the linear program. We use the exact algorithm for DAGs with no more than twenty candidate edges, and the approximation algorithm for larger DAGs.

3.2   Phase 2: Estimating Gaps

Let $s = \{c_i | i = 1, \ldots, h\}$ be one of the scaffold generated in Phase 1 where each c_i is a contig. In Phase 2, we estimate the length l_i of the gap between each pair c_i and c_{i+1} of adjacent contigs. We estimate all gap lengths $L = \{l_i | i = 1, \ldots, h-1\}$ at the same time using the distances between the contigs provided by the alignments and the corresponding order subgraphs. We assume that each l_i is chi-square distributed with α_i degrees of freedom. The choice of chi-square distribution is due to its additive properties, namely the sum of independent chi-squared variables is also chi-squared distributed. Recall that each order subgraph O_k provides an unique ordering $x_k = \{c_j | j = 1, \ldots, r\}$ of the contigs aligned to molecule o_k, while the coordinates of the alignment provide the distances between all pairs of adjacent contigs c_j and c_{j+1} as $y_k = \{d_j | j = 1, \ldots, r-1\}$. We use the distances d_j as samples to estimate gap lengths l_i. If edge (c_j, c_{j+1}) in O_k is removed in the order graph O when solving MIN-EUL in Phase 1, d_j will be considered not reliable and removed from y_k.

In the ideal case, d_j should be a sample of a single l_i (i.e., $c_j c_{j+1}$ in x_k corresponds to $c_p c_{p+1}$ in s). In practice however, $c_j c_{j+1}$ in x_k will corresponds to a different pair $c_p c_q$ in s where $q > p+1$ (i.e., $c_{p+1} \ldots c_{q-1}$ are missing from the order subgraph because some alignments with low confidence were removed in Step 1 of Phase 1). In this situation, after subtracting the length of missing contigs from d_j, $d_j - \sum_{c=c_{p+1}}^{c_{q-1}} |c|$ is a sample of $\sum_{i=p}^{q-1} l_i$ where $|c|$ represents the length of contig c. Since $l_p, \ldots, l_{q-1}$ are independent chi-square random variables, $\sum_{i=p}^{q-1} l_i$ is chi-square distributed with degree of freedom $\sum_{i=p}^{q-1} \alpha_i$, so that the log likelihood of this sample is

$$\log l = (\beta - 1) \log \gamma - \frac{\gamma}{2} - \beta \log 2 - \log \Gamma(\beta).$$

where $\beta = \sum_{i=p}^{q-1} \frac{\alpha_i}{2}$, $\gamma = d_j - \sum_{c=c_{p+1}}^{c_{q-1}} |c|$ and Γ is the gamma function (additional details can be found in Appendix, Sect. D). The total log likelihood is the sum of the log likelihoods across all samples. To find the α_i maximizing the total log likelihood, we use the Broyden-Fletcher-Goldfarb-Shanno (BFGS) algorithm [1]. Since the mean of a chi-square distribution equals its degree of freedom, we obtain the estimated gaps $\hat{l}_i = \hat{\alpha}_i$. For the case in which the l_i are pre-estimated as negative in the first step, the second and third steps are ignored and the pre-estimated distances are used as final estimates.

Finally, we add $\lceil \hat{l}_i \rceil$ nucleotides (represented by Ns) between each pair of contigs c_i and c_{i+1}. When $\hat{l}_i < 0$, we add exactly 100 Ns between c_i and c_{i+1}, which is the convention for a gap of unknown length.

4   Experimental Results

We compared OMGS against KSU SEWINGMACHINE (version 1.0.6, released in 2015) and Bionano HYBRIDSCAFFOLD (version 4741, released in 2016) which, to the best of our knowledge, are the only available scaffolding tools for Bionano Genomics optical maps. All tools were run with default parameters, unless otherwise specified. We collected experimental results on scaffolds of (i) cowpea (*Vigna unguiculata*) and (ii) fruit fly (*Drosophila melanogaster*).

4.1   Experimental Results on Cowpea

Cowpea is a diploid with a chromosome number $2n = 22$ and an estimated genome size of 620 Mb. We sequenced the cowpea genome using single-molecule real-time sequencing (Pacific Biosciences RSII). A total of 87 SMRT cells yielded about 6M reads for a total of 56.84 Gbp (91.7x genome equivalent). We tested the three scaffolding tool on a high-quality assembly produced by CANU [3,18] with parameters `corMhapSensitivity=high` and `corOutCoverage=100`, then polished it with QUIVER. We used CHIMERICOGNIZER to detect and break chimeric contigs, using seven other assemblies generated by CANU, FALCON [6] and ABRUIJN [20] as explained in [26].

In addition to standard contiguity statistics (N50[1], L50[2]), total assembled size and scaffold length distribution, we determined incorrect/chimeric scaffolds by comparing them against the high-density genetic map available from [24]. We BLASTed 121bp-long design sequence for the 51,128 genome-wide SNPs described in [24] against each assembly, then we identified which contigs had SNPs mapped to them, and what linkage group (chromosome) of the genetic map those mapped SNPs belonged to. Chimeric contigs were revealed when their mapped SNPs belonged to more than one linkage group. The last line of Tables 1 and 2 report the total size of contigs in each assembly for which (i) they have at least one SNPs mapped to it and (ii) all SNPs belong to the same linkage group (i.e., likely to be non-chimeric).

As said, the three scaffolding tools were run on a chimera-free assembly of cowpea described above using two available Bionano Genomics optical maps (the first obtained using the BspQI nicking enzyme, and the second obtained with the BssSI nicking enzyme). Since SEWINGMACHINE can only use a single optical map, we alternated the optical maps in input (BspQI map first, then BssSI and vice versa). SEWINGMACHINE provides two outputs depending on the minimum allowed alignment confidence, namely 'default' and 'relax'. Mode 'relax' considers more alignments than 'default', but it has a higher chance of introducing mis-joins. HYBRIDSCAFFOLD failed on the BssSI map, so we could not test it on alternating maps.

Table 1 shows that when using a single optical map, OMGS can generate comparable or better scaffolds than SEWINGMACHINE and HYBRIDSCAFFOLD.

---

[1] Length for which the set of contigs/scaffolds of that length or longer accounts for at least half of the assembly size.

[2] Minimum number of contigs/scaffolds accounting for at least half of the assembly.

With two optical maps, OMGS' correctness ("contigs/scaffolds with 100% consistent LG") and contiguity (N50) are significantly better than other two tools. Observe that OMGS' correctness ("contigs/scaffolds with 100% consistent LG") is even better than the input assembly. This can happen when contigs with SNPs belonging to same linkage group are scaffolded with contigs that have no SNP.

We also compared the performance of OMGS, SEWINGMACHINE and HYBRIDSCAFFOLD when using optical maps corrected by CHIMERICOGNIZER (on the same cowpea assembly). Observe in Table 2 that OMGS, SEWINGMACHINE and HYBRIDSCAFFOLD increased the correctness but decreased the contiguity when the corrected BspQI optical map was used. The results on the corrected BssSI optical map or both corrected optical maps did not change significantly. But again, OMGS produced better scaffolds than SEWINGMACHINE and HYBRIDSCAFFOLD.

4.2   Experimental Results on *D. Melanogaster*

D. melanogaster has four pairs of chromosomes: three autosomes, and one pair of sex chromosomes. The fruit fly's genome is about 139.5 Mb. We downloaded three *D. melanogaster* assemblies generated in [36] (https://github.com/danrdanny/ Nanopore_ISO1). The first assembly (295 contigs, total size 141 Mb, N50 = 3 Mb) was generated using CANU [3,18] on Oxford Nanopore (ONT) reads longer than 1 kb. The second assembly (208 contigs, total size 132 Mb, N50 = 3.9 Mb) was generated using MINIMAP and MINIASM [19] using only ONT reads. The third assembly (339 contigs, total size 134 Mb, N50 = 10 Mb) was generated by PLATANUS [16] and DBG2OLC [39] using 67.4x of Illumina paired-end reads and the longest 30x ONT reads. The first and third assemblies were polished using NANOPOLISH [21] and PILON [38]. The Bionano optical for *D. melanogaster* map was provided by the authors of [36]. This BspQI optical map (363 molecules, total size = 246 Mb, N50 = 841 kb) was created using IRYSSOLVE 2.1 from 78,397 raw Bionano molecules (19.9 Gb of data with a mean read length 253 kb).

As said, all tools were run with default parameters, with the exception of OMGS' minimum confidence, which was set at 20 (default is 15). To evaluate the performance of OMGS, HYBRIDSCAFFOLD and SEWINGMACHINE, we compared their output scaffolds to the high-quality reference genome of *D. melanogaster* (release 6.21, downloaded from FlyBase). We reported the total length of correct/non-chimeric scaffolds as a measure of the overall correctness. To determine which scaffolds were incorrect/chimeric we first selected BLAST alignments of the scaffolds against the reference genome which had an e-value lower than 1e-50 and an alignment length higher than 30 kbp. We defined a scaffold S to be *chimeric* if S had at least two high-quality alignments which satisfied one or more of the following conditions: (i) S aligned to different chromosomes; (ii) the orientation of S's alignments were different; or (iii) the difference between the distance of alignments on the scaffold and the distance of alignments on the reference sequence was larger than 100 Kbp.

Table 1. Comparing OMGS, SEWINGMACHINE (SM) and HYBRIDSCAFFOLD (HS) on a cowpea assembly using one or two optical maps. Numbers in boldface highlight the best N50 and scaffold consistency with the genetic map for one map (BspQI and BssSI) or two maps ('A + B' refers to the use of map A followed by map B, 'A&B' refers to the use of both maps at the same time).

| | | ONE OPTICAL MAP | | | | | | | |
| | | BspQI | | | | BssSI | | | |
	Input	SM (default)	SM (relax)	HS	OMGS	SM (default)	SM (relax)	HS	OMGS
contig/scaffold N50 (bp)	5,633,882	13,154,336	13,154,336	12,211,658	**14,339,314**	10,620,326	10,886,079	N/A	**11,536,649**
contig/scaffold L50	28	15	15	17	14	18	17	N/A	15
total assembled (bp)	511,101,122	521,209,608	521,210,640	516,455,893	518,265,608	518,987,660	518,945,404	N/A	518,252,638
# contigs/scaffolds	948	863	863	877	847	849	846	N/A	832
# contigs/scaffolds ≥100kbp	269	185	185	198	170	177	174	N/A	165
# contigs/scaffolds ≥1Mbp	94	59	59	63	56	63	62	N/A	59
# contigs/scaffolds ≥10Mbp	10	20	20	21	20	18	18	N/A	17
contigs/scaffolds with consistent LG (bp)	425,812,490	404,408,642	404,409,674	381,974,417	**410,552,582**	**425,572,265**	425,530,009	N/A	424,143,108

| | | TWO OPTICAL MAPS | | | | |
| | | BspQI+BssSI | BspQI+BssSI | BssSI+BspQI | BssSI+BspQI | BspQI&BssSI |
	Input	SM (default)	SM (relax)	SM (default)	SM (relax)	OMGS
contig/scaffold N50 (bp)	5,633,882	14,892,230	14,892,230	13,527,997	14,892,235	**16,364,046**
contig/scaffold L50	28	13	13	14	13	12
total assembled (bp)	511,101,122	525,577,823	525,198,231	525,827,900	525,105,345	521,324,385
# contigs/scaffolds	948	822	823	816	814	802
# contigs/scaffolds ≥100kbp	269	149	150	145	143	137
# contigs/scaffolds ≥1Mbp	94	46	46	48	46	44
# contigs/scaffolds ≥10Mbp	10	21	21	22	22	21
contigs/scaffolds with consistent LG (bp)	425,812,490	385,449,577	385,069,985	425,678,421	403,637,207	**432,639,234**

Table 2. Comparing OMGS, SEWINGMACHINE (SM) and HYBRIDSCAFFOLD (HS) on a cowpea assembly using optical maps corrected by CHIMERICOGNIZER. Numbers in boldface highlight the best N50 and scaffold consistency with the genetic map for one map (BspQI and BssSI) or two maps ('A + B' refers to the use of map A followed by map B, 'A&B' refers to the use of both maps at the same time).

| | | ONE OPTICAL MAP | | | | | | | |
| | | BspQI | | | | BssSI | | | |
	Input	SM (default)	SM (relax)	HS	OMGS	SM (default)	SM (relax)	HS	OMGS
contig/scaffold N50 (bp)	5,633,882	12,487,373	12,487,373	12,495,655	**13,505,314**	9,420,899	10,886,079	N/A	**11,256,770**
contig/scaffold L50	28	16	16	15	14	19	17	N/A	16
total assembled (bp)	511,101,122	519,785,777	519,785,777	515,519,585	518,405,022	517,678,278	517,636,022	N/A	517,318,151
# contigs/scaffolds	948	863	863	871	849	854	851	N/A	837
# contigs/scaffolds ≥100kbp	269	185	185	192	172	182	179	N/A	169
# contigs/scaffolds ≥1Mbp	94	60	60	60	58	66	65	N/A	62
# contigs/scaffolds ≥10Mbp	10	19	19	19	19	17	17	N/A	17
contigs/scaffolds with consistent LG (bp)	425,812,490	413,819,557	413,819,557	402,840,302	**421,466,164**	**424,262,883**	424,220,627	N/A	423,117,331

| | | TWO OPTICAL MAPS | | | | |
| | | BspQI+BssSI | BspQI+BssSI | BssSI+BspQI | BssSI+BspQI | BspQI&BssSI |
	Input	SM (default)	SM (relax)	SM (default)	SM (relax)	OMGS
contig/scaffold N50 (bp)	5,633,882	14,354,752	14,354,752	13,527,997	14,892,235	**16,364,046**
contig/scaffold L50	28	14	14	14	13	12
total assembled (bp)	511,101,122	523,520,329	523,139,705	521,540,185	525,105,345	520,697,623
# contigs/scaffolds	948	823	824	817	814	805
# contigs/scaffolds ≥100kbp	269	150	151	146	143	139
# contigs/scaffolds ≥1Mbp	94	48	48	48	46	46
# contigs/scaffolds ≥10Mbp	10	21	21	21	22	21
contigs/scaffolds with consistent LG (bp)	425,812,490	402,344,751	401,964,127	420,269,616	403,637,207	**431,921,182**

Table 3 reports the main statistics for the three *D. melanogaster* scaffolded assemblies. Even with one map, OMGS' scaffolds are better than SEWINGMACHINE and HYBRIDSCAFFOLD.

Table 3. Comparing OMGS, SewingMachine (SM) and HybridScaffold (HS) on three *D. melanogaster* assemblies (produced by MiniAsm, Canu, and Dbg2Olc) using the BspQI optical map. Numbers in boldface highlight the best N50 and the best scaffold consistency with the reference genome

	MiniAsm assembly				
	Input	SM (default)	SM (relax)	HS	OMGS
contig/scaffold N50 (bp)	3,866,686	4,494,241	**4,906,224**	3,866,686	**4,906,224**
contig/scaffold L50	9	8	8	9	8
total assembled (bp)	131,856,353	132,480,826	133,233,999	132,138,056	132,838,677
# contigs/scaffolds	208	205	203	206	206
# contigs/scaffolds ≥100kbp	85	82	80	83	83
# contigs/scaffolds ≥1Mbp	26	26	25	26	25
# contigs/scaffolds ≥10Mbp	2	2	2	2	2
non-chimeric contigs/scaffolds (bp)	131,317,873	125,305,638	**132,695,519**	131,174,201	132,300,197

	Canu assembly				
	Input	SM (default)	SM (relax)	HS	OMGS
contig/scaffold N50 (bp)	3,004,953	3,004,953	3,004,953	3,918,649	**5,336,340**
contig/scaffold L50	11	11	11	10	7
total assembled (bp)	140,720,404	140,923,974	140,923,974	140,867,226	140,960,395
# contigs/scaffolds	295	291	291	286	280
# contigs/scaffolds ≥100kbp	111	107	107	102	96
# contigs/scaffolds ≥1Mbp	31	31	31	29	27
# contigs/scaffolds ≥10Mbp	1	1	1	1	5
non-chimeric contigs/scaffolds (bp)	140,720,404	140,923,974	140,923,974	140,867,226	**140,960,395**

	Dbg2Olc assembly				
	Input	SM (default)	SM (relax)	HS	OMGS
contig/scaffold N50 (bp)	10,113,899	11,223,142	11,223,142	12,785,467	**12,928,771**
contig/scaffold L50	6	5	5	5	4
total assembled (bp)	134,109,164	134,164,629	134,164,629	134,162,857	134,208,377
# contigs/scaffolds	339	337	337	331	327
# contigs/scaffolds ≥100kbp	78	76	76	70	66
# contigs/scaffolds ≥1Mbp	22	22	22	17	16
# contigs/scaffolds ≥10Mbp	6	6	6	5	7
non-chimeric contigs/scaffolds (bp)	134,109,164	134,164,629	134,164,629	134,162,857	**134,208,377**

5   Conclusions

We presented a scaffolding tool called OMGS for improving the contiguity of *de novo* genome assembly using one or multiple optical maps. OMGS solves several optimization problems to generate scaffolds with optimal contiguity and correctness. Experimental results on *V. unguiculata* and *D. melanogaster* clearly demonstrate that OMGS outperforms SewingMachine and HybridScaffold both in contiguity and correctness using multiple optical maps.

Acknowledgements. This work was supported in part by National Science Foundation grants IIS-1814359, IOS-1543963, IIS-1526742 and IIS-1646333, the Natural Science Foundation of China grant 61772197 and the National Key Research and Development Program of China grant 2018YFC0910404.

Appendix

A   DAG Unique Ordering

Algorithm 1. Sketch of the algorithm for checking whether a DAG provides an unique ordering

---

```
1:  procedure ORDER_UNIQUENESS_CHECK(G = (V, E))
2:      S = nodes with no incoming edges
3:      while S ≠ ∅ do
4:          if |S| > 1 then
5:              return False
6:          remove a node n from S
7:          for each node m with an edge e = (n, m) do
8:              remove edge e from the E
9:              if m has no other incoming edges then
10:                 insert m into S
11:     return True
```

B   Statistical Test for Repetitive Regions

Here we provide additional details for the estimation of σ^2 during the analysis of repetitive regions. Recall that we collect all estimated repetitive lists in set $R = \{D_p$ is estimated repetitive$|p = 1, \ldots, P\}$ and the estimated mean $\hat{\mu}_p$ for each distance list D_p in the set R, where P is the total number of estimated repetitive lists. For each D_p, the distances d_i's are distributed as a Gaussian with mean $\hat{\mu}_p$ and variance σ^2. According to the density function of Gaussian distribution, the log likelihood of one D_p is

$$-\frac{|D_p|}{2} \log(2\pi) - \frac{|D_p|}{2} \log \sigma^2 - \frac{1}{2\sigma^2} \sum_{d_i \in D_p} (d_i - \hat{\mu}_p)^2.$$

The total log likelihood is the sum of the log likelihoods across all D_p's in R, which is

$$\log L(\sigma^2) = -\frac{\sum_{p=1}^{P} |D_p|}{2} \log \sigma^2 - \frac{1}{2\sigma^2} \sum_{p=1}^{P} \sum_{d_i \in D_p} (d_i - \hat{\mu}_p)^2,$$

after ignoring all terms not related to σ^2. To maximize $\log L(\sigma^2)$, we require that the derivative of total log likelihood

$$\frac{\partial \log L(\sigma^2)}{\partial \sigma^2} = 0,$$

that is,

$$-\frac{\sum_{p=1}^{P}|D_p|}{2\sigma^2} + \frac{1}{2(\sigma^2)^2}\sum_{p=1}^{P}\sum_{d_i\in D_p}(d_i - \hat{\mu}_p)^2 = 0.$$

After some simplification, the estimator for variance becomes

$$\hat{\sigma}^2 = \frac{\sum_{p=1}^{P}\sum_{d_i\in D_p}(d_i - \hat{\mu}_p)^2}{\sum_{p=1}^{P}|D_p|}.$$

C   Density Function of $d_{max} - d_{min}$

Here we provide additional details for calculating the density function of $d_{max} - d_{min}$. It is well-known that the joint density function of order statistics is

$$f_{X(i),X(j)}(u, v) = \frac{n!}{(i-1)!(j-1-i)!(n-j)!}f_x(u)f_x(v)[F_x(u)]^{i-1}[F_x(v) - F_x(u)]^{j-1-i}[1 - F_x(v)]^{n-j}$$

$$(1)$$

for $-\infty < u < v < +\infty$, where $X(i)$ and $X(j)$ are the i-th and j-th order statistics in $X_1,\ldots,X_n$ and F_x and f_x are the distribution function and density function of each X_i, respectively. Using (1), the joint density function of (d_{max}, d_{min}) can be expressed as

$$f_{d_{max},d_{min}}(u, v) = n(n-1)f_{d_i}(u)f_{d_i}(v)[F_{d_i}(v) - F_{d_i}(u)]^{n-2}$$

for $-\infty < u < v < +\infty$, where F_{d_i} and f_{d_i} are the distribution function and density function of $d_i \sim N(\hat{\mu}_{j,q}, \hat{\sigma}^2)$, respectively.

Now, let $X = d_{max} - d_{min}$ and $Y = d_{min}$. Then $d_{max} = X + Y$ and $d_{min} = Y$, and the corresponding Jacobian determinant is

$$J = \begin{vmatrix} \partial d_{\max}/\partial X & \partial d_{\max}/\partial Y \\ \partial d_{\min}/\partial X & \partial d_{\min}/\partial Y \end{vmatrix} = \begin{vmatrix} 1 & 1 \\ 0 & 1 \end{vmatrix} = 1.$$

Thus, the joint density function of (X, Y) is given by

$$f_{X,Y}(x, y) = f_{d_{max},d_{min}}(x+y, y)|J| = n(n-1)f_{d_i}(y)f_{d_i}(x+y)[F_{d_i}(x+y) - F_{d_i}(y)]^{n-2},$$

where $x \geq 0$ and $-\infty < y < +\infty$. By integrating over Y, the density function of $X = d_{max} - d_{min}$ becomes

$$f_{d_{max}-d_{min}}(x) = \int_{-\infty}^{+\infty} n(n-1)f_{d_i}(y)f_{d_i}(x+y)[F_{d_i}(x+y) - F_{d_i}(y)]^{n-2}dy, x \geq 0.$$

D Gap Estimation

Here we provide additional details for calculating the log likelihood function when estimating gaps. Recall that $l_p, \ldots, l_{q-1}$ are independent chi-square random variables, and $\sum_{i=p}^{q-1} l_i$ is chi-square distributed with degree of freedom $\sum_{i=p}^{q-1} \alpha_i$. Since the density function of a chi-square random variable X with degree of freedom k is

$$f_X(x) = \frac{1}{2^{k/2}\Gamma(k/2)} x^{k/2-1} e^{-x/2}$$

where Γ is the gamma function, the likelihood of $\sum_{i=p}^{q-1} l_i$ with observation

$$\gamma = d_j - \sum_{c=c_{p+1}}^{c_{q-1}} |c|$$

is

$$\frac{1}{2^\beta \Gamma(\beta)} \gamma^{\beta-1} e^{-\gamma/2},$$

where $\beta = \sum_{i=p}^{q-1} \frac{\alpha_i}{2}$. Therefore, the log likelihood function for one sample is

$$\log l = (\beta - 1)\log\gamma - \frac{\gamma}{2} - \beta\log 2 - \log\Gamma(\beta).$$

The total log likelihood is the sum of the log likelihoods across all samples.

References

1. Avriel, M.: Nonlinear Programming: Analysis and Methods. Courier Corporation, New York (2003)
2. Baharev, A., Schichl, H., Neumaier, A., Achterberg, T.: An exact method for the minimum feedback arc set problem, vol. 10, pp. 35–60. University of Vienna (2015)
3. Berlin, K., Koren, S., Chin, C.-S., Drake, J.P., Landolin, J.M., Phillippy, A.M.: Assembling large genomes with single-molecule sequencing and locality-sensitive hashing. Nature Biotechnol. **33**(6), 623 (2015)
4. Bickhart, D.M., et al.: Single-molecule sequencing and chromatin conformation capture enable de novo reference assembly of the domestic goat genome. Nature Genet. **49**(4), 643 (2017)
5. Boetzer, M., Henkel, C.V., Jansen, H.J., Butler, D., Pirovano, W.: Scaffolding pre-assembled contigs using sspace. Bioinformatics **27**(4), 578–579 (2010)
6. Chin, C.-S., et al.: Phased diploid genome assembly with single-molecule real-time sequencing. Nature Methods **13**(12), 1050 (2016)
7. Daccord, N., et al.: High-quality de novo assembly of the apple genome and methylome dynamics of early fruit development. Nature Genet. **49**(7), 1099 (2017)
8. Dayarian, A., Michael, T.P., Sengupta, A.M.: SOPRA: scaffolding algorithm for paired reads via statistical optimization. BMC Bioinform. **11**(1), 345 (2010)

9. Demaine, E.D., Immorlica, N.: Correlation clustering with partial information. In: Arora, S., Jansen, K., Rolim, J.D.P., Sahai, A. (eds.) APPROX/RANDOM -2003. LNCS, vol. 2764, pp. 1–13. Springer, Heidelberg (2003). https://doi.org/10.1007/978-3-540-45198-3_1

10. Dessmark, A., Jansson, J., Lingas, A., Lundell, E.-M., Persson, M.: On the approximability of maximum and minimum edge clique partition problems. Int. J. Found. Comput. Sci. 18(02), 217–226 (2007)

11. Donmez, N., Brudno, M.: SCARPA: scaffolding reads with practical algorithms. Bioinformatics 29(4), 428–434 (2012)

12. Gao, S., Nagarajan, N., Sung, W.-K.: Opera: reconstructing optimal genomic scaffolds with high-throughput paired-end sequences. In: Bafna, V., Sahinalp, S.C. (eds.) RECOMB 2011. LNCS, vol. 6577, pp. 437–451. Springer, Heidelberg (2011). https://doi.org/10.1007/978-3-642-20036-6_40

13. Gritsenko, A.A., Nijkamp, J.F., Reinders, M.J.T., de Ridder, D.: GRASS: a generic algorithm for scaffolding next-generation sequencing assemblies. Bioinformatics 28(11), 1429–1437 (2012)

14. Hunt, M., Newbold, C., Berriman, M., Otto, T.D.: A comprehensive evaluation of assembly scaffolding tools. Genome Biol. 15(3), R42 (2014)

15. Jiao, W.-B., et al.: Improving and correcting the contiguity of long-read genome assemblies of three plant species using optical mapping and chromosome conformation capture data. Genome Res. 27(5), 778–786 (2017)

16. Kajitani, R., et al.: Efficient de novo assembly of highly heterozygous genomes from whole-genome shotgun short reads. Genome Res. 24(8), 1384–1395 (2014). https://doi.org/10.1101/gr.170720.113

17. Koren, S., Treangen, T.J., Pop, M.: Bambus 2: scaffolding metagenomes. Bioinformatics 27(21), 2964–2971 (2011)

18. Koren, S., Walenz, B.P., Berlin, K., Miller, J.R., Bergman, N.H., Phillippy, A.M.: Canu: scalable and accurate long-read assembly via adaptive k-mer weighting and repeat separation. Genome Res. 27(5), 722–736 (2017). https://doi.org/10.1101/gr.215087.116

19. Li, H.: Minimap and miniasm: fast mapping and de novo assembly for noisy long sequences. Bioinformatics 32(14), 2103–2110 (2016)

20. Lin, Y., Yuan, J., Kolmogorov, M., Shen, M.W., Chaisson, M., Pevzner, P.A.: Assembly of long error-prone reads using de Bruijn graphs. Proc. National Acad. Sci. 113(52), E8396–E8405 (2016)

21. Loman, N.J., Quick, J., Simpson, J.T.: A complete bacterial genome assembled de novo using only nanopore sequencing data. Nature Methods 12(8), 733 (2015)

22. Luo, R., et al.: SOAPdenovo2: an empirically improved memory-efficient short-read de novo assembler. GigaScience 1(1), 18 (2012)

23. Mascher, M., et al.: A chromosome conformation capture ordered sequence of the barley genome. Nature 544(7651), 427 (2017)

24. Muñoz-Amatriaín, M., et al.: Genome resources for climate-resilient cowpea, an essential crop for food security. Plant J. 89(5), 1042–1054 (2017)

25. Nagarajan, N., Read, T.D., Pop, M.: Scaffolding and validation of bacterial genome assemblies using optical restriction maps. Bioinformatics 24(10), 1229–1235 (2008)

26. Pan, W., Lonardi, S.: Accurate detection of chimeric contigs via bionano optical maps. Bioinformatics (2018, in press)

27. Pan, W., Wanamaker, S.I., Ah-Fong, A.M.V., Judelson, H.S., Lonardi, S.: Novo&stitch: accurate reconciliation of genome assemblies via optical maps. Bioinformatics 34(13), i43–i51 (2018)

28. Pendleton, M., et al.: Assembly and diploid architecture of an individual human genome via single-molecule technologies. Nature Methods **12**(8), 780 (2015)
29. Pop, M., Kosack, D.S., Salzberg, S.L.: Hierarchical scaffolding with Bambus. Genome Res. **14**(1), 149–159 (2004)
30. Saha, S., Rajasekaran, S.: Efficient and scalable scaffolding using optical restriction maps. BMC Genomics **15**(5), S5 (2014)
31. Salmela, L., Mäkinen, V., Välimäki, N., Ylinen, J., Ukkonen, E.: Fast scaffolding with small independent mixed integer programs. Bioinformatics **27**(23), 3259–3265 (2011)
32. Samad, A., Huff, E.F., Cai, W., Schwartz, D.C.: Optical mapping: a novel, single-molecule approach to genomic analysis. Genome Res. **5**(1), 1–4 (1995)
33. Shelton, J.M., et al.: Tools and pipelines for BioNano data: molecule assembly pipeline and FASTA super scaffolding tool. BMC Genomics **16**(1), 734 (2015)
34. Simpson, J.T., Durbin, R.: Efficient de novo assembly of large genomes using compressed data structures. Genome Res. **22**(3), 549–556 (2012)
35. Simpson, J.T., Wong, K., Jackman, S.D., Schein, J.E., Jones, S.J.M., Birol, I.: ABySS: a parallel assembler for short read sequence data. Genome Res. **19**(6), 1117–1123 (2009). https://doi.org/10.1101/gr.089532.108
36. Solares, E.A., et al.: Rapid low-cost assembly of the Drosophila melanogaster reference genome using low-coverage, long-read sequencing. G3: Genes Genomes Genet. **8**(10), 3143–3154 (2018)
37. Tang, H., et al.: ALLMAPS: robust scaffold ordering based on multiple maps. Genome Biol. **16**(1), 3 (2015)
38. Walker, B.J., et al.: Pilon: an integrated tool for comprehensive microbial variant detection and genome assembly improvement. PloS One **9**(11), e112963 (2014)
39. Ye, C., Hill, C.M., Wu, S., Ruan, J., Ma, Z.S.: DBG2OLC: efficient assembly of large genomes using long erroneous reads of the third generation sequencing technologies. Sci. Rep. **6** (2016). Article number: 31900
40. Zheng, J., Lonardi, S.: Discovery of repetitive patterns in DNA with accurate boundaries. In: Fifth IEEE Symposium on Bioinformatics and Bioengineering (BIBE 2005), pp. 105–112, October 2005

Fast Approximation of Frequent k-mers and Applications to Metagenomics

Leonardo Pellegrina, Cinzia Pizzi, and Fabio Vandin$^{(\boxtimes)}$

Department of Information Engineering, University of Padova, Padova, Italy
{pellegri,cinzia.pizzi}@dei.unipd.it, fabio.vandin@unipd.it

Abstract. Estimating the abundances of all k-mers in a set of biological sequences is a fundamental and challenging problem with many applications in biological analysis. While several methods have been designed for the exact or approximate solution of this problem, they all require to process the entire dataset, that can be extremely expensive for high-throughput sequencing datasets. While in some applications it is crucial to estimate all k-mers and their abundances, in other situations reporting only *frequent* k-mers, that appear with relatively high frequency in a dataset, may suffice. This is the case, for example, in the computation of k-mers' abundance-based distances among datasets of reads, commonly used in metagenomic analyses.

In this work, we develop, analyze, and test, a sampling-based approach, called SAKEIMA, to approximate the frequent k-mers and their frequencies in a high-throughput sequencing dataset while providing rigorous guarantees on the quality of the approximation. SAKEIMA employs an advanced sampling scheme and we show how the characterization of the VC dimension, a core concept from statistical learning theory, of a properly defined set of functions leads to practical bounds on the sample size required for a rigorous approximation. Our experimental evaluation shows that SAKEIMA allows to rigorously approximate frequent k-mers by processing only a fraction of a dataset and that the frequencies estimated by SAKEIMA lead to accurate estimates of k-mer based distances between high-throughput sequencing datasets. Overall, SAKEIMA is an efficient and rigorous tool to estimate k-mers abundances providing significant speed-ups in the analysis of large sequencing datasets.

Keywords: k-mer analysis · Sampling algorithm · VC dimension · Metagenomics

1 Introduction

The analysis of substrings of length k, called *k-mers*, is ubiquitous in biological sequence analysis and is among the first steps of processing pipelines for a

This work is supported, in part, by the University of Padova grants *SID2017* and *STARS: Algorithms for Inferential Data Mining*.

L. J. Cowen (Ed.): RECOMB 2019, LNBI 11467, pp. 208–226, 2019.
https://doi.org/10.1007/978-3-030-17083-7_13

wide spectrum of applications, including: de novo assembly [21,31], error correction [9,24], repeat detection [11], genome comparison [25], digital normalization [3], RNA-seq quantification [20,33], metagenomic reads classification [30] and binning [7], fast search-by-sequence over large high-throughput sequencing repositories [27]. A fundamental task in k-mer analysis is to compute the frequency of all k-mers, with the goal to distinguish frequent k-mers from infrequent k-mers [13,15]. For example, this task is relevant in the analysis of high-throughput sequencing data, since infrequent k-mers are often assumed to result from sequencing errors. For several applications, the computation of k-mers frequencies is among the most computationally demanding steps of the analysis.

Many algorithms have been proposed for computing the exact frequency of all k-mers, such as Jellyfish [13], DSK [22], KMC 3 [10] and Squeakr-exact [19]. These methods typically perform a linear scan of the sequence to analyze, and use a combination of parallelism and efficient data structures (such as Bloom filters and Hash tables) to maintain membership and counting information associated to all k-mers. Since the computation of exact k-mer frequencies is computationally demanding, in particular for large sequence analysis or for high-throughput sequence datasets, recent methods have focused on providing approximate solution to the problem, improving the time and memory requirements. KmerStream [14], khmer [32], Kmerlight [26] and ntCard [17] proposed streaming approaches for the approximation of the k-mer frequencies histogram. Of these, only Kmerlight and ntCard provide analytical bounds on their accuracy guarantee. KmerGenie [4] performs a linear scan of the input to compute the frequencies of a (random) subset of the k-mers that appear in the input, and uses these frequencies to approximate the abundance histogram. The recently proposed Squeakr [19] relies on a probabilistic data structure to approximate the counts of individual k-mers. Turtle [23] focuses on finding k-mers that appear at least twice in the dataset, but still processes all the k-mer occurrences in the input dataset, as all the other aforementioned methods do.

All the methods cited above try to estimate the frequency of *all* k-mers or of all k-mers that appear at least few times (e.g., twice) in the dataset. While this is crucial in some applications (e.g., in genome assembly k-mers that occur exactly once often represents sequencing errors and it is therefore important to estimate the count of all observed k-mers), in other applications this is less justified. For example, in the comparison of high-throughput sequencing metagenomic datasets, *abundance-based distances or dissimilarities* (e.g., the Bray-Curtis dissimilarity) between k-mer counts of two datasets are often used [1,5,6] to assess the distance between the corresponding datasets. In contrast to *presence-based distances* [18] (e.g., Jaccard distance), abundance-based distances take into account the frequency of each k-mer, with frequent k-mers contributing more to the distance than k-mers that appear with low frequency, but still more than a handful of times, in the dataset. Thus, two natural questions are (i) whether the results obtained considering all k-mers can be estimated by considering the abundances of frequent k-mers only, and (ii) if the abundances of frequent k-mers can be computed more efficiently than the counts of all

k-mers. Recently, preliminary work [8] has shown that, for the cosine distance and $k = 12$, the answer to the first question is positive, and in Sect. 4 we show that this indeed the case for larger values of k and other abundance-based distances as well as presence-based distances (e.g., the Jaccard distance). To the best of our knowledge, the second question is hitherto unexplored. In addition, considering only frequent k-mers allows to focus on the most reliable information in a metagenomic dataset, since a high stochastic variability in low frequency k-mers is to be expected due to the sampling process inherent in sequencing.

A natural approach to reduce time and memory requirements for frequency estimation problems is to process only a portion of the data, for example by *sampling* some parts of a dataset. Sampling approaches are appealing because infrequent k-mers naturally tend to appear with lower probability in a sample, allowing to directly focus on frequent k-mers in subsequent steps. However, major challenges in sampling approaches are (i) to provide rigorous guarantees relating the results obtained by processing the sample and the results that would be obtained from the whole dataset, and (ii) to provide effective bounds on the size of the sample required to achieve such guarantees. The application of sampling to k-mers is even more challenging than in other scenarios since, for values of k in the typical range of interest to applications (e.g., 20–60), even the most frequent k-mers have relatively low frequency in the data. To the best of our knowledge, no approach based on sampling a portion of the input dataset has been proposed to approximate frequent k-mers and their frequencies while providing rigorous guarantees (Fig. 1).

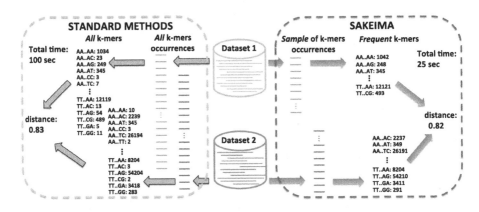

Fig. 1. SAKEIMA computes a fast and rigorous approximation of the *frequent* k-mers in a high-throughput sequencing dataset by sampling a fraction of all k-mer occurrences in a dataset, providing a significant speed-up for the computation of k-mer's abundance-based distances between datasets of reads (e.g., in metagenomic).

Our Contribution. We study the problem of approximating frequent k-mers, i.e., k-mers that appear with frequency above a user-defined threshold θ in a high-throughput sequencing dataset. In these regards, our contributions are fourfold.

First, we define a rigorous definition of approximation, governed by an accuracy parameter ε. Second, we propose a new method, $\underline{S}$ampling $\underline{A}$lgorithm for $\underline{K}$-m$\underline{E}$rs approx$\underline{IMA}$tion (SAKEIMA), to obtain an approximation to the set of frequent k-mers using *sampling*. SAKEIMA is based on a sampling scheme that goes beyond naïve sampling of k-mers and allows to estimate low frequency k-mers considering only a fraction of all k-mers occurrences in the dataset. Third, we provide analytical bounds to the sample size needed to obtain rigorous guarantees on the accuracy of the estimated k-mer frequencies, with respect to the ones measured on the entire dataset. Our bounds are based on the notion of VC dimension, a fundamental concept from statistical learning theory. To our knowledge, ours is the first method that applies concepts from *statistical learning* to provide a rigorous approximation of the k-mers frequencies. Fourth, we use SAKEIMA to extract frequent k-mers from metagenomic datasets from the Human Microbiome Project (HMP) and to approximate abundance-based and presence-based distances among such datasets, showing that SAKEIMA allows to accurately estimate such distances by analyzing only a fraction of the entire dataset, resulting in a significant speed-up.

Our approach is orthogonal to previous work: any exact or approximate algorithm can be applied to the sample extracted by SAKEIMA, that can therefore be used *before* applying previously proposed methods, thus reducing their computational requirements while providing rigorous guarantees on the results w.r.t. to the entire dataset. While we present our methodology in the case of finding frequent k-mers from a set of sequences representing a high-throughput sequencing dataset of short reads, our results can be applied to datasets of long reads and to whole-genome sequences as well.

2   Preliminaries

Let a dataset $\mathcal{D}$ be a bag of n reads $\mathcal{D} = \{r_0, \ldots, r_{n-1}\}$, where each read r_i, $0 \le i \le n-1$, is a string of length n_i from an alphabet Σ of cardinality $|\Sigma| = \sigma$. For $j \in \{0, \ldots, n_i - 1\}$, let $r_i[j]$ be the j-th character of r_i. For a given integer $k \le \min_i\{n_i : r_i \in \mathcal{D}\}$, we define a k-mer A as a string of length k from Σ, that is $A \in \Sigma^k$. We say that a k-mer A *appears* in r_i at position $j \in \{0, \ldots, n_i - k\}$ if $r_i[j + h] = A[h], \forall h \in \{0, \ldots, k-1\}$. For every $i, 0 \le i \le n-1$, and every $j \in \{0, \ldots, n_i - k\}$, we define the indicator function $\phi_{r_i,A}(j)$ that is 1 if the k-mer A appears in r_i at position j, while $\phi_{r_i,A}(j) = 0$ otherwise. The total number of k-mers in $\mathcal{D}$ is $t_{\mathcal{D},k} = \sum_{i=0}^{n-1}(n_i - k + 1)$. We define the *support* $o_{\mathcal{D}}(A)$ of a k-mer A as the number of distinct positions in $\mathcal{D}$ where A appears: $o_{\mathcal{D}}(A) = \sum_{i=0}^{n-1} \sum_{j=0}^{n_i-k} \phi_{r_i,A}(j)$. We define the *frequency* $f_{\mathcal{D}}(A)$ of A in $\mathcal{D}$ as the ratio between the number of distinct positions where A appears in $\mathcal{D}$ and the total number of k-mers in $\mathcal{D}$: $f_{\mathcal{D}}(A) = o_{\mathcal{D}}(A)/t_{\mathcal{D},k}$.

2.1 Frequent k-mers and Approximations

We are interested in obtaining the set $FK(\mathcal{D}, k, \theta)$ of frequent k-mers in a dataset $\mathcal{D}$ with respect to a minimum frequency threshold θ, defined as follows.

Definition 1. *Given a dataset $\mathcal{D}$, an integer $k > 0$, and a frequency threshold $\theta \in (0,1]$, the set $FK(\mathcal{D}, k, \theta)$ of Frequent k-Mers in $\mathcal{D}$ w.r.t. θ is the collection of all k-mers with frequency at least θ in $\mathcal{D}$ and of their corresponding frequencies in $\mathcal{D}$:*

$$FK(\mathcal{D}, k, \theta) = \{(A, f_{\mathcal{D}}(A)) : f_{\mathcal{D}}(A) \geq \theta\}. \tag{1}$$

$FK(\mathcal{D}, k, \theta)$ can be computed with a single scan of all the k-mers occurrences in $\mathcal{D}$ maintaining the k-mers supports in an appropriate data structure; however, when $\mathcal{D}$ is extremely large and k is not small, the exact computation of $FK(\mathcal{D}, k, \theta)$ is extremely demanding in terms of time and memory, since the number of k-mers grows exponentially with k. In this case, a fast to compute *approximation* of the set $FK(\mathcal{D}, k, \theta)$ may be preferable, provided it ensures rigorous guarantees on its quality. In this work, we focus on the following approximation.

Definition 2. *Given a dataset $\mathcal{D}$, an integer $k > 0$, a frequency threshold $\theta \in (0,1]$, and a constant $\varepsilon \in (0, \theta)$, an ε-approximation of $FK(\mathcal{D}, k, \theta)$ is a collection $C = \{(A, f_A) : f_A \in (0,1]\}$ such that:*

- *for any $(A, f_{\mathcal{D}}(A)) \in FK(\mathcal{D}, k, \theta)$ there is a pair $(A, f_A) \in C$;*
- *for any $(A, f_A) \in C$ it holds that $f_{\mathcal{D}}(A) \geq \theta - \varepsilon$;*
- *for any $(A, f_A) \in C$ it holds that $|f_{\mathcal{D}}(A) - f_A| \leq \varepsilon/2$.*

The definition above guarantees that every frequent k-mer of $\mathcal{D}$ is in the approximation and that no k-mer with frequency $< \theta - \varepsilon$ is in the approximation. The third condition guarantees that the estimated frequency f_A of A in the approximation is close (i.e., within $\varepsilon/2$) to the frequency $f_{\mathcal{D}}(A)$ of A in $\mathcal{D}$. It is easy to show that obtaining a ε-approximation of $FK(\mathcal{D}, k, \theta)$ with absolute certainty requires to process all k-mers in $\mathcal{D}$.

2.2 Simple Sampling-Based Algorithms and Bounds

We aim to provide an approximation to $FK(\mathcal{D}, k, \theta)$ with *sampling*, by processing only *randomly selected portions* of $\mathcal{D}$. The simplest sampling scheme is the one in which a random sample is a bag P of m positions taken uniformly at random, with replacement, from the set $P_{\mathcal{D},k} = \{(i,j) : i \in [0, n-1], j \in [0, n_i - k]\}$ (note that $|P_{\mathcal{D},k}| = t_{\mathcal{D},k}$) of all positions where k-mers occurs in the dataset $\mathcal{D}$, corresponding to m occurrences of k-mers (with repetitions) taken uniformly at random. Given such sample P, an integer $k > 0$, and a minimum frequency threshold $\theta \in (0,1]$ one can define the set of frequent k-mers (and their frequencies) in the sample P as $FK(P, k, \theta) = \{(A, f_P(A)) : f_P(A) \geq \theta\}$, where $f_P(A)$ is the frequency of k-mer A in the sample.

Obtaining a ε-approximation from a random sample with absolute certainty is impossible, thus we focus on obtaining a ε-approximation with probability $1 - \delta > 0$, where $\delta \in (0,1)$ is a *confidence* parameter, whose value is provided by the user. Intuitively, the set $FK(\mathcal{D}, k, \theta)$ of frequent k-mers is well approximated by the set of frequent k-mers in a random sample P when P is sufficiently large. One natural question regards how many samples are needed to obtain the desired ε-approximation. By using Hoeffding's inequality [16] to bound the deviation of the frequency of a k-mer A in the sample from $f_{\mathcal{D}}(A)$ and a union bound on the maximum number σ^k of k-mers, where $\sigma = |\Sigma|$, we have the following result that provides a first such bound, and a corresponding first algorithm to obtain a ε-approximation to $FK(\mathcal{D}, k, \theta)$. (Due to space constraints proofs are omitted and will be provided in the full version of this extended abstract.)

Proposition 1. *Consider a sample P of size m of $\mathcal{D}$. If $m \geq \frac{2}{\varepsilon^2} \left(\ln\left(2\sigma^k\right) + \ln\left(\frac{1}{\delta}\right)\right)$ for fixed $\varepsilon \in (0, \theta), \delta \in (0,1)$, then, with probability $\geq 1 - \delta$, $FK(P, k, \theta - \varepsilon/2)$ is a ε-approximation of $FK(\mathcal{D}, k, \theta)$.*

In addition, by using known results in statistical learning theory [16,29] relating the VC dimension (see Sect. 3 for its definition) of a family of functions and a novelly derived bound on the family of functions $\{f_{\mathcal{D}}(A)\}$, we obtain the following improved bound and algorithm. (The derivation will be provided in the full version.)

Proposition 2. *Let P be a sample of size m of $\mathcal{D}$. For fixed $\varepsilon \in (0, \theta), \delta \in (0,1)$, if $m \geq \frac{2}{\varepsilon^2} \left(1 + \ln\left(\frac{1}{\delta}\right)\right)$ then $FK(P, k, \theta - \varepsilon/2)$ is an ε-approximation for $FK(\mathcal{D}, k, \theta)$ with probability $\geq 1 - \delta$.*

3   Advanced and Practical Bounds and Algorithms for k-mer Approximations

While the bound of Proposition 2 significantly improves the simple bounds of Sect. 1, since the factor $\ln(2\sigma^k)$ has been reduced to 1, it still has an inverse quadratic dependency with respect to the accuracy parameter ε, that is problematic when the quantities to estimate are small. In these cases, one needs a small ε to produce a meaningful approximation (since $\varepsilon < \theta$), and the inverse quadratic dependence of the sample size from ε often results in a sample size larger than the entire input, defeating the purpose of sampling. The case of k-mers is particularly challenging, since the sum $\sum_{A \in \Sigma^k} f_{\mathcal{D}}(A)$ of all k-mers frequencies is exactly 1. Therefore the higher the number of distinct k-mers appearing in the input, the lower their frequencies will be, with the consequence that θ (and therefore ε) typically needs to be set to a very low value. For example, a typical dataset from the Human Microbiome Project (HMP) has $n \approx 10^8$ reads of (average) length ≈ 100: therefore if we are interested in k-mers for $k = 31$, by setting $\delta = 0.05$ the bound of Sect. 2.2 gives $\varepsilon \approx 10^{-5}$, that is only k-mers with frequency $\geq 10^{-5}$ could be reliably reported by sampling. However, in datasets we considered, no or a very small number (≤ 30) of k-mers have frequency $\geq 10^{-5}$, therefore according to the result from Sect. 2.2 we cannot obtain a meaningful approximation of k-mers and their frequencies. In the remaining of this section we develop more refined sampling schemes and estimation techniques leading to a practical sampling-based algorithm.

3.1   Sampling Bags of Positions and VC Dimension Bound

We propose a method to provide an efficiently computable approximation to $FK(\mathcal{D}, k, \theta)$ when the minimum frequency θ is low, by properly defining samples so that any k-mer A will appear in a sample with probability higher than $f_{\mathcal{D}}(A)$, thus lessening the the dependence of the sample size from $1/\varepsilon^2$. For this to be achievable, we need to relax the notion of approximation defined in Sect. 2. In particular, the guarantees, provided by our method, in such relaxed approximation are that *all* k-mers with frequency above θ', with θ' slightly higher than θ, are reported in output, and that no k-mer having frequency below $\theta - \varepsilon$ is reported in output. (See Proposition 5 for the definition of θ'.) Our experiments show that the fraction of k-mers having frequency $\in [\theta, \theta')$ which are non reported is very small. Our method works by sampling *bags of positions* instead than single positions. In particular, an element of the sample is now a set of ℓ positions chosen independently at random from the set $P_{\mathcal{D},k}$ of all positions.

Let $I_\ell = \{(i_1, j_1), (i_2, j_2), \ldots, (i_\ell, j_\ell)\}$ be a bag of ℓ positions for k-mers in $\mathcal{D}$, chosen uniformly at random from the set $P_{\mathcal{D},k}$. We define the indicator functions $\hat{\phi}_A(I_\ell)$ that, for a given bag I_ℓ of ℓ positions, is equal to 1 if k-mer A appears in *at least* one of the ℓ positions in I_ℓ and is equal to 0 otherwise. That is $\hat{\phi}_A(I_\ell) = \min\left\{1, \sum_{(i,j) \in I_\ell} \phi_{r_i, A}(j)\right\}$. We define the ℓ-*positions sample* P_ℓ as a bag of m bags $\{I_{\ell,0}, I_{\ell,1}, \ldots, I_{\ell,m-1}\}$, where each $I_{\ell,j}, 0 \leq j \leq m-1$ is a bag of ℓ positions, sampled independently, and

$$\hat{f}_{P_\ell}(A) = \frac{1}{m} \sum_{I_{\ell,i} \in P_\ell} \frac{\hat{\phi}_A(I_{\ell,i})}{\ell}. \tag{2}$$

Intuitively, $\hat{f}_{P_\ell}(A)$ is the biased version of the unbiased estimator $f_{P_\ell}(A) = \frac{1}{m} \sum_{I_{\ell,i} \in P_\ell} \frac{\sum_{(i,j) \in I_{\ell,i}} \phi_{r_i, A}(j)}{\ell}$ of $f_{\mathcal{D}}(A)$, where the bias arises from considering a value of 1 every time $\sum_{(i,j) \in I_{\ell,i}} \phi_{r_i, A}(j) > 1$.

In our analysis we use the Vapnik-Chervonenkis (VC) dimension [28,29], a statistical learning concept that measures the expressivity of a family of binary functions. We define a range space Q as a pair $Q = (X, R_X)$ where X is a finite or infinite set and R_X is a finite or infinite family of subsets of X. The members of R_X are called *ranges*. Given $D \subset X$, the *projection* of R_X on D

is defined as $proj_{R_X}(D) = \{r \cap D : r \in R_X\}$. We say that D is *shattered by* R_X if $proj_{R_X}(D) = 2^{|D|}$. The *VC dimension of* Q, denoted as $VC(Q)$, is the maximum cardinality of a subset of X shattered by R_X. If there are arbitrary large shattered subsets of X shattered by R_X, then $VC(Q) = \infty$.

A finite bound on the VC dimension of a range space Q implies a bound on the number of random samples required to obtain a good approximation of its ranges, defined as follows.

Definition 3. *Let $Q = (X, R_X)$ be a range space and let D be a finite subset of X. For $\varepsilon \in (0, 1]$, a subset B of D is an ε-approximation of D if for all $r \in R_X$ we have:* $\left| \frac{|D \cap r|}{|D|} - \frac{|B \cap r|}{|B|} \right| \leq \varepsilon/2.$

The following result [16] relates ε and the probability that a random sample of size m is an ε-approximation for a range space of VC dimension at most v.

Proposition 3 ([16]). *There is an absolute positive constant c such that if (X, R_X) is a range-space of VC dimension at most v, D is a finite subset of X, and $0 < \varepsilon, \delta < 1$, then a random subset $B \subset D$ of cardinality m with $m \geq \frac{4c}{\varepsilon^2}\left(v + \ln\left(\frac{1}{\delta}\right)\right)$ is a ε-approximation of D with probability at least $1 - \delta$.*

The universal constant c has been experimentally estimated to be at most 0.5 [12].

We now prove an upper bound to the VC dimension $VC(Q)$ of the range space Q associated to the class of functions $\hat{\phi}_A$ that grows sub-linearly with respect to ℓ. To this aim, we first define the range space associated to bags of ℓ positions of k-mers.

Definition 4. *Let $\mathcal{D}$ be a dataset of n reads and let k and ℓ be two integers ≥ 1. We define $Q = (X_{\mathcal{D},k,\ell}, R_{\mathcal{D},k,\ell})$ to be the following range space:*

- *$X_{\mathcal{D},k,\ell}$ is the set of all bags of ℓ positions of k-mers in $\mathcal{D}$, that is the set of all possible subsets, with repetitions, of size ℓ from from $P_{\mathcal{D},k}$;*
- *$R_{\mathcal{D},k,\ell} = \{P_{\mathcal{D},\ell}(A)|A \in \Sigma^k\}$ is the family of sets of starting positions of k-mers, such that for each k-mer A, the set $P_{\mathcal{D},\ell}(A)$ is the set of all bags of ℓ starting positions in $\mathcal{D}$ where A appears at least once.*

We prove the following results on the VC dimension of the above range space.

Proposition 4. *Let Q the range space from Definition 4. Then: $VC(Q) \leq \lfloor \log_2(\ell) \rfloor + 1$.*

Using the result above, we prove the following.

Proposition 5. *Let $\ell \geq 1$ be an integer and P_ℓ be a bag of m bags of ℓ positions of $\mathcal{D}$ with*

$$m \geq \frac{2}{(\ell\varepsilon)^2}\left(\lfloor \log_2 \min(2\ell, \sigma^k) \rfloor + \ln\left(\frac{1}{\delta}\right)\right). \tag{3}$$

Then, with probability at least $1 - \delta$:

– for any k-mer $A \in FK(\mathcal{D}, k, \theta)$ such that $f_{\mathcal{D}}(A) \geq \theta' = 1 - (1 - \ell\theta)^{1/\ell}$ it holds $\hat{f}_{P_\ell}(A) \geq \theta - \varepsilon/2$;
– for any k-mer A with $\hat{f}_{P_\ell}(A) \geq \theta - \varepsilon/2$ it holds $f_{\mathcal{D}}(A) \geq \theta - \varepsilon$;
– for any k-mer $A \in FK(\mathcal{D}, k, \theta)$ it holds $f_{\mathcal{D}}(A) \geq \hat{f}_{P_\ell}(A) - \varepsilon/2$;
– for any k-mer A with $\hat{f}_{P_\ell}(A) - \varepsilon/2 \geq 0$, it holds $f_{\mathcal{D}}(A) \geq 1 - (1 - \ell(\hat{f}_{P_\ell}(A) - \varepsilon/2))^{1/\ell}$;
– for any k-mer A with $\ell(\hat{f}_{P_\ell}(A) + \varepsilon/2) \leq 1$ it holds $f_{\mathcal{D}}(A) \leq 1 - (1 - \ell(\hat{f}_{P_\ell}(A) + \varepsilon/2))^{1/\ell}$.

Note that from Proposition 5 the set $\{(A, f_{P_\ell}(A)) : \hat{f}_{P_\ell}(A) \geq \theta - \varepsilon/2\}$ is *almost* a ε-approximation to $FK(\mathcal{D}, k, \theta)$: in particular, there may be k-mers A for which $\mathbb{E}[\hat{f}_{P_\ell}(A)] = (1 - (1 - f_{\mathcal{D}}(A))^\ell)/\ell < \theta$ while $f_{\mathcal{D}}(A) = \mathbb{E}[f_{P_\ell}(A)] \geq \theta$ and such that for the given sample P_ℓ we have $\hat{f}_{P_\ell}(A) \approx \mathbb{E}[\hat{f}_{P_\ell}(A)] - \varepsilon/2$. While this can happen, we can limit the probability of this happening by appropriately choosing ℓ, and still enjoy the reduction in sample size of the order of $\frac{\log_2 \ell}{\ell^2}$ w.r.t. Proposition 2 obtained by considering bags of bags of ℓ positions. In particular, this result allows the user to set θ, ε, δ, and ℓ to effectively find, with probability at least $1 - \delta$, *all* frequent k-mers A for which $f_{\mathcal{D}}(A) \geq \theta'$ and do not report any k-mer with frequency below $\theta - \varepsilon$, while still being able to report in output almost all k-mers with frequency $\in [\theta, \theta')$. Our experimental analysis (Sect. 4) shows that in practice choosing ℓ close from below to $1/\theta$ is very effective to obtain such result. Then, the third, fourth, and fifth guarantees from Proposition 5 state that we can use the biased estimates $\hat{f}_{P_\ell}(A)$ to derive *guaranteed upper and lower bounds* to $f_{\mathcal{D}}(A)$ that will be much tighter than the one obtained using the bounds of Sect. 2.2. We will show how to obtain further improved upper and lower bounds to $f_{\mathcal{D}}(A)$ in Sect. 3.3. Such lower bounds lb_A can be used, for example, to prove that the set $\{(A, f_{P_\ell}(A)) : lb_A \geq \theta - \varepsilon\}$ enjoys the same last four guarantees from Proposition 5 while the first one holds for a $\theta' < 1 - (1 - \ell\theta)^{1/\ell}$; therefore, when false negatives are problematic, the set $\{(A, f_{P_\ell}(A)) : lb_A \geq \theta - \varepsilon\}$ can be used to obtain a different approximation of $FK(\mathcal{D}, k, \theta)$ with fewer false negatives.

3.2 SAKEIMA: An Efficient Algorithm to Approximate Frequent k-mers

We now present our <u>Sa</u>mpling <u>A</u>lgorithm for <u>K</u>-m<u>E</u>rs approx<u>IMA</u>tion (SAKEIMA), that builds on Proposition 5 and efficiently samples a bag P_ℓ of bags of ℓ-positions from $\mathcal{D}$ to obtain an approximation of the set $FK(\mathcal{D}, k, \theta)$ with probability $1 - \delta$, where δ is a parameter provided by the user.

Algorithm 1. SAKEIMA

Input: dataset $\mathcal{D}$, total number of k-mers $t_{\mathcal{D},k}$ in $\mathcal{D}$,
frequency threshold θ, accuracy parameter $\varepsilon \in (0, \theta)$,
confidence parameter $\delta \in (0, 1)$, integer $\ell \geq 1$.
Output: approximation $\{(A, f_A)\}$ of $FK(\mathcal{D}, k, \theta)$ with probability $\geq 1 - \delta$

1 $m \leftarrow \left\lceil \frac{2}{(\ell\varepsilon)^2} \left(\lfloor \log_2 \min(2\ell, \sigma^k) \rfloor + \ln\left(\frac{2}{\delta}\right) \right) \right\rceil$; $\lambda \leftarrow \frac{m\ell}{t_{\mathcal{D},k}}$;

2 $T \leftarrow$ empty hash table;

3 **forall** *reads* $r_i \in \mathcal{D}$ **do**

4   **forall** $j \in [0, n_i - k]$ **do**

5      $A \leftarrow k$-mer in position j of read r_i;

6      $a \leftarrow Poisson(\lambda)$;

7   **if** $a > 0$ **then** $T[A] \leftarrow T[A] + a$;

8 $\mathcal{O} \leftarrow \emptyset$; $t \leftarrow \sum_{A \in T} T[A]$;

9 $P_\ell \leftarrow$ random partition of t occurrences in T into m bags;

10 **forall** k-*mers* $A \in T$ **do**

11   $f_A \leftarrow T[A]/t$;

12   $\mathcal{P}_A \leftarrow$ bags of P_ℓ where A appears at least once;

13   $\hat{f}_A \leftarrow |\mathcal{P}_A|/m$;

14   **if** $\hat{f}_A \geq \theta - \varepsilon/2$ **then** $\mathcal{O} \leftarrow \mathcal{O} \cup (A, f_A)$;

15 **return** $\mathcal{O}$;

SAKEIMA is described in Algorithm 1. SAKEIMA performs a pass on the stream of k-mers appearing in $\mathcal{D}$, and for each position in the stream it samples the number a of times that the position appears in the sample P_ℓ independently at random from the Poisson distribution $Poisson(\lambda)$ of parameter $\lambda = m\ell/t_{\mathcal{D},k}$. SAKEIMA stores such values in a counting structure T (lines 3–7) that keeps, for each k-mer A, the total number of occurrences of A in the sample P_ℓ. (Note that $t_{\mathcal{D},k}$, that can be computed with a very quick linear scan of the dataset, where n_i is computed for every $r_i \in \mathcal{D}$ without extracting and processing (e.g., inserting or updating information for) k-mers; in alternative a lower bound to $t_{\mathcal{D},k}$ can be used, simply resulting in a number of samples higher than needed). Then, such occurrences are partitioned into the m bags $I_{\ell,0}, \ldots, I_{\ell,m-1}$ (line 9); this can be efficiently implemented by assigning each occurrence to a random bag while keeping the difference between the final size of the bags ≤ 1. For each k-mer A appearing at least once in the sample (line 10), the unbiased estimate f_A is computed as the number $T[A]$ of occurrences of A in the sample P_ℓ (line 11) divided by the total number of positions in the sample, while the biased estimate $\hat{f}_A$ is computed as the number $|\mathcal{P}_A|$ of distinct bags of P_ℓ where A appears at least once divided by the number m of bags (lines 12–13). Then SAKEIMA flags A as frequent if $\hat{f}_A \geq \theta - \varepsilon/2$ (line 14) and, in this case, the couple (A, f_A) is added to the output set $\mathcal{O}$ (line 14), since f_A is the best (and unbiased) estimate to $f_{\mathcal{D}}(A)$. Note that bags for different values of ℓ (on the same sampled positions) can be obtained by maintaining a table T_ℓ and a set $\mathcal{P}_{A,\ell}$ for each value ℓ of interest.

Note that SAKEIMA does not sample m bags of *exactly* ℓ positions each, since the number of occurrences of each position in $\mathcal{D}$ in the sample P_ℓ is sampled independently from a Poisson distribution, even if the expected number of total occurrences sampled from the algorithm is $m\ell$. However, the independent Poisson distributions used by SAKEIMA provide an accurate approximation of the random sampling of *exactly* $m\ell$ positions used in the analysis of Sect. 3.1. In particular, this holds when one focuses on the events of interests for our approximation of Sect. 3.1 (e.g., the event "there exists a k-mer A such that $|\mathbb{E}[\hat{f}_{P_\ell}(A)] - \hat{f}_{P_\ell}(A)| > \varepsilon/2$"). In fact, a simple adaptation of a known result (Corollary 5.11 of [16]) on the relation between sampling with replacement and the use of independent Poisson distributions gives the following.

Proposition 6. *Let E be an event whose probability is either monotonically increasing or monotonically decreasing in the number of sampled positions. If E has probability p when the independent Poisson distributions are used, then E has probability at most $2p$ when the sampling with replacement is used.*

As a simple corollary, the output $\mathcal{O}$ features the guarantees of Proposition 5 with probability $\geq 1 - \delta'$, with $\delta' = 2\delta$.

3.3   Improved Lower and Upper Bounds to k-mers Frequencies

Note that Proposition 5 guarantees that we can obtain upper and lower bounds to $f_\mathcal{D}(A)$ for every $A \in FK(\mathcal{D}, k, \theta)$ from the sample of bags of ℓ positions. These bounds are meaningful only in specific ranges of the frequencies; for example, the lower bound from the third guarantee in Proposition 5 is meaningful when the frequency of A is fairly low, i.e. $f_\mathcal{D}(A) \approx 1/\ell$, while for very frequent k-mers they could be a multiplicative factor $1/\ell$ away from than the correct value. For example, if a k-mer is very frequent and appears in all bags of ℓ k-mers in a sample S, its corresponding lower bound is still only $1/\ell - \varepsilon/2$.

However, Proposition 5 can be generalized to obtain tighter upper and lower bounds to the frequency of all k-mers. For given ℓ, ε, and δ, let m as given in Proposition 5. Note that the total number of k-mer's positions in the sample P_ℓ is $m\ell$. Let $\mathcal{L}$ be a set of integer values $\mathcal{L} = \{\ell_i\}$ with $\ell_i \in [1, m\ell], \forall i = 0, \ldots, |\mathcal{L}| - 1$. Now, for every $\ell_i \in \mathcal{L}$, we can partition the *same* $m\ell$ k-mers that are in P_ℓ into $m_i = m\ell/\ell_i$ partitions having size ℓ_i. Let P_{ℓ_i} be such a random partition of such positions into m_i bags of ℓ_i positions each. Note that each P_{ℓ_i} is a "valid" sample (i.e., a sample of independent bags of positions, each obtained by uniform sampling with replacement) for Proposition 5, even if the P_{ℓ_i}'s are not independent. From each P_{ℓ_i}, we define a maximum deviation ε_i from Proposition 5 as $\varepsilon_i = \frac{1}{\ell_i}\sqrt{\frac{2}{m_i}\left(\lfloor \log_2(\min(2\ell_i, \sigma^k)) \rfloor + \ln\left(|\mathcal{L}|/\delta\right)\right)}$. We have the following result.

Proposition 7. *With probability at least $1 - \delta$, for all k-mers A simultaneously and for all the random partitions induced by $\mathcal{L}$ it holds*

$$- f_\mathcal{D}(A) \geq \max\{\hat{f}_{P_{\ell_i}}(A) - \varepsilon_i/2 : i = 0, \ldots, |\mathcal{L}| - 1\};$$

- $f_{\mathcal{D}}(A) \geq \max\{1 - (1 - \ell(\hat{f}_{P_{\ell_i}}(A) - \varepsilon_i/2))^{1/\ell} : i = 0, \ldots, |\mathcal{L}| - 1 \text{ and } \hat{f}_{P_\ell}(A) - \varepsilon_i/2 \geq 0\}$;
- $f_{\mathcal{D}}(A) \leq \min\{1 - (1 - \ell(\hat{f}_{P_{\ell_i}}(A) + \varepsilon_i/2))^{1/\ell} : i = 0, \ldots, |\mathcal{L}| - 1 \text{ and } \hat{f}_{P_\ell}(A) + \varepsilon_i/2 \leq 1/\ell\}$.

In our experiments, we use $\mathcal{L} = \{\ell_i\}$ with $\ell_i = \ell/2^i, \forall i \in [0, \lfloor \log_2 \ell \rfloor - 1]$; in this case, note that $P_{\ell_0} = P_\ell$. Using this scheme, we can compute upper and lower bounds for k-mers having frequencies of many different orders of magnitude, but any (application dependent) distribution can be specified by the user. These upper and lower bounds can be used to obtain different approximations of $FK(\mathcal{D}, k, \theta)$ with different guarantees. For example, by reporting all k-mers (and their frequencies) that have an upper bound $\geq \theta$, we have an approximation that guarantees that all k-mers A with $f_{\mathcal{D}}(A) \geq \theta$ are in the approximation.

4   Experimental Results

In this section we present the results of our experimental evaluation for SAKEIMA. Sect. 4.1 describes the datasets, our implementation for SAKEIMA[1], and the baseline for comparisons. In Sect. 4.2, we report the results for computing the approximation of the frequent k-mers using SAKEIMA. Section 4.3 reports the results of using our approximation to compute abundance-based and presence-based distances between metagenomic datasets.

4.1   Datasets and Implementation

We considered six datasets from the Human Microbiome Project (HMP)[2], one of the largest publicly available collection of metagenomic datasets from high-throughput sequencing. In particular, we selected the three largest datasets of stool and the three largest of tongue dorsum (Table 1). These datasets constitute the most challenging instances, due to their size, and provide a test case with different degrees of similarities among datasets.

We implemented SAKEIMA in C++ as a modification of Jellyfish [13] (the version we used is 2.2.10[3]), a very popular and efficient algorithm for exact k-mer counting. Doing so, our algorithm enjoys the succinct counting data structure provided by Jellyfish publicly available implementation. We remark that our sampling-based approach can be used in combination with any other highly tuned method available for exact, approximate, and parallel k-mer counting. For this reason, we only compare SAKEIMA with the exact counting performed by Jellyfish, since they share the underlying characteristics, allowing us to evaluate the impact of SAKEIMA sampling strategy. We did not include the time to compute $t_{\mathcal{D},k}$ in our experiments since it was always negligible (i.e., less than 2 min) w.r.t. the time for counting k-mers.

---

[1] Available at https://github.com/VandinLab/SAKEIMA.
[2] https://hmpdacc.org/HMASM.
[3] https://github.com/gmarcais/Jellyfish.

Table 1. Datasets for our experimental evaluation. For each dataset $\mathcal{D}$ the table shows: the dataset name and site ((s) for stool, (t) for tongue dorsum); the total number $t_{\mathcal{D},k}$ of k-mers ($k = 31$) in $\mathcal{D}$; the number $|\mathcal{D}|$ of reads it contains; the maximum read length $\max_{n_i} = \max_i\{n_i|r_i \in \mathcal{D}\}$; the average read length $\mathrm{avg}_{n_i} = \sum_{i=0}^{n-1} n_i/n$.

Dataset	$t_{\mathcal{D},k}$	$|\mathcal{D}|$	$\max_{n_i}$	avg_{n_i}
SRS024388(s)	$7.92 \cdot 10^9$	$1.20 \cdot 10^8$	102	97.21
SRS011239(s)	$8.13 \cdot 10^9$	$1.24 \cdot 10^8$	102	96.69
SRS024075(s)	$8.82 \cdot 10^9$	$1.38 \cdot 10^8$	96	94.88
SRS075404(t)	$7.75 \cdot 10^9$	$1.22 \cdot 10^8$	102	94.51
SRS062761(t)	$8.26 \cdot 10^9$	$1.18 \cdot 10^8$	101	101.00
SRS043663(t)	$9.15 \cdot 10^9$	$1.31 \cdot 10^8$	101	101.00

For the computation of the abundance-based distances from the k-mer counts of two dataset, we implemented in C++ a simple algorithm that loads the counts of one dataset in main memory and then performs one pass on the counts of the other dataset, producing the distances in output. We executed all our experiments on the same machine with 512 GB of RAM and a 2.30 GHz Intel Xeon CPU, compiling both implementations with g++ 4.9.4. SAKEIMA can be used in combination with more efficient algorithms and implementations for the computation of these (and other) distances [1], resulting in speed-ups analogous to the ones we present below. For all the experiments of SAKEIMA, given θ and a dataset $\mathcal{D}$, we fixed the parameters $\delta = 0.1$, $\varepsilon = \theta - 2/t_{\mathcal{D},k}$, $m = 100$, and we fix ℓ to the minimum value satisfying the ε-approximation. For all the experiments we have ℓ close from below to $1/\theta$. For all the metrics we considered, we report the results for one random run.

4.2 Approximation of the Frequent k-mers

We fixed $k = 31$, and we compared SAKEIMA with the exact counting of all k-mers (from Jellyfish) in terms of: (i) running time[4], including, for both algorithms, the time required to write the output on disk; (ii) memory requirement. We also assessed the accuracy of the output of SAKEIMA.

Figure 2 shows the running times and the peak memory as function of θ. Note that for the exact counting algorithm these metrics do not depend on θ, since it always counts all k-mers. SAKEIMA is always faster than the exact counting, with a difference that increases when θ increases and a speed-up around 2 even for $\theta = 2 \cdot 10^{-8}$. The memory requirement of SAKEIMA reduces when θ increases, and for $\theta = 2 \cdot 10^{-8}$ it is half of the memory required by the exact counting.

---

[4] Every instance of SAKEIMA and Jellyfish was executed with 1 worker, i.e., sequentially. Note that the Poisson approximation employed by SAKEIMA allows multiple workers to independently process the input k-mers, therefore SAKEIMA can be used in a parallel scenario. We will investigate the impact of parallelism in the extended version of this work.

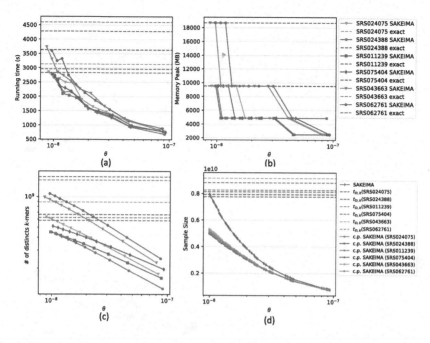

Fig. 2. Running time, memory requirements, and number of distinct k-mers counted, for SAKEIMA and exact counting as function of θ. (a) Running time. (b) Memory requirement. (c) Number of distinct k-mers counted. (d) Sample sizes of SAKEIMA, total size $t_{\mathcal{D},k}$ of the datasets, and number (c.p.) of dataset's distinct covered positions (i.e., included in SAKEIMA's sample), as function of θ.

This is due to SAKEIMA's sample size being much smaller than the dataset size (Fig. 2(d)), therefore a large portion of extremely low frequency k-mers are naturally left out from the random sample and do not need to be accounted for in the counting data structure, as confirmed by counting the number of *distinct* k-mers that are inserted in the counting data structure by the two algorithms (Fig. 2(c)). (The difference between the memory requirement and the number of distinct k-mers is given by Jellyfish's strategy to doubles the size of the counting data structure when it is full.)

In terms of quality of the approximation, the output of SAKEIMA satisfied the guarantees given by Proposition 5 for all runs of our experiments, therefore with probability higher than $1 - \delta$. While SAKEIMA may incur in false negatives, its false negative ratio (i.e., the fraction of k-mers in $FK(\mathcal{D}, k, \theta)$ not reported by SAKEIMA) is always $\leq 3 \cdot 10^{-4}$ (Fig. 3(a)), even if the sampling technique of Sect. 3.1 does not provide rigorous guarantees on such quantity. Therefore SAKEIMA is very effective in reporting almost all frequent k-mers. As mentioned in Sect. 3.3, SAKEIMA can be easily modified so to report all frequent k-mers in output, even if at the cost of reporting also more k-mers with frequency between $\theta - \varepsilon$ and θ. In addition, the estimated frequencies f_A reported by SAKEIMA

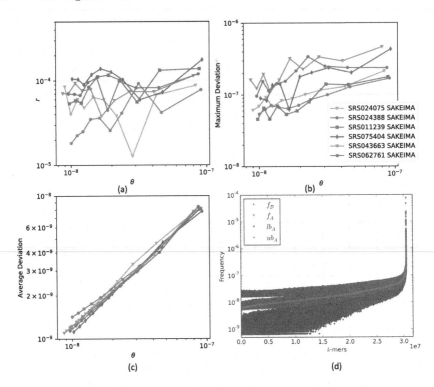

Fig. 3. Quality of the approximation of $FK(\mathcal{D}, k, \theta)$ produced by SAKEIMA. (a) False negative rate, i.e., the fraction r of k-mers in $FK(\mathcal{D}, k, \theta)$ not reported by SAKEIMA. (b) Maximum deviation $|f_A - f_{\mathcal{D}}(A)|$ of the estimates reported by SAKEIMA for various θ. (c) Average value of $|f_A - f_{\mathcal{D}}(A)|$ for the k-mers A reported by SAKEIMA for various θ. (d) Frequencies and bounds for dataset SRS062761 and $\theta = 10^{-8}$ shown for k-mers sorted in increasing order of exact frequencies. Red: exact frequencies $f_{\mathcal{D}}(A)$. Green: estimate f_A of $f_{\mathcal{D}}(A)$ from SAKEIMA. Blue: lower bound lb_A to $f_{\mathcal{D}}(A)$ from SAKEIMA. Brown: upper bound ub_A to $f_{\mathcal{D}}(A)$ from SAKEIMA. (Color figure online)

are always close to the true values $f_{\mathcal{D}}(A)$, with a small maximum deviation $|f_A - f_{\mathcal{D}}(A)|$ (Fig. 3(b)), and an even smaller average deviation (Fig. 3(c)). In addition, the upper and lower bounds computed as in Sect. 3.3 provide small confidence intervals always containing the value $f_{\mathcal{D}}(A)$ (e.g., Fig. 3(d) for dataset SRS062761), and could be used to obtain sets of k-mers with various guarantees from the sample used by SAKEIMA.

4.3   Application to Metagenomics: Computation of Ecological Distances

We evaluate the use of SAKEIMA to speed up the computation of commonly used k-mer based ecological distances [1] between datasets of Next-Generation Sequencing (NGS) reads. We present results for the Bray-Curtis distance;

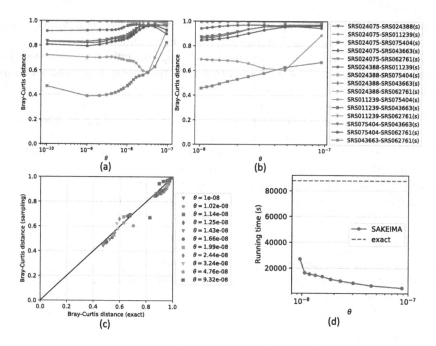

Fig. 4. Results for Bray-Curtis (BC) distances of metagenomic datasets. (a) BC distance computed using k-mers with frequency $\geq \theta$. (b) BC distances computed using the approximation of k-mers with frequency $\geq \theta$ from SAKEIMA. (c) Comparison of the BC distance using all k-mers with exact counts and the approximation of frequent k-mers by SAKEIMA. (d) Total time required by SAKEIMA and the exact approach to find frequent k-mers and compute all distances between datasets as a function of θ.

analogous results hold for other distances and will be presented in the full version of this extended abstract.

We first investigated how the distances change when those are computed considering only the *frequent* k-mers (w.r.t. a frequency threshold θ) instead that the full spectrum of k-mers appearing in the data. Therefore, given a pair of datasets $\mathcal{D}_1$ and $\mathcal{D}_2$ and θ, we computed the sets $\mathcal{O}_1 = FK(\mathcal{D}_1, k, \theta)$ and $\mathcal{O}_2 = FK(\mathcal{D}_2, k, \theta)$ using Jellyfish and then computed a generalized version of the distances for all pairs of datasets we used for our experiments. For the Bray-Curtis distance, this generalization is defined as: $BC(\mathcal{D}_1, \mathcal{D}_2, \mathcal{O}_1, \mathcal{O}_2)$
$= 1 - 2 \frac{\sum_{A \in \mathcal{O}_1 \cap \mathcal{O}_2} \min\{o_{\mathcal{D}_1}(A), o_{\mathcal{D}_2}(A)\}}{\sum_{A \in \mathcal{O}_1} o_{\mathcal{D}_1}(A) + \sum_{A \in \mathcal{O}_2} o_{\mathcal{D}_2}(A)}.$

Note that when $\theta \leq 10^{-10}$ then $FK(\mathcal{D}, k, \theta)$ coincides with the set of *all* k-mers, for any of the datasets we tested. The results (Fig. 4(a)) show that for θ up to 5×10^{-8} the values of the distances are fairly stable and therefore one can use only frequent k-mers for such values of θ to compute the distances, and that for θ up to 10^{-7} the relation between distances of different pairs of datasets are almost always conserved. We underline that the exact counting approach needs to count *all* the k-mers and only afterwards can filter the infrequent ones before writing them to disk to compute $FK(\mathcal{D}, k, \theta)$. We then used SAKEIMA to extract approximations (of k-mers and their frequencies) of $FK(\mathcal{D}_1, k, \theta)$ and $FK(\mathcal{D}_2, k, \theta)$ and used such approximations to compute the distances among datasets (Fig. 4(b)). Strikingly, the distances computed from the output of SAKEIMA are very close to their exact variant (Fig. 4(c)). Interestingly this holds also for the Jaccard distance, a presence-based distance that does not depend neither on k-mer abundances nor on k-mer ranking by frequencies (detailed results will be provided in the full version of this extended abstract).

We then compared, for different values of θ, the total running time required to compute the approximations of the frequent k-mers using SAKEIMA for all datasets in Table 1 and all distances among such datasets using SAKEIMA approximations with the running time required when the exact counting algorithm is used for the same tasks. SAKEIMA reduces the computing time by more than 75% (Fig. 4(d)). This result comes from both the efficiency of SAKEIMA and from the fact that by focusing on the most frequent k-mers we greatly reduce the number of distinct k-mers that need to be processed for computing the distances. Therefore SAKEIMA can be used for a very fast comparison of metagenomic datasets while preserving the ability of distinguishing similar datasets from different ones.

5   Conclusion

We presented SAKEIMA, a sampling-based algorithm to approximate frequent k-mers and their frequencies with rigorous guarantees on the quality of the approximation. We show that SAKEIMA can be used to speed up the analysis of large high-throughput sequencing metagenomic datasets, in particular to compute abundance-based distances among such datasets. Interestingly SAKEIMA allows to compute accurate approximations also for presence-based distances (e.g., the Jaccard distance), even if for such distances other, potentially faster, tools [18] are available. SAKEIMA can be combined with any highly optimized method that counts all k-mers in a set of strings, including recent parallel methods designed for comparative metagenomics [1]. While we presented results for k-mers from datasets of short reads, SAKEIMA can also be used for the analysis of spaced seeds [2], large datasets of long reads, and whole genome sequences.

References

1. Benoit, G., Peterlongo, P., et al.: Multiple comparative metagenomics using multiset k-mer counting. PeerJ Comput. Sci. **2**, e94 (2016)
2. Břinda, K., Sykulski, M., Kucherov, G.: Spaced seeds improve k-mer-based metagenomic classification. Bioinformatics **31**(22), 3584–3592 (2015)
3. Brown, C.T., Howe, A., et al.: A reference-free algorithm for computational normalization of shotgun sequencing data. arXiv preprint arXiv:1203.4802 (2012)
4. Chikhi, R., Medvedev, P.: Informed and automated k-mer size selection for genome assembly. Bioinformatics **30**(1), 31–37 (2013)
5. Danovaro, R., Canals, M., et al.: A submarine volcanic eruption leads to a novel microbial habitat. Nat. Ecol. Evol. **1**(6), 0144 (2017)

6. Dickson, L.B., Jiolle, D., et al.: Carryover effects of larval exposure to different environmental bacteria drive adult trait variation in a mosquito vector. Sci. Adv. **3**(8), e1700585 (2017)
7. Girotto, S., Pizzi, C., Comin, M.: MetaProb: accurate metagenomic reads binning based on probabilistic sequence signatures. Bioinformatics **32**(17), i567–i575 (2016)
8. Hrytsenko, Y., Daniels, N.M., Schwartz, R.S.: Efficient distance calculations between genomes using mathematical approximation. In: Proceedings of the ACM-BCB, p. 546 (2018)
9. Kelley, D.R., Schatz, M.C., Salzberg, S.L.: Quake: quality-aware detection and correction of sequencing errors. Genome Biol. **11**(11), R116 (2010)
10. Kokot, M., Długosz, M., Deorowicz, S.: KMC 3: counting and manipulating k-mer statistics. Bioinformatics **33**(17), 2759–2761 (2017)
11. Li, X., Waterman, M.S.: Estimating the repeat structure and length of DNA sequences using ℓ-tuples. Genome Res. **13**(8), 1916–1922 (2003)
12. Löffler, M., Phillips, J.M.: Shape fitting on point sets with probability distributions. In: Fiat, A., Sanders, P. (eds.) ESA 2009. LNCS, vol. 5757, pp. 313–324. Springer, Heidelberg (2009). https://doi.org/10.1007/978-3-642-04128-0_29
13. Marçais, G., Kingsford, C.: A fast, lock-free approach for efficient parallel counting of occurrences of k-mers. Bioinformatics **27**(6), 764–770 (2011)
14. Melsted, P., Halldórsson, B.V.: KmerStream: streaming algorithms for k-mer abundance estimation. Bioinformatics **30**(24), 3541–3547 (2014)
15. Melsted, P., Pritchard, J.K.: Efficient counting of k-mers in DNA sequences using a Bloom filter. BMC Bioinform. **12**(1), 333 (2011)
16. Mitzenmacher, M., Upfal, E.: Probability and Computing: Randomization and Probabilistic Techniques in Algorithms and Data Analysis. Cambridge University Press, Cambridge (2017)
17. Mohamadi, H., Khan, H., Birol, I.: ntCard: a streaming algorithm for cardinality estimation in genomics data. Bioinformatics **33**(9), 1324–1330 (2017)
18. Ondov, B.D., Treangen, T.J., et al.: Mash: fast genome and metagenome distance estimation using MinHash. Genome Biol. **17**(1), 132 (2016)
19. Pandey, P., Bender, M.A., Johnson, R., Patro, R.: Squeakr: an exact and approximate k-mer counting system. Bioinformatics **34**(14), 568–575 (2017)
20. Patro, R., Mount, S.M., Kingsford, C.: Sailfish enables alignment-free isoform quantification from RNA-seq reads using lightweight algorithms. Nat. Biotechnol. **32**(5), 462 (2014)
21. Pevzner, P.A., Tang, H., Waterman, M.S.: An Eulerian path approach to DNA fragment assembly. Proc. National Acad. Sci. **98**(17), 9748–9753 (2001)
22. Rizk, G., Lavenier, D., Chikhi, R.: DSK: k-mer counting with very low memory usage. Bioinformatics **29**(5), 652–653 (2013)
23. Roy, R.S., Bhattacharya, D., Schliep, A.: Turtle: Identifying frequent k-mers with cache-efficient algorithms. Bioinformatics **30**(14), 1950–1957 (2014)
24. Salmela, L., Walve, R., Rivals, E., Ukkonen, E.: Accurate self-correction of errors in long reads using de Bruijn graphs. Bioinformatics **33**(6), 799–806 (2016)
25. Sims, G.E., Jun, S.-R., Wu, G.A., Kim, S.-H.: Alignment-free genome comparison with feature frequency profiles (FFP) and optimal resolutions. Proc. National Acad. Sci. **106**(8), 2677–2682 (2009)
26. Sivadasan, N., Srinivasan, R., Goyal, K.: Kmerlight: fast and accurate k-mer abundance estimation. arXiv preprint arXiv:1609.05626 (2016)
27. Solomon, B., Kingsford, C.: Fast search of thousands of short-read sequencing experiments. Nat. Biotechnol. **34**(3), 300 (2016)

28. Vapnik, V.: Statistical Learning Theory. Wiley, New York (1998)
29. Vapnik, V., Chervonenkis, A.: On the uniform convergence of relative frequencies of events to their probabilities. Theory Prob. Appl. **16**(2), 264 (1971)
30. Wood, D.E., Salzberg, S.L.: Kraken: ultrafast metagenomic sequence classification using exact alignments. Genome Biol. **15**(3), R46 (2014)
31. Zerbino, D.R., Birney, E.: Velvet: algorithms for de novo short read assembly using de Bruijn graphs. Genome Res. **18**(5), 821–829 (2008)
32. Zhang, Q., Pell, J., et al.: These are not the k-mers you are looking for: efficient online k-mer counting using a probabilistic data structure. PloS One **9**(7), e101271 (2014)
33. Zhang, Z., Wang, W.: RNA-Skim: a rapid method for RNA-Seq quantification at transcript level. Bioinformatics **30**(12), i283–i292 (2014)

De Novo Clustering of Long-Read Transcriptome Data Using a Greedy, Quality-Value Based Algorithm

Kristoffer Sahlin[1(✉)] and Paul Medvedev[1,2,3]

[1] Department of Computer Science and Engineering, Pennsylvania State University,
State College, USA
kxs624@psu.edu
[2] Department of Biochemistry and Molecular Biology, Pennsylvania State University,
State College, USA
[3] Center for Computational Biology and Bioinformatics,
Pennsylvania State University, State College, USA

Abstract. Long-read sequencing of transcripts with PacBio Iso-Seq and Oxford Nanopore Technologies has proven to be central to the study of complex isoform landscapes in many organisms. However, current *de novo* transcript reconstruction algorithms from long-read data are limited, leaving the potential of these technologies unfulfilled. A common bottleneck is the dearth of scalable and accurate algorithms for clustering long reads according to their gene family of origin. To address this challenge, we develop isONclust, a clustering algorithm that is greedy (in order to scale) and makes use of quality values (in order to handle variable error rates). We test isONclust on three simulated and five biological datasets, across a breadth of organisms, technologies, and read depths. Our results demonstrate that isONclust is a substantial improvement over previous approaches, both in terms of overall accuracy and/or scalability to large datasets. Our tool is available at https://github.com/ksahlin/isONclust.

1 Introduction

Long-read sequencing of transcripts with Pacific Biosciences (PacBio) Iso-Seq and Oxford Nanopore Technologies (ONT) has proven to be central to the study of complex isoform landscapes in, *e.g.*, humans [1–4], animals [5], plants [6], fungi [7] and viruses [8]. Long reads can reconstruct more complex regions than can short RNA-seq reads because the often complex assembly step is not required. However, they suffer from higher error rates, which present different challenges. Using a reference genome can help alleviate these challenges, but, for non-model organisms or for complex gene families, *de novo* transcript reconstruction methods are required [2,9].

For non-targeted Iso-Seq data, the commonly used tool for *de novo* transcript reconstruction is the ToFU pipeline from PacBio [7]. However, ToFU generates

© Springer Nature Switzerland AG 2019
L. J. Cowen (Ed.): RECOMB 2019, LNBI 11467, pp. 227–242, 2019.
https://doi.org/10.1007/978-3-030-17083-7_14

a large number of redundant transcripts [7,10,11], and most studies have had to resort to additional short-read sequencing [11,12] or relying on a draft reference [5]. For ONT data, there are, to the best of our knowledge, no methods yet for *de novo* transcript reconstruction. Therefore, we believe that the full potential of long-read technologies for *de novo* transcript reconstruction from non-targeted data has yet to be fully realized.

Algorithms for this problem are roughly composed of two steps [2,7]. Since most transcripts are captured in their entirety by some read, there is no need to detect dovetail overlaps between reads (i.e. a suffix of one read matching the prefix of another), and to perform subsequent graph construction and traversal (as in RNA-seq assembly). Instead, the first step is to group reads together into clusters according to their origin, and the second is to error-correct the reads using the information within each cluster. This is the approach taken by ToFU, but it clusters reads according to their isoform (rather than gene) of origin. This separates reads that share exons into different clusters – reducing the effective coverage for downstream error correction. For genes with multiple isoforms, this significantly fragments the clustering and, we suspect, causes many of the redundant transcripts that have been reported. For ONT data, there exists a tool to cluster reads into their gene family of origin (CARNAC-LR [9]), but it performed sub-optimally in our experiments and scales poorly for larger datasets. Thus, better clustering methods are needed to realize the full potential of long reads in this setting.

There exists a plethora of algorithms for *de novo* clustering of generic nucleotide-[13–16], and protein-sequences [14,17–19]. Several algorithms have also been proposed for clustering of specific nucleotide data such as barcode sequences [20], EST sequences [21–23], full-length cDNA [24], RAD-seq [25], genomic or metagenomic short reads [26–31], UMI-tagged reads [32], full genomes and metagenomes [33], and contigs from RNA-seq assemblies [34,35]. However, our clustering problem has unique distinguishing characteristics: transcripts from the same gene have large indels due to alternative splicing, and the error rate and profile differs both between [2] and within [36] reads. Furthermore, the large number of reads limits the scalability of algorithms that require an all-to-all similarity computation. *De novo* clustering of long-read transcript sequences has to our knowledge only been studied in [2,7,37] for Iso-Seq data and in [9] for ONT data. However, neither IsoCon [2] nor Cogent [37] scale to large datasets, and the limitations of ToFu [7] and CARNAC-LR [9] have already been described above. In [30], the authors demonstrated that using quality values can significantly improve clustering accuracy, especially for higher error rates, but their method was not designed for long reads.

Motivated by the shortcomings of the existing tools, we develop isONclust, an algorithm for clustering long reads according to their gene family of origin. isONclust is available at https://github.com/ksahlin/isONclust. isONclust is greedy (in order to scale) and makes use of quality values (in order to handle variable error rates). We test isONclust on three simulated and five biological datasets, across a breadth of organisms, technologies, and read depths. Our results demonstrate that ISONCLUST is a substantial improvement over previous approaches, both in terms of overall accuracy and/or scalability to large datasets. ISONCLUST opens the door to the development of more scalable and more accurate methods for *de novo* transcript reconstruction from long-read datasets.

2   Methods

2.1   Definitions

Let r be a string of nucleotides that we refer to as a read. Let $q(r_i)$ be the probability of base call error at position $1 \leq i \leq |r|$. This value can be derived from the Phred quality value Q at position i as $q(r_i) = 10^{-(Q/10)}$. Let ϵ_r be the average base error rate, $\epsilon_r = \sum_{i=1}^{|r|} q(r_i)/|r|$. Given two integers w and k such that $1 \leq k \leq w \leq |r|$, the *minimizer at position i* is the lexicographically smallest substring of r of length k that starts at a position in the interval of $[i, i+w)$ [38]. Let $M(r)$ be the set of ordered pairs containing all the minimizers of r and their start positions on the read. For example, for $r = ACGCCGATC, k = 2, w = 4$, we have $M(r) = \{(AC, 0), (CC, 3), (AT, 6)\}$. All the strings of $M(r)$ are referred to as the *minimizers* of r.

2.2   ISONCLUST **Overview**

ISONCLUST is a greedy clustering algorithm. Initially, we sort the reads so that sequences that are longer and have higher quality scores appear earlier (details in Sect. 2.3). We then process the reads one by one, in this sorted order. At any point, we maintain a clustering of the reads processed so far, and, for each cluster, we maintain one of the reads as the *representative* of the cluster. We also maintain a hash-table H such that for any k-mer x, $H(x)$ returns all representatives that have x as a minimizer.

At each point that a new read is processed, ISONCLUST consists of three steps. In the first step, we find the number of minimizers shared with each of the current representatives, by querying H for each minimizer in the read and maintaining an array of representative counts. We refer to any representative that shares at least one minimizer with the read as a *candidate*. In the second step, we use a minimizer-based method to try to assign a read to one of the candidate representative's cluster. To do this, we process the candidate representatives in the order of most shared minimizers to the least. For each representative, we estimate the fraction of the read's sequence that is shared with it (details in Sect. 2.3). If the fraction is above 0.7, then we assign the read to that representative's cluster; if not, we proceed to the next candidate. However, if the number of shared minimizers with the current representative drops below 70% of the top-sharing representative, or below an absolute value of 5, we terminate the search and proceed to the third step.

In the third step, we fall back on a slower but exact Smith-Waterman alignment algorithm. If two transcripts have highly variable exon structure or contain many mutations (e.g., multi-copy genes or highly-mutated allele), then it could create long regions of no shared minimizers and prevent the minimizer matching approach from detecting the similarity. An alignment approach is more sensitive and can detect that they should be clustered together. To control the run-time, we align the read only to the representative with the most shared minimizers (several in the case of a tie). Similar to the second step, we estimate the fraction of the read's sequence that is shared with the representative (details in Sect. 2.3), and if the quality is above a threshold (details in Sect. 2.3), the read is assigned to that representative's cluster. Otherwise, the read is assigned to a new cluster and made its representative.

2.3   ISONCLUST In-Depth

Homopolymer Compression: An important aspect of ISONCLUST is that reads are homopolymer compressed, i.e. all consecutive appearances of a base are replaced by a single occurrence. For example, the sequence GCCTGGG is replaced by GCTG. When a homopolymer is compressed, the base quality assigned to the compressed base is taken as the highest quality of the bases in the original homopolymer region. The reason for using the highest quality value is that it is a lower bound on the presence of at least one nucleotide in that run. The homopolymer compression removes a large amount of homopolymer indel errors during minimizer matching or alignment and at the same time removes repetitive minimizers (from, e.g., poly-A tails). We borrowed this idea from [39,40], where it was used to improve the sensitivity of PacBio read alignment.

Sorting Order: Prior to greedily traversing the reads, we sort them in decreasing order of their score. We define the score $s(r)$ of a read r as the expected number of error-free k-mers in r. Let X_i be a binary indicator variable modelling the event that the k-mer starting at position i of r has no sequencing error ($X_i = 1$). Then, we have

$$s(r) = E\left[\sum_{i=1}^{|r|-k+1} X_i\right] = \sum_{i=1}^{|r|-k+1} E[X_i] = \sum_{i=1}^{|r|-k+1} \prod_{j=0}^{k-1}(1 - q(r_{i+j}))$$

The score of a read can be quickly computed in a linear scan by maintaining a running product over a sliding window of k quality scores.

The ordering produced by this score function is crucial for the accuracy of our greedy approach. Observe that our algorithm never re-computes which read in a cluster is the representative, and all future reads are compared only to a cluster's representative and not to other reads in the cluster. This is done for the sake of efficiency, but, as a downside, once a read initiates a new cluster, it becomes its representative forever. However, our score function guarantees that it will have the largest expected number of error-free k-mers of any future read in the cluster. In the case of alternatively spliced genes, this means that the representative likely contains the most complete exon repertoire of the gene. This allows us to make the assumption that all exon differences during minimizer matching or alignment are encountered as deletions with respect to the representative. In both the matching and alignment parts, we will therefore not penalize for long deletions in the read. We do penalize for insertions in reads with respect to representatives because we assume that they cannot be due to exon differences. We note that in the cases that our assumption does not hold (e.g. when several exons are not present in the longest isoform), we may miss some matches and/or alignments. However, we do tolerate some fraction of unmatched sequence in later steps.

Estimating Fraction of Shared Sequence, Based on a Minimizer Match: Consider a read r, a representative c, the set $M(r)$, and the minimizers of r that are shared with c. We would like to quickly estimate the fraction f of r's sequence that would align to c, if an alignment were to have been performed. When two consecutive minimizers in $M(r)$ match c, we simply count the sequence spanned between their positions towards f. For the harder case, consider a sequence of i consecutive unmatched minimizers in r that are flanked on both sides by either a matched minimizer or the end of the read. We must decide if this is due to the region being unalignable or due to true sequencing errors. Let $p(\epsilon_r, \epsilon_c)$ be the probability that a minimizer in a read is not matched to another read, given that they are both generated from the same transcript with respective error rates ϵ_r and ϵ_c. Then, the probability that i consecutive minimizers of r are unmatched as a result of sequencing error can be estimated as $p(\epsilon_r, \epsilon_c)^i$. If this probability is above 0.1, we count the whole region towards f, otherwise we do not.

Estimating Minimizer Mismatch Rate: Deriving an analytical formula for $p(\epsilon_r, \epsilon_c)$ is a challenge, as the probability of observing a spurious minimizer in a window is a complex function depending on, e.g., the sequence of the true minimizer in the window, the sequence in the window, the error profile, and the properties of homopolymer compression. Instead, we use simulations to create a lookup table for p. We randomly generate a transcript of length 1 kbp and, from that transcript, two reads r and c with error rates ϵ_r and ϵ_c, respectively. The errors are equally distributed between insertion and deletion errors since Iso-Seq and ONT errors are dominated by indels. Further customization of the error profile to more accurately reflect the technology is possible, but we found that it had little effect. We then homopolymer compress the reads and count the fraction of r's minimizers that do not match c. We repeat the process 1,000 times, each time starting with a new transcript. The average fraction of non-matching minimizers is used as the estimate for $p(\epsilon_r, \epsilon_c)$. We pre-computed the lookup table for a range of ϵ_r and ϵ_c values that we observe in practice, but it can also be computed on the fly for datasets outside of these ranges.

Estimating Fraction of Shared Sequence, Based on the Alignment:
When the minimizer matching approach fails to find a match, ISONCLUST aligns
the read r to the most promising representative c using the parasail [41] semi-
global implementation of Smith-Waterman (start or end insertions in either
sequence are not penalized). Let $\epsilon = \epsilon_r + \epsilon_c$ be the combined error rate of r
and c. The parameters to Smith-Waterman are described in the experimental
appendix [42] and are a function of ϵ. Based on this alignment, we would like
to estimate the fraction of r whose alignment to c is consistent with having
the same underlying sequence but allowing sequencing errors (i.e. the same goal
as we had during minimizer matching). We aim to tolerate a mismatch rate of
ϵ. Consider the pairwise alignment A, represented as a matrix where the two
rows correspond to r and c, and each cell contains a symbol indicating either
a match, mismatch, or a gap. Consider a window of k columns in A starting
at position i of r. Let $W_i = 1$ if the number of columns in the window that
are not matches is $\leq \lceil \epsilon k \rceil$; otherwise, let $W_i = 0$. We let the shared fraction
$f = (\sum_{i=1}^{|r|-k+1} W_i)/|r - k + 1|$ and add r to the cluster of c if f is above 0.4.

Time Complexity: Our tool is a greedy heuristic, hence it is challenging
to derive a worst-case run-time that is informative. We attempt to do so by
parametrizing our analysis and fixing the number of representatives identified
as candidates for a read as d. The initial sorting step takes $\mathcal{O}(n \log n)$ time.
Then for each read, the identification of minimizers takes $\mathcal{O}(\ell)$ time, where ℓ
is the read length. Here, we treat w and k as constants. There are at most ℓ
minimizers, and each one hits at most d representatives, hence identifying candidate representatives takes $\mathcal{O}(\ell d)$ time. Ranking the candidate representatives
can be done using counting sort in $\mathcal{O}(d)$ time. For minimizer matching, each
of the at most d candidates can be processed using a linear scan through the
read, leading to a total of $\mathcal{O}(\ell d)$ time. The alignment step is done only once and
is dominated by the $\mathcal{O}(\ell^2)$ Smith-Waterman time. Hence, the total run-time is
$\mathcal{O}(n \log n + n\ell d + n\ell^2)$. In the worst-case, d can be $\Omega(n)$, but it is much less in
practice.

Parameters and Thresholds: The only parameters to ISONCLUST are the
window size w and the k-mer size k. We found through trial-and-error that
$k = 15$ and $w = 50$ work well for Iso-Seq data, and $k = 13$ and $w = 20$ work well
for ONT data. Note that these lengths are applied for homopolymer compressed
reads, thus a 13-mer is likely to be much longer in the original read. There are
also several other hard thresholds used by ISONCLUST, as described above. We
set these through a mix of intuition and testing on simulated data; nevertheless,
we found that ISONCLUST is robust to these thresholds. In particular, we did
not vary them for any of our experiments, which included a diverse collection of
real datasets. We therefore do not recommend users to change these thresholds.

3 Results

3.1 Experimental Setup

Datasets: We used eight datasets, in order to test the robustness of ISONCLUST with respect to different technologies, organisms, and read depths (Table 1). We first simulated three Iso-Seq read datasets from 107,844 unique cDNA sequences from ENSEMBL using SiMLOrD [43]. The datasets contained 100,000, 500,000, and 1,000,000 reads that were simulated with uniform distribution over the cDNA fragments. Next, we included a semi-biological Iso-Seq dataset (denoted RC0) where the transcripts are synthetically produced but then sequenced with Iso-Seq using the PacBio Sequel system. Then, we added three fully biological Iso-Seq datasets: PacBio Sequel datasets from a zebrafinch (ZEB) and a hummingbird (HUM), and a PacBio RSII system dataset from human brain tissue (ALZ). Finally, we included a ONT dataset of human cDNA sequenced with a MinION, which exhibits a different error profile and higher error rates than Iso-Seq. The non-simulated Iso-Seq and ONT datasets are publically available at [44] and [45], respectively.

Tools: The authors of CARNAC-LR [9] observed the inability of most clustering tools designed for other purposes [14,20,25,27] to run on long-read transcriptomic data. However, we did consider four additional such tools: qCluster [30], LINCLUST [18], DNACLUST [16], and MeShClust [15]. We also considered four tools specifically designed for long-read transcriptome data (CARNAC-LR, Iso-Con, isoseq3-cluster, and Cogent). Isoseq3-cluster (which we will refer to simply as ISOSEQ3) is the clustering tool used in the most recent version of PacBio's *de novo* transcript reconstruction pipeline. Out of these eight tools, we found that only three (CARNAC-LR, ISOSEQ3, and LINCLUST) could process our two smallest datasets (SIM-100k and RC0). We therefore only include these tools in our final evaluations. Command lines and parameter settings for the nine tools we tried are described in the experimental appendix [42].

Ground Truth: Since the true clustering is not known, we use a clustering based on alignments to the reference genome as a proxy. We first align the reads with minimap2 [40] to the reference genome (hg38 for human, Tgut_diploid_1.0 for zebrafinch [46], and Canna_diploid_1.0 for hummingbird [46]), with different parameters for Iso-Seq and ONT data (for details, see [42]). The aligned reads are then clustered greedily by merging the clusters of any two reads whose alignments overlap. We refer to the cluster of a read obtained via this alignment to the reference as the *class* of the read. Reads that could not be aligned and hence could not be assigned to a class were excluded from all downstream accuracy evaluations. Some class metrics for the datasets are shown in Table 1.

Using alignments to the reference to define classes is an imperfect proxy of the true clustering. There are likely systemic misalignments due to gene sequence content, artifacts of the aligner, or chimeric reads due to *e.g.* reverse transcription errors. Thus our approach does not yield a reliable estimate for the absolute performance of a tool, but we believe it is a reasonable proxy to access the relative performance between different tools.

Table 1. Datasets used for evaluation. The error rate of a read is estimated by summing the probability of a base call error (obtained from the quality value) over all bases in a read, divided by the read length. The error rate is estimated on original reads (without homopolymer compression). The average error rate per dataset is computed by averaging the estimated error rate over all reads in the dataset. A singleton class (S) refers to a class that contains only one read and a non-singleton class (NS) refers to a class with more than one read. †many of these originated from the synthetic spike-in non-human transcripts.

Dataset	Avg. error rate (%)	n. classes		n. reads		% reads in NS classes
		NS	S	Total	Unaligned	
ALZ	1.7	13,350	10,187	814,667	98	98.7
RC0	1.2	11,052	9,119	185,790	11,423†	88.9
HUM	1.8	13,683	4,450	288,699	3,882	97.1
ZEB	1.9	12,891	4,936	309,749	129	98.4
SIM-100k	1.9	9,106	3,351	100,000	4	96.6
SIM-500k	1.9	14,792	2,152	500,000	4	99.6
SIM-1000k	1.9	16,510	1,594	1,000,000	4	99.8
ONT	12.9	14,863	13,665	890,503	38,061	94.2

Evaluation Metrics: There exists several metrics to measure quality of clustering. We mainly use the V-measure and its two components completeness and homogeneity [47]. Let X be an array of n integers, where n is the number of reads and the ith value is the cluster id given by a clustering algorithm. Similarly, let Y be an array with the assigned ground truth class ids of the reads, ordered as in X. *Homogeneity* is defined as $h = 1 - H(Y|X)/H(Y)$ and *completeness* as $c = 1 - H(X|Y)/H(X)$. Here, $H(*)$ and $H(*|*)$ refer to the entropy and conditional entropy functions, respectively [47]. Intuitively, homogeneity penalizes over-clustering, i.e. wrongly clustering together reads, while completeness penalizes under-clustering, i.e. mistakenly keeping reads in different clusters. The *V-measure* is then defined as the harmonic mean of homogeneity and completeness. These are analogous to precision, recall, and F-score measures for binary classification problems. We chose the V-measure metric as it is independent of the number of classes, the number of clusters, and the size of the dataset—and can therefore be compared across different tools [47]. Moreover, it can be decomposed in terms of homogeneity and completeness for a better understanding of the algorithm behavior.

In order to avoid bias with respect to a single accuracy measure, we also included the commonly used adjusted Rand index (ARI) [48]. Intuitively, ARI measures the percentage of read pairs correctly clustered, normalized so that a perfect clustering achieves an ARI of 1 and a random cluster assignment achieves an ARI of 0. The formal definition is more involved [48] and, since it is standard, we omit it here for brevity.

In addition, we measure the percent of reads that are in non-singleton clusters. Since the coverage per gene is sufficiently high in all our datasets, the percentage of reads that are in non-singleton classes is high (89–100%, Table 1). Thus, any reads in singleton clusters in excess of this amount are indicative of reads that likely could have been clustered by the algorithm, but did not. Finally, we measure the runtime (Table 2) and memory usage (Table 3) of all the experiments.

Table 2. Run-time for the clustering algorithms. isONclust was run on 1 core. The other tools were run with 8 cores specified. Runtime for CARNAC-LR includes mapping time with minimap. ([†]the run was terminated after 10 days.)

Dataset	Run-time (minutes)			
	isONclust	isoseq3	carnac-lr	linclust
ALZ	173	194	>14,400[†]	**132**
RC0	40	11	**7**	8
HUM	105	53	105	**33**
ZEB	130	58	689	**35**
SIM-100k	26	5	4	**4**
SIM-500k	111	58	187	**28**
SIM-1000k	185	223	1,271	**67**
ONT	1,630	N/A	5,053	**39**

3.2 Comparison Against Other Tools

The most direct comparison of our tool is to CARNAC-LR, which solves the same problem we do. One of its stated limitations is a worst-case cubic runtime [9], and we indeed observe that it does not scale well with growing sizes of our datasets (Table 2). For the largest Iso-Seq dataset (ALZ, 814k reads), CARNAC-LR did not complete within 10 days. For the other two large datasets (SIM-1000k and ONT), CARNAC-LR was >6x and >3x slower than isONclust, respectively. In terms of accuracy, CARNAC-LR performed reasonably well but always had a lower V-measure and ARI than isONclust. CARNAC-LR also placed less reads in non-singleton clusters than isONclust. For the ONT data, in particular, it was only able to place 54% of the reads into non-singleton clusters (compared to 94.5% for isONclust), even though 94.2% of the reads were in non-singleton classes (Table 1).

Table 3. Peak memory usage for the clustering algorithms. ISONCLUST was run on 1 core. The other tools were run with 8 cores specified.

Dataset	Memory (Gb)			
	ISONCLUST	ISOSEQ3	CARNAC-LR	LINCLUST
ALZ	**1.9**	5.3	N/A	9.8
RC0	**0.4**	1	0.9	1.6
HUM	**0.8**	2.8	6.2	4.6
ZEB	**0.9**	3	3.4	5
SIM-100k	**0.3**	0.6	0.5	0.9
SIM-500k	**0.8**	2.7	2.3	4.4
SIM-1000k	**1.8**	5.1	5.9	8.9
ONT	**1.6**	N/A	3.9	2.9

ISOSEQ3 solves a slightly different problem than ISONCLUST: its objective is to cluster reads together from the same isoform of a gene, rather than from the same gene family (*i.e.* in the case of alternative splicing, it will have separate clusters for each isoform). Thus, completeness, V-measure, and ARI with respect to our ground truth are not fair metrics by which to evaluate ISOSEQ3. Nevertheless, ISOSEQ3 leaves many reads unclustered: 26–36% of the reads from the real datasets and 53–89% of the reads from the simulated datasets (Table 4). In some cases, this could be caused by low coverage per isoform; however, SIM-1000k contains on average 9 reads per isoform, which should enable an algorithm to cluster substantially more than 53% of the reads. In terms of homogeneity, ISOSEQ3 slightly outperforms ISONCLUST, indicating that ISOSEQ3 is the right tool if the goal is a conservative clustering. ISOSEQ3 is designed for only Iso-Seq data and is thus not run on the ONT dataset.

Finally, we compare against LINCLUST, which has a generic objective to cluster any sequences above a given sequence similarity and coverage. We explored several combinations of parameters to achieve the best results (more details in [42]). While LINCLUST was the fastest tool, it has substantially worse accuracy on Iso-Seq data than other tools and was able to cluster only 0.1% of the ONT reads. This is not surprising, given that it was not designed for transcriptome data.

3.3   Performance Observations

Scalability: For Iso-Seq, we can use the simulated data, which only varies in read depth, to conclude that ISONCLUST has linear scaling with respect to the number of reads (Table 2). The absolute run-time is 3.1 h for the largest Iso-Seq dataset, which is acceptable but could be further improved through parallelization or code optimization. For ONT data, the dearth of mature transcriptomic read simulators makes a controlled evaluation of scalability challenging. Though

Table 4. Performance and accuracy of the tools on our datasets. %NS is the percentage of reads in non-singleton clusters. The number of clusters is split between NS (non-singleton clusters) and S (singleton clusters).

Dataset	Tool	Accuracy				%NS reads	n. clusters	
		V	c	h	ARI		NS	S
ALZ	ISONCLUST	**0.944**	**0.899**	0.993	**0.630**	**96.1**	23265	32169
	ISOSEQ3	0.813	0.686	**0.998**	0.423	73.9	63512	212246
	LINCLUST	0.839	0.725	0.996	0.518	80.8	57942	156476
RC0	ISONCLUST	**0.977**	**0.961**	0.994	**0.804**	**90.1**	12513	18459
	ISOSEQ3	0.923	0.859	**0.997**	0.640	66.6	14025	62085
	CARNAC-LR	0.94	0.904	0.98	0.346	82.4	11002	32778
	LINCLUST	0.933	0.877	0.996	0.566	77.7	18116	41363
HUM	ISONCLUST	**0.958**	**0.971**	0.947	**0.716**	**97.3**	12140	7773
	ISOSEQ3	0.88	0.805	**0.97**	0.486	67.2	24171	94558
	CARNAC-LR	0.934	0.944	0.924	0.489	93.3	9565	19323
	LINCLUST	0.888	0.825	0.962	0.462	78.9	28066	61046
ZEB	ISONCLUST	**0.965**	**0.965**	0.965	**0.809**	**97.1**	12767	8949
	ISOSEQ3	0.878	0.79	**0.986**	0.476	64.5	24097	110028
	CARNAC-LR	0.93	0.94	0.92	0.401	93.4	9315	20555
	LINCLUST	0.881	0.801	0.979	0.455	76.1	31119	74119
SIM-100k	ISONCLUST	**0.984**	**0.987**	0.981	**0.829**	**96.7**	8931	3346
	ISOSEQ3	0.863	0.76	**0.998**	0.007	10.9	5013	89114
	CARNAC-LR	0.979	0.99	0.969	0.734	96.1	8165	3945
	LINCLUST	0.911	0.845	0.988	0.258	76.5	17856	23478
SIM-500k	ISONCLUST	**0.984**	**0.988**	0.98	**0.831**	**99.5**	13996	2274
	ISOSEQ3	0.809	0.681	**0.995**	0.006	33.1	68704	334547
	CARNAC-LR	0.971	0.974	0.967	0.695	97.1	12761	14527
	LINCLUST	0.895	0.82	0.985	0.263	89.8	48608	51026
SIM-1000k	ISONCLUST	**0.984**	**0.988**	0.98	**0.832**	**99.8**	15590	1945
	ISOSEQ3	0.788	0.654	**0.993**	0.006	46.8	180629	532410
	CARNAC-LR	0.958	0.949	0.967	0.674	94.3	14423	56502
	LINCLUST	0.89	0.813	0.984	0.264	91.8	68752	81641
ONT	ISONCLUST	**0.886**	**0.825**	0.957	**0.353**	**94.5**	39464	48935
	CARNAC-LR	0.797	0.669	**0.984**	0.095	54.2	27483	408270
	LINCLUST	0.72	0.563	1	<0.001	0.1	516	889346

we are 3x faster than CARNAC-LR on our dataset, the absolute run-time is still fairly high (27.2 h) and improving it is an immediate future goal. We expect that parallelization will yield significant speed-ups, keeping in mind that other tools were run on eight cores compared to only one core for ISONCLUST (Table 2). Memory consumption was relatively low for all tools, with ISONCLUST comsuming the least memory (Table 3).

Role of Class Size: We investigated if ISONCLUST's clustering accuracy is affected by the class size (*i.e.*, the number of reads present in a class). We binned the reads according to ranges of class size and computed the completeness and homogeneity with respect to each bin (Fig. 1). The completeness clearly decreases with increased class size, indicating that ISONCLUST tends to have more fragmented clusters as the class size increases. Homogeneity has no clear trend for class sizes up to 50, but decreases after that.

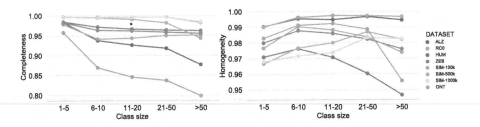

Fig. 1. Completeness and homogeneity of ISONCLUST across various class sizes.

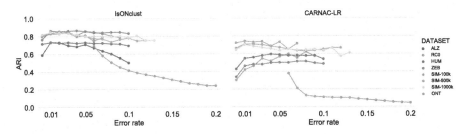

Fig. 2. Accuracy (measured by the ARI) of ISONCLUST and CARNAC-LR as a function of error rates. The read error rate is inferred by ISONCLUST. Reads are binned according to their error rate, rounded to the nearest two decimal points. Datapoints for where there are at least 1,000 reads are shown.

Role of Read Error Rates: Base errors pose a challenge to any clustering algorithm, so we measured how they affected our accuracy. We batch reads with respect to their error rate and measure the ARI within each batch (Fig. 2, left panel). For Iso-Seq, ISONCLUST has relatively stable ARI across different error rates (with ALZ being the exception), which we believe is due to our algorithm's use of quality values. This is not true for ONT, where error rates of 7–20% have a detrimental effect on ISONCLUST. Nevertheless, compared to CARNAC-LR, ISONCLUST has substantially higher ARI across error rates, datasets, and technologies (Fig. 2, right panel); e.g. for the ONT dataset, ISONCLUST does better at 20% error rate than CARNAC-LR does at 7%.

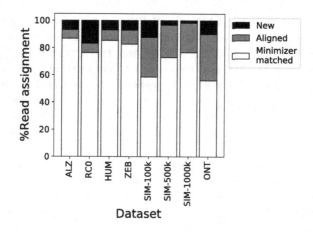

Fig. 3. Distribution of the stages of our algorithm. A read is either minimizer-matcher or aligned to an existing cluster, or a new cluster is formed.

Breakdown of Algorithm Stages: For each read, ISONCLUST either assigns it to a new cluster or to an existing cluster. If the read goes to an existing cluster, then it is either by minimizer matching or by alignment. We measure the distribution of reads into these three cases for all our datasets (Fig. 3). For the non-simulated Iso-Seq data, alignment was invoked only 6–10% of the time. However, for the ONT and simulated Iso-Seq data, alignment was invoked more frequently (22–34%), indicating room for future run-time improvement.

4   Conclusion

In this paper, we presented ISONCLUST, a clustering tool for long-read transcriptome data. The design choices of our algorithm are mostly driven by scaling and the desire to use quality values. In order to scale, we made the algorithm greedy so that it can avoid doing an all-to-all similarity comparison. We avoid the natural but time consuming step of recomputing the best representative within a cluster after each update. Our initial sorting step mitigates the potentially negative effects of this by making sure that the representative is guaranteed to have the largest expected number of error-free k-mers among all reads in the cluster. Furthermore, we avoid the expensive alignment step whenever possible by using minimizer matching. In terms of quality values, we use them throughout the algorithm, including in the initial sorting step, in deciding whether mismatched minimizers are the result of sequencing error, and in computing pairwise alignment. The use of quality values is critical to the success of our algorithm and to its ability to handle both PacBio and Nanopore data.

Our results indicate that ISONCLUST is a substantial improvement over existing methods, with higher accuracy and/or better scaling than other comparable tools. We also demonstrated that ISONCLUST performs well across a breadth of instruments (PacBio's Sequel, PacBio's RSII, and Oxford Nanopore), organisms (human, zebrafinch, and hummingbird), with each also having a different quality of reference for estimating the ground truth, and read depths (from 100k to 1mil reads). In all these scenarios, ISONCLUST outperforms others on all relevant accuracy metrics, with the exception that ISOSEQ3 produces a more homogeneous clustering (though at the cost of clustering much fewer reads).

Ultimately, we would like to combine ISONCLUST with a post-clustering error-correcting module in order to reconstruct transcripts *de novo* from non-targeted Iso-Seq and ONT data. We have previously taken this approach in our IsoCon tool [2] for targeted Iso-Seq data. IsoCon, however, is not able to scale to the much larger non-targeted datasets and to the higher error rates of ONT. With the development of ISONCLUST, we are now able to overcome these challenges in the clustering step. Our next step is to tackle the error correction problem within each cluster. The ultimate goal is to develop a tool for *de novo* transcript reconstruction, which will be the first such tool for ONT data and an improvement over other methods for Iso-Seq data.

Acknowledgements. This work has been supported in part by NSF awards DBI-1356529, CCF-551439057, IIS-1453527, and IIS-1421908 to PM.

References

1. Byrne, A., et al.: Nanopore long-read RNAseq reveals widespread transcriptional variation among the surface receptors of individual B cells. Nature Commun. **8**, 16027 (2017)
2. Sahlin, K., Tomaszkiewicz, M., Makova, K.D., Medvedev, P.: Deciphering highly similar multigene family transcripts from Iso-Seq data with IsoCon. Nature Commun. **9**(1), 4601 (2018)
3. Tseng, E., Tang, H.T., AlOlaby, R.R., Hickey, L., Tassone, F.: Altered expression of the FMR1 splicing variants landscape in premutation carriers. Biochimica et Biophys. Acta (BBA)-Gene Regul. Mech. **1860**(11), 1117–1126 (2017)
4. Nattestad, M., et al.: Complex rearrangements and oncogene amplifications revealed by long-read DNA and RNA sequencing of a breast cancer cell line. Genome Res. **28**(8), 1126–1135 (2018)
5. Kuo, R.I., Tseng, E., Eory, L., Paton, I.R., Archibald, A.L., Burt, D.W.: Normalized long read RNA sequencing in chicken reveals transcriptome complexity similar to human. BMC Genomics **18**(1), 323 (2017)
6. Hoang, N.V., et al.: A survey of the complex transcriptome from the highly polyploid sugarcane genome using full-length isoform sequencing and de novo assembly from short read sequencing. BMC Genomics **18**(1), 395 (2017)
7. Gordon, S.P., et al.: Widespread polycistronic transcripts in fungi revealed by single-molecule mRNA sequencing. PloS One **10**(7), e0132628 (2015)
8. Tombácz, D., et al.: Long-read isoform sequencing reveals a hidden complexity of the transcriptional landscape of herpes simplex virus type 1. Front. Microbiol. **8**, 1079 (2017)
9. Marchet, C., et al.: De novo clustering of long reads by gene from transcriptomics data. Nucleic Acids Res. **47**(1), e1 (2018). https://doi.org/10.1093/nar/gky834

10. Workman, R.E., Myrka, A.M., Wong, G.W., Tseng, E., Welch Jr., K.C., Timp, W.: Single-molecule, full-length transcript sequencing provides insight into the extreme metabolism of the ruby-throated hummingbird Archilochus colubris. GigaScience **7**(3), 1–12 (2018). https://doi.org/10.1093/gigascience/giy009

11. Li, J., et al.: Long read reference genome-free reconstruction of a full-length transcriptome from astragalus membranaceus reveals transcript variants involved in bioactive compound biosynthesis. Cell Discov. **3**, 17031 (2017)

12. Liu, X., Mei, W., Soltis, P.S., Soltis, D.E., Barbazuk, W.B.: Detecting alternatively spliced transcript isoforms from single-molecule long-read sequences without a reference genome. Mol. Ecol. Resour. **17**(6), 1243–1256 (2017)

13. Edgar, R.C.: Search and clustering orders of magnitude faster than BLAST. Bioinformatics **26**(19), 2460–2461 (2010)

14. Li, W., Godzik, A.: Cd-hit: a fast program for clustering and comparing large sets of protein or nucleotide sequences. Bioinformatics **22**(13), 1658–1659 (2006)

15. James, B.T., Luczak, B.B., Girgis, H.Z.: MeShClust: an intelligent tool for clustering DNA sequences. Nucleic Acids Res. **46**(14), e83 (2018)

16. Ghodsi, M., Liu, B., Pop, M.: DNACLUST: accurate and efficient clustering of phylogenetic marker genes. BMC Bioinform. **12**(1), 271 (2011)

17. Paccanaro, A., Casbon, J.A., Saqi, M.A.: Spectral clustering of protein sequences. Nucleic Acids Res. **34**(5), 1571–1580 (2006)

18. Steinegger, M., Söding, J.: Clustering huge protein sequence sets in linear time. Nat. Commun. **9**(1), 2542 (2018)

19. Steinegger, M., Söding, J.: MMseqs2 enables sensitive protein sequence searching for the analysis of massive data sets. Nat. Biotechnol. **35**(11), 1026–1028 (2017)

20. Zorita, E., Cusco, P., Filion, G.J.: Starcode: sequence clustering based on all-pairs search. Bioinformatics **31**(12), 1913–1919 (2015)

21. Bevilacqua, V., et al.: EasyCluster2: an improved tool for clustering and assembling long transcriptome reads. BMC Bioinform. **15**, S7 (2014)

22. Dost, B., Wu, C., Su, A., Bafna, V.: TCLUST: a fast method for clustering genome-scale expression data. IEEE/ACM Trans. Comput. Biol. Bioinform. (TCBB) **8**(3), 808–818 (2011)

23. Christoffels, A., Gelder, A.V., Greyling, G., Miller, R., Hide, T., Hide, W.: STACK: sequence tag alignment and consensus knowledgebase. Nucleic Acids Res. **29**(1), 234–238 (2001)

24. Burke, J., Davison, D., Hide, W.: d2_cluster: a validated method for clustering EST and full-length cDNA sequences. Genome Res. **9**(11), 1135 (1999)

25. Chong, Z., Ruan, J., Wu, C.I.: Rainbow: an integrated tool for efficient clustering and assembling RAD-seq reads. Bioinformatics **28**(21), 2732–2737 (2012)

26. Solovyov, A., Lipkin, W.I.: Centroid based clustering of high throughput sequencing reads based on n-mer counts. BMC Bioinform. **14**(1), 268 (2013)

27. Bao, E., Jiang, T., Kaloshian, I., Girke, T.: SEED: efficient clustering of next-generation sequences. Bioinformatics **27**(18), 2502–2509 (2011)

28. Fu, L., Niu, B., Zhu, Z., Wu, S., Li, W.: CD-HIT: accelerated for clustering the next-generation sequencing data. Bioinformatics **28**(23), 3150–3152 (2012)

29. Shimizu, K., Tsuda, K.: SlideSort: all pairs similarity search for short reads. Bioinformatics **27**(4), 464–470 (2010)

30. Comin, M., Leoni, A., Schimd, M.: Clustering of reads with alignment-free measures and quality values. Algorithms Mol. Biol. **10**(1), 4 (2015)

31. Alanko, J., Cunial, F., Belazzougui, D., Mäkinen, V.: A framework for space-efficient read clustering in metagenomic samples. BMC Bioinform. **18**(3), 59 (2017)

32. Orabi, B., et al.: Alignment-free clustering of UMI tagged DNA molecules. Bioinformatics (2018). https://doi.org/10.1093/bioinformatics/bty888

33. Ondov, B.D., et al.: Mash: fast genome and metagenome distance estimation using MinHash. Genome Biol. **17**(1), 132 (2016)

34. Davidson, N.M., Oshlack, A.: Corset: enabling differential gene expression analysis for de novoassembled transcriptomes. Genome Biol. **15**(7), 410 (2014). https://doi.org/10.1186/s13059-014-0410-6

35. Malik, L., Almodaresi, F., Patro, R.: Grouper: graph-based clustering and annotation for improved de novo transcriptome analysis. Bioinformatics **34**(19), 3265–3272 (2018)

36. Krishnakumar, R., et al.: Systematic and stochastic influences on the performance of the MinION nanopore sequencer across a range of nucleotide bias. Sci. Rep. **8**(1), 3159 (2018)

37. Tseng, E.: Cogent: coding genome reconstruction using Iso-Seq data (2018). https://github.com/Magdoll/Cogent

38. Roberts, M., Hayes, W., Hunt, B.R., Mount, S.M., Yorke, J.A.: Reducing storage requirements for biological sequence comparison. Bioinformatics **20**(18), 3363–3369 (2004)

39. Au, K.F., Underwood, J.G., Lee, L., Wong, W.H.: Improving PacBio long read accuracy by short read alignment. PloS one **7**(10), e46679 (2012)

40. Li, H.: Minimap2: pairwise alignment for nucleotide sequences. Bioinformatics **34**(18), 3094–3100 (2018)

41. Daily, J.: Parasail: SIMD C library for global, semi-global, and local pairwise sequence alignments. BMC Bioinform. **17**(1), 81 (2016)

42. Sahlin, K., Medvedev, P.: Exprimental details appendix to "de novo clustering of long-read transcriptome data using a greedy, quality-value based algorithm" (2019). https://github.com/ksahlin/isONclust/wiki/Paper-Appendix

43. Stöcker, B.K., Köster, J., Rahmann, S.: SimLoRD: simulation of long read data. Bioinformatics **32**(17), 2704–2706 (2016)

44. Iso-Seq in house datasets. https://github.com/PacificBiosciences/IsoSeq-SA3nUP/wiki/Iso-Seq-in-house-datasets. Accessed 24 Oct 2018

45. Direct RNA and cDNA sequencing of a human transcriptome on Oxford Nanopore MinION and GridION. https://github.com/nanopore-wgs-consortium/NA12878/blob/master/RNA.md. Accessed 24 Oct 2018

46. Korlach, J., et al.: De novo PacBio long-read and phased avian genome assemblies correct and add to reference genes generated with intermediate and short reads. GigaScience **6**(10) (2017). https://doi.org/10.1093/gigascience/gix085

47. Rosenberg, A., Hirschberg, J.: V-measure: a conditional entropy-based external cluster evaluation measure. In: Proceedings of the 2007 Joint Conference on Empirical Methods in Natural Language Processing and Computational Natural Language Learning (EMNLP-CoNLL) (2007)

48. Hubert, L., Arabie, P.: Comparing partitions. J. Classif. **2**(1), 193–218 (1985)

A Sticky Multinomial Mixture Model of Strand-Coordinated Mutational Processes in Cancer

Itay Sason[1], Damian Wojtowicz[2], Welles Robinson[3], Mark D. M. Leiserson[3], Teresa M. Przytycka[2], and Roded Sharan[1(✉)]

[1] School of Computer Science, Tel Aviv University, 69978 Tel Aviv, Israel
roded@tau.ac.il
[2] National Center for Biotechnology Information, National Library of Medicine, National Institutes of Health, Bethesda, MD 20894, USA
[3] Center for Bioinformatics and Computational Biology, University of Maryland, College Park, MD 20742, USA

Abstract. The characterization of mutational processes in terms of their signatures of activity relies, to the most part, on the assumption that mutations in a given cancer genome are independent of one another. Recently, it was discovered that certain segments of mutations, termed processive groups, occur on the same DNA strand and are generated by a single process or signature. Here we provide a first probabilistic model of mutational signatures that accounts for their observed stickiness and strand-coordination. The model conditions on the observed strand for each mutation, and allows the same signature to generate a run of mutations. We show that this model provides a more accurate description of the properties of mutagenic processes than independent-mutation models or strand oblivous models, achieving substantially higher likelihood on held-out data. We apply this model to characterize the processivity of mutagenic processes across multiple types of cancer in terms of replication and transcriptional strand-coordination.

1 Introduction

Mutational processes are key factors in shaping cancer genomes [1,9,23] and their characterization has important implications for understanding the disease and choosing targeted therapies [3,4,15].

Multiple algebraic and statistical approaches have been suggested to the detection of mutational processes from somatic mutation data [2,5,11,17,19]. These methods, which focus on single-base substitions, are based on learning the pattern of mutations of each potential process as well as its activity (aka *exposure*) in any given tumor in a way that will best explain the observed mutation data.

State-of-the-art approaches for learning mutational signatures include non-negative matrix factorization (NMF) methods [2,5,11,17] that aim to explain the mutation counts as a sum over all signatures of the probability of a specific

© Springer Nature Switzerland AG 2019
L. J. Cowen (Ed.): RECOMB 2019, LNBI 11467, pp. 243–255, 2019.
https://doi.org/10.1007/978-3-030-17083-7_15

mutation to be generated by the respective signature times its exposure. Other approaches that borrow from the world of topic modeling, aim to provide a probabilistic model of the data so as to maximize the model's likelihood [7,19]. However, most of these methods assume that mutations are independent of one another and cannot capture processes that create dependencies among them.

Recently, it was observed that APOBEC-related signatures operate in a strand-coordinated manner where pairs of consecutive mutations tend to mutate from the same reference allele and occur on the same strand [14]. Morganella et al. [12] generalized these observations and found segments of such mutations (i.e., same reference allele and same strand) that they termed processive groups. The significance and abundance of these processive groups suggested that certain mutational processes display stickiness and strand-coordination properties.

The biological reasons for this strand coordination are related, at least in part, to the asymmetric role that the two strands play in many cellular processes that operate on DNA. For example, it is well known that APOBEC enzymes act on single stranded DNA [16]. In cellular processes that require strand separation, including replication and transcription, one of the strands is often more exposed than the other, leading to strand coordination of APOBEC mutations. Due to the differences in which the leading and lagging strands are processed by the replication apparatus, APOBEC asymmetry in these two strands is particularly strong [8,12,18,22]. Since leading and lagging strand are also processed by different polymerases, among other differences, additional signatures may have replication related strand coordination. Transcription-coupled repair is another source of strand-specific mutagenesis, and multiple signatures have been found to have mutation strand bias in template versus non-template strands [1,8,12,22].

Here we provide a first probabilistic model, sticky multinomial mixture model (sMMM), of cancer mutation data that accounts for the stickiness and strand-coordination of mutational signatures. The model captures independent mutations as well as processive groups in one probabilistic framework. In cross validation tests on multiple datasets, sMMM outperforms independent-mutation models or sticky models that do not account for the strand information. We apply our model to gain new insights about the stickiness and strand preferences of known signatures.

2   Methods

2.1   Preliminaries

We follow previous work and assume that somatic mutations in cancer fall into $M = 96$ categories (denoting the mutation identity and its flanking bases). These mutations are assumed to be the result of the activity of K mutational processes, each of which is associated with a signature $S_i = (e_i(1), \ldots, e_i(M))$ of probabilities to emit each of the mutation categories. Henceforth, we denote the mutation categories observed in a given tumor by $o_1, \ldots, o_T$. We assume that o_i was emitted by signature s_i (whose identity is hidden from us).

A *multinomial mixture model (MMM)* assumes the following generative process for the mutation data. For each mutation, independently of all others, a signature $s \in S$ is drawn from a multinomial $\pi = (\pi_1, \ldots, \pi_K)$; subsequently, the mutation category is drawn from S_i. The model parameters can be learned using the Expectation-Maximization (EM) algorithm.

2.2 Model Specification

Following Morganella et al. [12] and subsequent work [21], we assume that mutations are either formed one at a time or belong to a processive group of mutations that are the result of a single mutational process. To capture such dependencies we propose a sticky MMM (sMMM) framework in which every signature is active along a contiguous segment of mutations with hidden indicator variables r_i denoting whether the signature stays active when moving from s_{i-1} to s_i (with $r_1 = 0$). Segmenting the mutations in this way is analogous to speaker diarization in audio analysis, the process of partitioning an input audio stream into homogeneous segments according to the speaker identity. Importantly, we also define a strand-coordinated variant of sMMM where we condition r_i on the identities of the strands on which these two mutations occurred. In this way, mutation segments in which the indicator is positive model processive groups. The basic model is sketched in Fig. 1.

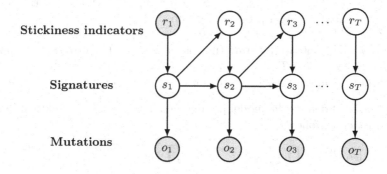

Fig. 1. A probabilistic model of processive groups in mutation data. For clarity we omit the dependency of the r_i indicators on strand identity.

Formally, the model we propose, sMMM, is parameterized by a $K \times M$ matrix e of signature emission probabilities, signature start probabilities $\pi = (\pi_1, \ldots, \pi_K)$ that are assumed to be sample-specific, and signature stickiness values $\alpha = (\alpha_1, \ldots, \alpha_K)$ that are shared across samples. For simplicity, we omit sample indices below and focus the description on a single sample. The model can be described by the following conditional probability distributions:

- $\Pr\left[o_t = m | s_t = S_i\right] = e_i(m)$

- $\Pr\left[r_{t+1} = 1 | s_t = S_i\right] = \begin{cases} 0 & id_{t+1} \neq id_t \\ \alpha_i & id_{t+1} = id_t \end{cases}$

- $\Pr\left[s_{t+1} = S_j | r_{t+1} = 1, s_t = S_i\right] = \begin{cases} 0, & i \neq j \\ 1, & i = j \end{cases}$

- $\Pr\left[s_{t+1} = S_i | r_{t+1} = 0\right] = \pi_i$

Here, t is a mutation index that ranges from 1 to T, b is a mutation category, and id_t is the identity of the strand at mutation t (for convenience, $id_0 := id_1$). In the sequel we refer to any sequence of at least two consecutive mutations of the same signature (according to the model) that are on the same strand and mutated from the same nucleotide as a *processive group*.

2.3   Model Training

On input mutations $O = (o_1, \ldots, o_T)$ and corresponding hidden signatures and activity indicators $Z = (s_1, \ldots, s_T, r_1 \ldots, r_T)$, we define the following sufficient statistics:

- $E_i(m, Z) := |\{1 \leq t \leq T | s_t = S_i, o_t = m\}|$, the number of times signature i emitted mutation category m.
- $A_i(Z) := |\{1 \leq t < T | s_t = s_{t+1} = S_i, r_{t+1} = 1\}|$, the number of times signature i repeated within a processive group.
- $B_i(Z) := |\{1 \leq t < T | s_t = S_i, r_t = 0, id_t = id_{t+1}\}|$, the number of times signature i could potentially start (not observing r_{t+1}) a processive group.
- $C_i(Z) := |\{1 < t \leq T | s_t = S_i, r_{t+1} = 0, id_t = id_{t-1}\}|$, the number of times signature i could potentially end (not observing r_t) a processive group.

Let $\theta = (\pi, \alpha, e)$ denote the model parameters. The complete likelihood function can be written as follows:

$$\Pr\left[Z, O | \theta\right] = \prod_i \prod_m e_i(m)^{E_i(m,Z)} \cdot \prod_i \alpha_i^{A_i(Z)} \pi_i^{B_i(Z)} (1 - \alpha_i)^{C_i(Z)}$$

We optimize θ using the EM algorithm. In order to describe the algorithm we make a simplifying assumption that every input sequence spans the same strand (i.e., $id_t = id_t + 1$ for all $t < T$) which can be achieved by splitting the input data to same-strand segments. We start by deriving forward and backward algorithm variants for our model. The forward algorithm computes for any i, t the joint probability $f_i(t) := \Pr\left[o_1 \ldots o_t, s_t = S_i\right]$ recursively:

$$f_i(1) = \Pr\left[o_1, s_1 = S_i\right] = \pi_i e_i(o_1)$$

$$f_i(t+1) = \Pr\left[o_1...o_{t+1}, s_{t+1} = S_i\right] = \sum_j \Pr\left[o_1, \ldots, o_{t+1}, s_t = S_j, s_{t+1} = S_i\right]$$

$$= \sum_j \Pr\left[o_1...o_t, s_t = S_j\right] \cdot \Pr\left[s_{t+1} = S_i, o_{t+1}|s_t = S_j\right]$$

$$= \sum_j f_j(t) \Pr\left[s_{t+1} = S_i|s_t = S_j\right] \Pr\left[o_{t+1}|s_{t+1} = S_i\right]$$

$$= e_i(o_{t+1}) \left(\alpha_i f_i(t) + \pi_i \sum_j (1 - \alpha_j) f_j(t) \right)$$

The backward algorithm computes the probability $b_i(t) := \Pr\left[o_{t+1}...o_T|s_t = S_i\right]$ for any i, t recursively:

$$b_i(T) = 1$$

$$b_i(t-1) = \Pr\left[o_t...o_T|s_{t-1} = S_i\right] = \sum_j \Pr\left[o_t, \ldots, o_T, s_t = S_j|s_{t-1} = S_i\right]$$

$$= \sum_j \Pr\left[o_{t+1}, \ldots, o_T|s_t = S_j\right] \Pr\left[s_t = S_j, o_t|s_{t-1} = S_i\right]$$

$$= \sum_j b_j(t) \Pr\left[s_t = S_j|s_{t-1} = S_i\right] \Pr\left[o_t|s_t = S_j\right]$$

$$= e_i(o_t) b_i(t) \alpha_i + (1 - \alpha_i) \sum_j e_j(o_t) b_j(t) \pi_j$$

The Q function for the EM algorithm is the expected complete log-likelihood:

$$Q(\theta|\theta^0) = \sum_Z \Pr\left[Z|O, \theta^0\right] \cdot \log \Pr\left[Z, O|\theta\right]$$

$$= \sum_i \sum_m E_i(m) \log(e_i(m)) + \sum_i A_i \log(\alpha_i) + B_i \log(\pi_i) + C_i \log(1 - \alpha_i)$$

where

$$E_i(m) = \sum_Z \Pr\left[Z|O, \theta^0\right] E_i(m, Z) \quad A_i = \sum_Z \Pr\left[Z|O, \theta^0\right] A_i(Z)$$

$$B_i = \sum_Z \Pr\left[Z|O, \theta^0\right] B_i(Z) \quad C_i = \sum_Z \Pr\left[Z|O, \theta^0\right] C_i(Z)$$

which is maximized for

$$e_i(m) = \frac{E_i(m)}{\sum_{\tilde{m}} E_i(\tilde{m})} \quad \alpha_i = \frac{A_i}{A_i + C_i} \quad \pi_i = \frac{B_i}{\sum_j B_j}$$

In the following we show how to compute $E_i(m), A_i, B_i, C_i$ under the current parameters θ^0 using the forward and backward algorithms.

$$E_i(m) = \sum_{t=1}^{T-1} \Pr[s_t = S_i, o_t = m|O] = \frac{1}{\Pr[O]} \sum_{t|o_t=m} f_i(t)f_i(t)$$

$$A_i = \sum_{t=2}^{T} \Pr[s_{t-1} = S_i, r_t = 1, s_t = S_i|O] = \frac{1}{\Pr[O]} \sum_{t=2}^{T} f_i(t-1)\alpha_i e_i(o_t)b_i(t)$$

$$B_i = \frac{1}{\Pr[O]} \sum_{t=1}^{T} \Pr[s_t = S_i, r_t = 0, O]$$

$$= \frac{1}{\Pr[O]} \left(\Pr[s_1 = S_i, O] + \sum_{t=2}^{T} \Pr[s_t = S_i, r_t = 0, O] \right)$$

$$= \frac{1}{\Pr[O]} \left(f_i(1)b_i(1) + \sum_{t=1}^{T-1} \pi_i e_i(o_{t+1})b_i(t+1) \cdot \left(\sum_j f_j(t)(1 - \alpha_j) \right) \right)$$

$$C_i = \sum_{t=1}^{T-1} \Pr[s_t = S_i, r_{t+1} = 0|O]$$

$$= \frac{1}{\Pr[O]} \sum_{t=1}^{T-1} \sum_j \Pr[s_t = S_i, r_{t+1} = 0, s_{t+1} = S_j, O]$$

$$= \frac{1}{\Pr[O]} \sum_{t=1}^{T-1} f_i(t)(1 - \alpha_i) \cdot \left(\sum_j \pi_j e_j(o_{t+1})b_j(t+1) \right)$$

$\Pr[O]$ can be easily computed using the forward algorithm:

$$\Pr[O] = \sum_i \Pr[O, s_T = S_i] = \sum_i f_i(T)$$

Lemma 1. *Each iteration of the EM algorithm can be computed in $O(Tn)$ time for T mutations and n signatures.*

Proof. Observe that the forward and backward algorithms can be computed in $O(Tn)$ time by computing the sums $\sum_j (1 - \alpha_j)f_j(t)$ and $\sum_j e_j(o_t)b_j(t)\pi_j$ only once for every mutation index t. Similarly, we can compute the Q function within the same time bounds.

3   Results

3.1   Data Description

We analyzed breast cancer (BRCA), colon ademocarcinoma (COAD), chronic lymphocytic leukemia (CLL), and malignant lymphoma (MALY) mutation datasets from whole-genome sequences from the International Cancer Genome Consortium (ICGC) (Table 1). We downloaded mutations from the ICGC data portal (https://docs.icgc.org/). For BRCA, we used release 22 to match Nik-Zainal et al. [13], while for the other datasets we used the most recent release (release 27). For each dataset, we followed the standard approach introduced by Alexandrov et al. [2] and classified mutations into 96 categories based on the 5' flanking base, substitution, and 3' flanking base, following the convention that substitutions are written with the pyrimidine first (i.e. C:G>T:A is written as C>T instead of G>A). For each dataset, we analyzed the COSMIC signatures (https://cancer.sanger.ac.uk/cosmic/signatures) [6] known to be active in the corresponding cancer type (enumerated in Table 1). We chose the four cancer types because they are known to have active signatures that were shown to display strand bias: Signatures 2 and 13 are active in BRCA, CLL, and MALY, and display replication strand bias [12], and Signature 10 is active in COAD and displays transcription strand bias [1].

Table 1. Datasets analyzed in this study: breast cancer (BRCA), colon ademocarcinoma (COAD), chronic lymphocytic leukemia (CLL), and malignant lymphoma (MALY).

Cancer type	#Samples	#Mutations	COSMIC signatures
BRCA	560	3,479,652	1, 2, 3, 5, 6, 8, 13, 17, 18, 20, 26, 30
COAD	44	52,827	1, 5, 6, 10
CLL	100	270,870	1, 2, 5, 9, 13
MALY	100	1,220,526	1, 2, 5, 9, 13, 17

We also classified mutations based on strand-specific information from Tomkova et al. [22] and Haradhvala et al. [8]. Specifically, we classified mutations as being on the leading or lagging strand during replication, or the template or non-template strand for transcription, and on whether transcription and replication occur in the same direction in that position. To do so, we used a dataset[1] from [22] (originally processed by [8]) that classifies each non-overlapping bin of 20,000 bases in the genome based on the inferred replication or transcription strand using six lymphoblastoid cell lines. For each strand annotation, we marked successive mutations in each genome based on whether they shared the same positive class (e.g., both are lagging). If a classification was not available for either of the successive mutations, or they spanned another class, we considered that pair to have a mismatch (i.e. $id_t \neq id_{t+1}$). Following [12], we also performed an additional classification where we marked successive mutations based on whether they shared the same reference allele (which also indicates that they are on the same strand).

---

[1] The `data/tableTerritories_Haradhvala_territories_50_bins.txt` file in https://bitbucket.org/bsblabludwig/replicationasymmetry.

3.2 Comparing the Sticky MMM to an Independent Mutation Model

We first compared the multinomial mixture model (MMM) with our basic sticky MMM (sMMM) using held-out data. MMM serves as a stand-in for state-of-the-art non-probabilistic mutation signature methods such as non-negative matrix factorization, as MMM is a probabilistic method that encodes the standard assumption that each mutation in a tumor is independent of all others. We performed an additional comparison with the sMMM by including the restriction that stickiness only occurs for pairs of mutations sharing the same reference allele, i.e., mutated from the same nucleotide and occurring on the same strand. Henceforth, we call this variant strand-coordinated sMMM.

For each model, we fixed the mutational signatures to the COSMIC signatures as described in the previous section and did not attempt to learn new signatures. We applied the EM algorithm with a tolerance of 0.01 and a maximum number of 100 iterations. We evaluated the algorithm's performance by computing log-likelihood using the Viterbi algorithm on held-out data. Specifically, we used a cross-validation scheme where in each iteration a complete chromosome is hidden.

The results are summarized in Table 2 and clearly show the superiority of the sMMM across the four cancer types analyzed. In each cancer type, the sMMM has higher held-out likelihood than the independent mutation MMM, demonstrating that mutation signatures have stickiness that is shared across samples and that modeling this stickiness provides greater predictive power for held-out data. Further, the difference between the sMMM and MMM becomes much larger and highly significant when the sMMM is restricted to only allow stickiness between mutations with the same reference allele.

Table 2. Comparing the multinomial mixture model (MMM) and two sticky MMMs on log-likelihood of held-out data, computed from cross-validation. P-values based on a Wilcoxon signed-rank test between paired MMM and strand-coordinated sMMM scores appear in the last column.

Dataset	Held-out log-likelihood			P-value
	MMM	sMMM	Strand-coordinated sMMM	
BRCA	−13742688	−13738247	**−13694138**	1.79e−86
COAD	−170052	−170050	**−170021**	4.99e−02
CLL	−1178031	−1177764	**−1173274**	3.90e−18
MALY	−5235054	−5232011	**−5205369**	3.90e−18

In addition to the Watson/Crick strand annotation, whose application is reported above, we also applied sMMM with strand annotations that characterize strand roles in replication and/or transcription. These included leading/lagging annotation, template/non-template annotation and whether transcription and replication proceed in the same or opposite directions. These annotations were available for only a fraction of the mutations, hence their likelihood cannot be directly compared to those in Table 2. Instead, focusing on the largest dataset (BRCA), we compare the leading-coordination to the lagging-coordination and compare the remaining annotations, all of which are applicable only in genic regions, to one another. The results are provided in Table 3. While differences in log-likelihood are smaller, they indicate that assuming the convention of associating the mutations with pyrimidines, mutational processes tend to be associated more with lagging vs. leading strand and with non-template vs. template strand, both with significant p-values. These observation align well with the previously observed preference of APOBEC-related processes to act on the lagging strand [8,12], and the bias of multiple signatures in breast cancer for mutations on the template versus non-template strand [12].

Table 3. A comparison of sMMM models with different strand coordinations for the breast cancer dataset. Displayed are the difference in log-likelihood in cross validation and a corresponding p-value based on a Wilcoxon signed-rank test.

Strand types	Log-likelihood ratio	P-value
Lagging/leading	780	1.27e−6
Non-template/template	645	4.70e−3
Opposite-direction/same-direction	91	6.54e−2

3.3 Strand-Coordinated sMMM Defines Processive Groups in Breast Cancers

Morganella et al. defined processive groups as sets of adjacent substitutions of the same mutational signature sharing the same reference allele [12]. Our model allows us to compute maximum likelihood estimates that sequences of mutations are generated in processive groups, hence we could apply it to characterize the processivity of the different signatures in breast cancer. In order to compare and contrast our findings with those of Morganella et al. [12], we likewise removed regions of localized hypermutation (*kataegis*), and used the same statistical test for the significance of a processive group of a given length.

Using the same length threshold of more than 10, we confirmed the association of processive groups with Signatures 2, 6, 13, 17 and 26 [12]. In addition, our strand-coordinated sMMM revealed that processivity is also a feature of Signature 18 (Fig. 2A). The number of processive groups of length more than 10 was particularly high for Signatures 2 and 13 (Fig. 2B).

Next, we tested whether using the mutation-independent MMM rather than the strand-coordinated sMMM was important for the accurate discovery of processive groups. Overall, the stand-coordinated model uncovered 134 groups in 43 patients while MMM model captured only 38 in 11 patients (Fig. 2B). These differences underscore the higher sensitivity of the strand-coordinated sMMM model for detecting processive groups, which may in part explain the observed differences in likelihood between MMM and strand-coordinated sMMM on held-out data.

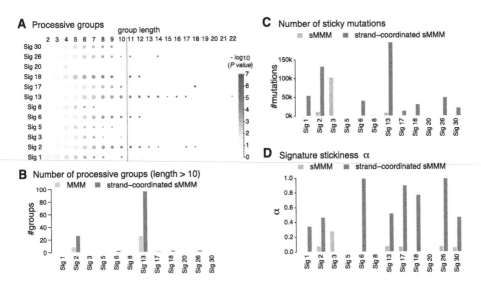

Fig. 2. (**A**) Relationship between processive group lengths (columns) and mutational signatures (rows) modeled by strand-coordinated sMMM in BRCA. The size of each circle represents the number of groups (log10) observed for the specified group length and for each signature. The color of each circle corresponds to the p-value of detecting a processive group of a given length in randomized data (-log10). (**B**) The number of processive groups of length above 10 for all signatures modeled by MMM (gray) and strand-coordinated sMMM (red) in BRCA. (**C**). The number of sticky mutations as modeled by sMMM (blue) and strand-coordinated sMMM (red) in BRCA. (**D**) Signature stickiness α as learned by sMMM (blue) and strand-coordinated sMMM (red) in BRCA. (Color figure online)

Processive groups, as summarized in Fig. 2A, B, capture statistically significant patterns in cancer genomes and are considered features of specific signatures. An alternative characterization could be provided by the model parameter α – the "stickiness" of a signature – which is learned by the strand-coordinated sMMM. Thus, we analyzed how these two views of strand-coordinated mutagenesis relate to each other. We considered only signatures for which there is a sufficient number of sticky mutations to properly learn this parameter (Fig. 2C). For comparison purposes, we also included stickiness values computed with the strand-oblivious sMMM. We found that in the strand-coordinated sMMM, the most sticky signatures were 1, 2, 6, 13, 17, 18, 26 and 30 (Fig. 2D). Signatures 2, 6, 13, 17, 18 and 26 are exactly the same signatures that were found to be associated with processive groups. While Signature 30 did not make the length

10 cutoff for processive groups, its processive segments are also relatively long. Interestingly Signature 5 was not found to be sticky despite the fact that its processive groups were also quite long. In contrast, there is some stickiness to Signature 1. The meaning of these intriguing findings is a subject for further investigations.

Notably, Signature 3 has the highest strand-oblivious sMMM stickiness and no stickiness in strand-coordinated sMMM, suggesting that mutations associated with this signature tend to neighbor each other but in a strand non-specific way.

4   Conclusions

We have designed a novel probabilistic model for mutation data that accounts for the processivity of mutational processes. In cross validation, the model was shown to outperform the standard independent-mutation model across multiple cancer datasets. Importantly, we obtained an even greater gain in predictive power when modeling processivity and strand-coordination together by only allowing stickiness for mutations on the same Watson/Crick strand with the same reference allele. By incorporating this additional genomic information, we obtained more accurate assignments of mutations to signatures (as evidenced by the increase in likelihood of held-out data), and also built a principled framework for testing the biological importance of different strand characteristics.

In that vein, a promising next step may be modeling multiple strand characteristics simultaneously, rather than considering them individually. For example, there is evidence in humans and other species that transcription and replication are co-oriented [10,20]. Experiments in bacteria suggest that one reason is to prevent replication fork collapse from collisions of replication and transcription machinery and, thus, we hypothesize that when replication and transcription are in opposite directions we may see stickiness of mutation signatures. In fact, this hypothesis has some support from our results in breast cancer (Table 3) where we show somewhat higher likelihood for the opposite-direction signal. It is not yet clear why this is the case, partially because it is difficult to separate the signal from leading/lagging and template/non-template strands. By simultaneously modeling these different strand characteristics, we may be able to shed light on these relationships.

We also demonstrated that the models provide an alternative and compelling way to capture processivity of mutational signatures. In particular, rather than using arbitrary cut-offs to assign mutations to signatures for calling processive groups, this property can be estimated in a principled way from our model.

Acknowledgements. This study was supported in part by a fellowship from the Edmond J. Safra Center for Bioinformatics at Tel-Aviv University. DW and TMP are supported by the Intramural Research Programs of the National Library of Medicine (NLM), National Institutes of Health, USA. RS was supported by Len Blavatnik and the Blavatnik Family foundation. We thank Mark Keller for his help in processing mutation datasets.

References

1. Alexandrov, L.B., Nik-Zainal, S., Wedge, D.C., Aparicio, S., Behjati, S., et al.: Signatures of mutational processes in human cancer. Nature **500**(7463), 415–421 (2013)
2. Alexandrov, L.B., Nik-Zainal, S., Wedge, D.C., Campbell, P.J., Stratton, M.R.: Deciphering signatures of mutational processes operative in human cancer. Cell Rep. **3**(1), 246–259 (2013)
3. Davies, H., Glodzik, D., Morganella, S., Yates, L.R., Staaf, J., et al.: HRDetect is a predictor of BRCA1 and BRCA2 deficiency based on mutational signatures. Nat. Med. **23**(4), 517–525 (2017)
4. Davies, H., Morganella, S., Purdie, C.A., Jang, S., Borgen, E., et al.: Whole-genome sequencing reveals breast cancers with mismatch repair deficiency. Cancer Res. **77**(18), 4755–4762 (2017)
5. Fischer, A., Illingworth, C.J., Campbell, P.J., Mustonen, V.: EMu: probabilistic inference of mutational processes and their localization in the cancer genome. Genome Biol. **14**(4), 1–10 (2013)
6. Forbes, S.A., Beare, D., Boutselakis, H., Bamford, S., Bindal, N., et al.: Cosmic: somatic cancer genetics at high-resolution. Nucleic Acids Res. **45**(D1), D777–D783 (2017)
7. Funnell, T., Zhang, A., Shiah, Y.-J., Grewal, D., Lesurf, R., et al.: Integrated single-nucleotide and structural variation signatures of DNA-repair deficient human cancers. bioRxiv, p. 267500 (2018)
8. Haradhvala, N., Polak, P., Stojanov, P., Covington, K., Shinbrot, E., et al.: Mutational strand asymmetries in cancer genomes reveal mechanisms of DNA damage and repair. Cell **164**(3), 538–549 (2016)
9. Helleday, T., Eshtad, S., Nik-Zainal, S.: Mechanisms underlying mutational signatures in human cancers. Nat. Rev. Genet. **15**(9), 585–598 (2014)
10. Huvet, M., Nicolay, S., Touchon, M., Audit, B., d'Aubenton Carafa, Y., Alain Arneodo, C.T.: Human gene organization driven by the coordination of replication and transcription. Genome Res. **17**(9), 1278–1285 (2007)
11. Kim, J., Mouw, K.W., Polak, P., Braunstein, L.Z., Kamburov, A., et al.: Somatic ERCC2 mutations are associated with a distinct genomic signature in urothelial tumors. Nat. Genet. **48**(6), 600–606 (2016)
12. Morganella, S., Alexandrov, L.B., Glodzik, D., Zou, X., Davies, H., et al.: The topography of mutational processes in breast cancer genomes. Nat. Commun. **7**, 11383 (2016)
13. Nik-Zainal, S., Davies, H., Staaf, J., Ramakrishna, M., Glodzik, D., et al.: Landscape of somatic mutations in 560 breast cancer whole-genome sequences. Nature **534**(7605), 47–54 (2016)
14. Nik-Zainal, S., Wedge, D.C., Alexandrov, L.B., Petljak, M., Butler, A.P., et al.: Association of a germline copy number polymorphism of APOBEC3A and APOBEC3B with burden of putative APOBEC-dependent mutations in breast cancer. Nat. Genet. **46**(5), 487–491 (2014)
15. Polak, P., Kim, J., Braunstein, L.Z., Karlic, R., Haradhavala, N.J., et al.: A mutational signature reveals alterations underlying deficient homologous recombination repair in breast cancer. Nat. Genet. **49**(10), 1476 (2017)
16. Refsland, E., Harris, R.: The APOBEC3 family of retroelement restriction factors. Curr. Top. Microbiol. Immunol. **371**, 1–27 (2013)

17. Rosales, R.A., Drummond, R.D., Valieris, R., Dias-Neto, E., da Silva, I.T.: signeR: an empirical Bayesian approach to mutational signature discovery. Bioinformatics **33**(1), 8–16 (2016)
18. Seplyarskiy, V., Soldatov, R., Popadin, K., Antonarakis, S., Bazykin, G., Nikolaev, S.: APOBEC-induced mutations in human cancers are strongly enriched on the lagging DNA strand during replication. Genome Res. **26**, 174–82 (2016)
19. Shiraishi, Y., Tremmel, G., Miyano, S., Stephens, M.: A simple model-based approach to inferring and visualizing cancer mutation signatures. PLOS Genet. **11**(12), e1005657 (2015)
20. Srivatsan, A., Tehranchi, A., MacAlpine, D.M., Wang, J.D.: Co-orientation of replication and transcription preserves genome integrity. PLoS Genet. **6**(1), e1000810 (2010)
21. Supek, F., Lehner, B.: Clustered mutation signatures reveal that error-prone DNA repair targets mutations to active genes. Cell **170**(3), 534–547.e23 (2017)
22. Tomkova, M., Tomek, J., Kriaucionis, S., Schuster-Böckler, B.: Mutational signature distribution varies with DNA replication timing and strand asymmetry. Genome Biol. **19**(1), 129 (2018)
23. Tubbs, A., Nussenzweig, A.: Endogenous DNA damage as a source of genomic instability in cancer. Cell **168**(4), 644–656 (2017)

Disentangled Representations
of Cellular Identity

Ziheng Wang[1](✉), Grace H. T. Yeo[1], Richard Sherwood[2,3,4],
and David Gifford[1,5](✉)

[1] Department of Electrical Engineering and Computer Science, M.I.T,
Cambridge, MA, USA
{ziheng,gifford}@mit.edu
[2] Division of Genetics, Brigham and Women's Hospital, Boston, MA, USA
[3] Department of Medicine, Harvard Medical School, Boston, MA, USA
[4] Hubrecht Institute, 3584 CT Utrecht, The Netherlands
[5] Department of Biological Engineering, M.I.T, Cambridge, MA, USA

Abstract. We introduce a disentangled representation for cellular identity that constructs a latent cellular state from a linear combination of condition specific basis vectors that are then decoded into gene expression levels. The basis vectors are learned with a deep autoencoder model from single-cell RNA-seq data. Linear arithmetic in the disentangled representation successfully predicts nonlinear gene expression interactions between biological pathways in unobserved treatment conditions. We are able to recover the mean gene expression profiles of unobserved conditions with an average Pearson $r = 0.73$, which outperforms two linear baselines, one with an average $r = 0.43$ and another with an average $r = 0.19$. Disentangled representations hold the promise to provide new explanatory power for the interaction of biological pathways and the prediction of effects of unobserved conditions for applications such as combinatorial therapy and cellular reprogramming. Our work is motivated by recent advances in deep generative models that have enabled synthesis of images and natural language with desired properties from interpolation in a "latent representation" of the data.

Keywords: Single-cell RNA seq · Gene expression ·
Generative modeling · Deep learning

1 Introduction

The gene expression profile of a cell represents the nonlinear composition of distinct biological processes [24]. Recent research has sought to discern "basis vectors" that constructively explain gene expression. The gene expression profile of a cell is postulated to be a composition of its basis vectors, which represent activities in different biological pathways or systems [4, 16, 17]. Inspired by this idea and recent advances in generative modelling, we hypothesized that we could

© Springer Nature Switzerland AG 2019
L. J. Cowen (Ed.): RECOMB 2019, LNBI 11467, pp. 256–271, 2019.
https://doi.org/10.1007/978-3-030-17083-7_16

find treatment-condition specific basis vectors that could be combined to predict the gene expression profile of a cell when it is exposed to a novel combination of conditions [11,19]. We depart from previous work by discovering basis vectors in a latent space instead of the gene expression space itself. We enforce a disentangled representation in our latent space, and represent the activity corresponding to a specific condition by a distinct portion of the latent code. After performing linear operations on basis vectors in latent space, we use a nonlinear decoder from latent to expression space to recover nonlinear interactions in expression space.

One application of disentangled representations is the prediction of gene expression levels of cells under unobserved combinations of treatment conditions. Cells are often perturbed by a combination of factors for disease treatment or cell type engineering [1,23]. However, the search for the correct combination of factors is plagued by the exponential number of potential combinations. Combinatorial cancer drug studies can not examine all 2^{1000} combinations of 1000 drugs, and stem cell induction studies can not try all 2^{50} combinations of 50 transcription factors. We propose to learn basis vectors from a limited number of combinations to inform us about the space we have not explored. Instead of experimentally testing all possible combinations of conditions, we train on a simplified set of combinations, and from these learn basis vectors that can be combined to predict gene expression in an arbitrary combination. Our model can thus guide cellular engineering efforts to produce cells with a desired gene expression pattern.

Inspired by recent work in predicting gene expression profiles from SNP information [27], histone modification [22] and protein binding profiles [6] using deep learning approaches, we use a deep autoencoder to learn the condition-specific basis vectors using training data from single-cell RNA-seq data (scRNA-seq). [9,18] It is not possible to train our autoencoder with a few bulk RNA-seq experiments. The thousands of observations from single cells in conditions of interest provided by single-cell RNA-seq is critical in allowing us to train our autoencoder to learn condition-specific basis vectors. In this study we use scRNA-seq data for 31 different cell populations, each receiving a distinct combination of five signalling pathway activators or inhibitors. After learning the basis vectors corresponding to these pathways in the disentangled latent space, we demonstrate that we can interpolate in the latent space to generate realistic looking profiles for treatment combinations unseen in the training data.

2   Methods

2.1   Disentangled Representations Encode Condition Specific Expression

We consider representing basis vectors either in gene expression space or in a latent space that needs to be decoded into gene expression space. In both cases our ultimate goal is to predict a gene expression profile for an unseen combination of conditions.

Let the training set consist of n cells indexed $\{1 \dots n\}$. Let x_i denote cell i's gene expression profile. Each profile will be represented by a vector of length d genes. Let $U \in \mathbb{R}^d$ denote the set of all possible gene expression profiles. We will consider S possible treatments, and use s_i to denote the set of treatments cell i receives. There are 2^S possible such sets (treatment combinations). Let $\mathbf{s}_i \in \mathbb{R}^S$ denote the vector of length S that represents the treatments seen by cell i. In this representation $\mathbf{s}_i$ has a nonzero entry at position j if and only if $j \in s_i$. Let $V \in \mathbb{R}^S$ denote the set of $\mathbf{s}_i$ we have in the training set. $\mathbf{s}_i$ has two nice properties:

1. **Orthogonality** $\mathbf{s}_i \cdot \mathbf{s}_j = 0$ if $s_i \cap s_j = \{\emptyset\}$.
2. **Linearity** $\mathbf{s}_i$ lives in a linear subspace $\sum_{a \in s_i} c_a \mathbf{s}_a$, where $\mathbf{s}_a$ is a possible latent code under single treatment a.

These properties suggest that the canonical basis of $\mathbb{R}^S$ is an obvious basis for V. A basis vector corresponds to a single treatment condition, and the single nonzero entry is 1. For any s_i, we can easily construct a possible $\mathbf{s}_i$ as a linear combination of these basis vectors.

As a control we will assume x_i is a linear function of cell i's treatment combination represented by $\mathbf{s}_i$. Thus, there exists a linear transformation: $L : \mathbb{R}^S \rightarrow U$ between the vector spaces $\mathbb{R}^S$ and U. In this case given $\mathbf{s}_i$ and $\mathbf{s}_j$, $L(\mathbf{s}_i + \mathbf{s}_j) = L(\mathbf{s}_i) + L(\mathbf{s}_j)$. Arithmetic in $\mathbb{R}^S$ correspond directly to arithmetic in U. Thus there is a simple set of basis vectors for U in expression space, and they correspond to gene expression profiles under a single treatment (the image of the canonical basis of $\mathbb{R}^S$). Every possible gene expression profile can then be expressed as a linear combination of these basis vectors. To generate gene expression profiles for an unseen treatment combination we can sample linear combinations of the corresponding single-treatment basis vectors. This is a natural baseline for this task.

In our method we assume that x_i is not a linear function of cell i's treatment combination. Instead, we assume there is a nonlinear function $T : \mathbb{R}^S \rightarrow U$ such that $T(\mathbf{s}_i + \mathbf{s}_j) \neq T(\mathbf{s}_i) + T(\mathbf{s}_j)$. In this case, arithmetic in $\mathbb{R}^S$ no longer correspond to arithmetic in U. Therefore, we can no longer find a set of linear basis vectors for U. However, arithmetic in $\mathbb{R}^S$ is still well-defined. If we learn T, we'll be able to compute $T(\mathbf{s}_i + \mathbf{s}_j)$ directly, without resorting to linear arithmetic in U. To generate gene expression profiles for an unseen treatment combination, we can sample linear combinations of basis vectors in $\mathbb{R}^S$ and "decode" them to U using T. In practice, we can observe the action of T restricted on V. Thus we can learn a function D that approximates T. There is a problem remaining: given a training example x_i and s_i, we don't know the corresponding $\mathbf{s}_i$. We know which elements in s_i are nonzero, but we don't know their values.

This can be worked around by learning another map E that maps from expression space U to $\mathbb{R}^S$. In practice, this should be trained in conjunction with D so that the learned values for $\mathbf{s}_i$ are relevant for decoding. This points to a deep autoencoder architecture with E as the encoder module and D as the decoder module. $\mathbb{R}^S$ corresponds to the latent space. In practice, we prescribe u nonzero units to a single treatment condition in the latent code to increase its expressiveness with limited training budget. Thus the latent code for cell i, $z_i \in \mathbb{R}^{S \times u}$, obeys a block structure: $z_i = z_{i,1}|z_{i,2}|\ldots|z_{i,|S|}$ where $z_{i,a} = 0$ if $a \notin s_i$. This structure is enforced through training via a regularization loss as described below.

2.2 Latent Space Discovery with a Constrained Autoencoder

We learn a decoder D of our latent space and a corresponding encoder E using the autoencoder architecture shown in Fig. 1. We first compress gene expression profiles into their top k principal component (PC) coefficients, where k is a tunable parameter. Attempts to use the autoencoder model without the PCA component was not successful because the inherent noise in the ESC dataset was preventing any meaningful training of the neural network. The set of PC coefficients is used as input to encoder E, which is a three-layer fully-connected neural network: $z_i = \sigma(\sigma(x_i W_1 + b_1)W_2 + b_2)W_3 + b_3$. σ denotes the standard relu activation function. All hidden layers have the same size h, a tunable parameter.

The block structure of the latent code is enforced by a combination of L1 and L2 regularization [28]. Let $\hat{z}_i$ denote the vector where the units that are allowed to be nonzero under s_i all have the value 1. For example, if $S = 3$, $u = 2$ (three possible treatment conditions, 2 latent units used for each treatment condition) and $s_i = \{1, 3\}$, then $\hat{z}_i = 110011$. Note that $\hat{z}_i$ is an expanded version of s_i defined above, where all nonzero values are set to 1. We define the L1 ratio loss on z_i to be:

$$L1 = \frac{\|z_i \cdot Not(\hat{z}_i)\|_1}{\|z_i \cdot \hat{z}_i\|_1} \tag{1}$$

The Not operator above turns 1 into 0 and 0 into 1 entry-wise. The minimization of this loss tries to force all blocks not corresponding to s_i to be 0 while encouraging the blocks that do correspond to s_i to take on non-zero values. A standard L2 loss is also imposed on the blocks that are supposed to be active to encourage more even weight spreading across nonzero units [28].

The regularized latent code is then used as input to a fully-connected neural network decoder to generate reconstructed PC coefficients. Instead of producing a single reconstructed vector, the decoder produces a Gaussian distribution with diagonal covariance matrix to represent its uncertainty about its predictions [10]. The loss is the negative log likelihood of the observed PC coefficients vector used as encoder input under this distribution. The stochastic decoder backpropagates the loss through the parameters of the distribution using reparameterization [11]. The mean of the Gaussian is represented by $\sigma(\sigma(\sigma(z_i W_4 + b_4)W_5 + b_5)W_6 + b_6$, whereas the diagonal of the covariance matrix is represented by $\sigma(\sigma(\sigma(z_i W_4 + b_4)W_5 + b_5)W_7 + b_7)$.

The loss associated with each data point thus has three components: L1 ratio loss, L2 loss and reconstruction loss. In practice, a scaling factor of 20 for the L1 ratio loss and L2 loss was necessary for them to have roughly the same starting magnitude as the reconstruction loss.

Deep learning models can be influenced by random initialization effects and stochasticity in the training process [12]. We mitigate these effects by training an ensemble of models, and pool the generated results from all those models.

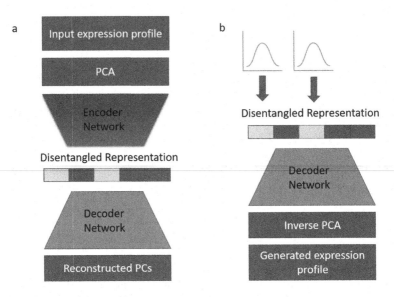

Fig. 1. (a) Schematic of neural network model used in training. (b) Schematic of neural network model used in generation.

2.3 Prediction of Gene Expression Profiles of Unseen Combinations

To predict the gene expression profile of an unseen combination of treatment conditions we generate a latent representation z for the unseen combination by using zero blocks for conditions not present in the combination and non-zero sample blocks for the active conditions as illustrated in Fig. 1b. This corresponds to adding together basis vectors for the selected conditions in latent space. We assume that the distribution of each latent block observed over the training set is representative of its corresponding condition in the test set and that different latent blocks are independent. Under these assumptions, for an active condition we generate its latent block by sampling from the block's observed distribution over the training data. The sampling is implemented through random selection from the list of its observed values in the training data.

In practice, we sample a population of the latent codes for the unseen combination to generate a population of gene expression profiles. Then we can decode the latent codes to a population of gene expression profiles using the decoder and inverse-PCA. From this population, the predicted mean gene expression profile of that treatment combination can be computed.

2.4   Experimental Design

The model is designed for the purpose of generating typical gene expression profiles for cells under treatment combinations not seen in the training data. In practice, we evaluate the model's performance by obtaining gene expression profiles for every possible treatment combination, then dividing the given data into a training set that contains certain treatment combinations and a test set that contains the rest. We can train the model on the training set and try to generate gene expression profiles for the treatment combinations in the test set. Then we compare the generated profiles with the actual experimental data in the test set to measure the model's performance. We can either compare the Pearson r across all the genes for each treatment combination or compare the Pearson r across all the combinations for each gene. We use two datasets described below to evaluate our model.

2.5   Synthetic Data

We generated synthetic data to test our model's performance on potential nonlinear interactions between pathways. Our simulated data consist of four treatment conditions. Condition one and condition two, when present in conjunction, exert a combinatorial effect. All other interactions between conditions are linear.

The "mean profile" for each combinatorial treatment condition X_s, where s is a set that could contain numbers 1 to 4 depending on what treatments are active, is calculated as follows:

$$X_1, X_2, X_3, X_4 \sim \mathcal{N}(500, 300)^{100} \tag{2}$$

$$X_c \sim \frac{1}{2}\mathcal{N}(500, 300)^{100} + \frac{1}{2}(X_1 + X_2) \tag{3}$$

$$otherwise, X_S = \begin{cases} \frac{1}{|S|-1}(X_c + \sum_{i \in S/\{1,2\}} X_i), & \text{if } \{1, 2\} \subset S \\ \frac{1}{|S|}\sum_{i \in S} X_i, & \text{otherwise} \end{cases} \tag{4}$$

300 individual "gene expression samples" are created from the "mean profile" for each treatment combination by adding a fixed set of 300 perturbations to the "mean profile". The perturbations are chosen to be on a five dimensional linear manifold.

2.6   Experimental Setup and Data Normalization

We generated an experimental dataset consisting of 32 sub-populations of epiblast-stage mESCs treated with all combinations of activators and inhibitors targeting five key developmental signaling pathways (Wnt, retinoic acid, Tgfb, Bmp and Fgf). Cells comprising these 32 conditions were labeled by condition and multiplexed into a single-cell RNA-sequencing run [8]. For preprocessing, UMI-unique counts were log-normalized in Seurat [21], which accounts for drop-out effects, and variable genes were identified by fitting an abundance-dependent trend to the genes [14]. Then, the data was scaled, removing unwanted sources of variation arising from number of genes and percentage of mitochondrial reads. [8] A mean of 109 and median of 105 cells (Min:39, Max:196) are present in each condition group. To balance the data from different treatment conditions, we upsample the data so each condition contains 300 gene expression profiles. 4635 highly variable genes are selected for analysis.

Combinatorially controlled genes were identified using a Bayesian regression framework as follows: Gene expression of cell i was modeled as $x_i \sim \mathcal{N}(y_i^T\beta, \sigma^2)$ where y_i is a vector consisting of linear terms indicating activation or inhibition of each pathway, and up to some k-th order interaction terms. Models containing varying orders of interaction terms were then fit using MCMC and compared using leave-one-out cross-validation. The framework was fit in pymc3. [20] Models were fit for the top 500 most variable genes [8].

2.7 Hyperparameter Selection

For the synthetic dataset, we select model hyperparameters based on performance in a validation generation task where data from all single and quadruple treatment combinations are used to train the model and generate expression profiles for triple treatment combinations. We use the hyperparameters which generated the best Pearson correlations across combinations for each gene on the test task, where data from all single, triple and quadruple combinations are used to generate expression profiles for double combinations. We select $k = 5$, $u = 4$ and $h = 80$. 11 different models are trained to form an ensemble.

For the ESC dataset, we select hyperparameters based on performance in a validation generation task where data from all single, quadruple and quintet combinations are used to train the model and generate expression profiles for double combinations. The hyperparameters which generated the best Pearson correlations across combinations for each gene is used on the test task, where data from all single, double, quadruple and quintet combinations are used to generate expression profiles for triple combinations. The selected hyperparameters are $k = 50$, $u = 4$ and $h = 40$. 21 different models are trained to form an ensemble.

Our Results section reports the performance on test tasks. For both simulated data and synthetic data, we find that the model performance is generally insensitive to the way the data is split into training and test sets. (Data not shown due to space).

2.8 Linear Baseline Methods

To examine the inherent noise present in our datasets, we introduce a supervised logistic regression classifier baseline. For each upsampled population of training data used to train an autoencoder model in the ensemble, a logistic regression classifier is trained to predict the on/off status of each single treatment condition. (e.g. five are trained for the ESC dataset). The logistic classifiers are used independently to predict the treatment combination of a cell from the PC coefficients of its gene expression profile.

For the generation task, we are not aware of any existing methods that reconstruct gene expression profiles for unseen treatment combinations. As a point of comparison we construct two baseline methods that make more linearity assumptions of the gene expression space than our model to reconstruct the mean gene expression profile of an unseen treatment combination.

First, we use a basis vector baseline method (control 1) that assumes that the interactions between treatment conditions are linear. Our baseline method uses a basis vector of gene expression for each of the five treatment conditions. The basis vector for a single condition is obtained by linear averaging over all gene expression profiles in the training set where the condition is active. We generate a predicted mean gene expression profile for a treatment combination by averaging the basis vectors for the individual conditions in the combination.

We also construct a marginalization linear baseline (control 2) to generate mean expression profile. It could be more resilient to the effects of combinatorial interactions in some cases. For example, to generate the mean profile for treatment combination $\{1, 3\}$ in the simulated data, we average all cells with treatment combinations $\{1, 2, 3\}$, $\{1, 3, 4\}$ and $\{1, 2, 3, 4\}$. In general, for treatment combination s in the test set, we generate a mean gene expression profile from all cells in the training set whose treatment combination contains s.

3   Results

3.1   Model Performance

Before using the model to generate gene expression profiles, we evaluate model training performance by measuring autoencoder reconstruction of the PC coefficients of held out data. The ESC dataset is extremely noisy. The first 50 PCs only capture 20% of the total variation. Most of the variation is captured by the first few PCs and the later PCs are exponentially less informative. We can examine the autoencoder's performance in reconstructing each PC coefficient by calculating the Pearson correlation coefficient between reconstructed values and observed values for that coefficient across the training and sets (Fig. 2). As expected, our model learns to reconstruct the first few PCs well and does less well for less significant principal components. The variations across models in the ensemble are indicated using error bars. We see limited variation across models in the ensemble. We can also examine the Pearson correlation coefficient across reconstructed PCs and observed PCs for each cell. All training and test set examples from all ensembles are pooled in the distribution plot. Results are shown in the bar charts in Fig. 2. Reconstruction performance is comparable across the training data and test data for both the synthetic and experimental datasets, suggesting that overfitting does not occur during training.

We also perform a control experiment where the regularization terms are turned off. The reconstruction performance is identical to that obtained with those terms active, suggesting that the regularization imposed does not hurt reconstruction.

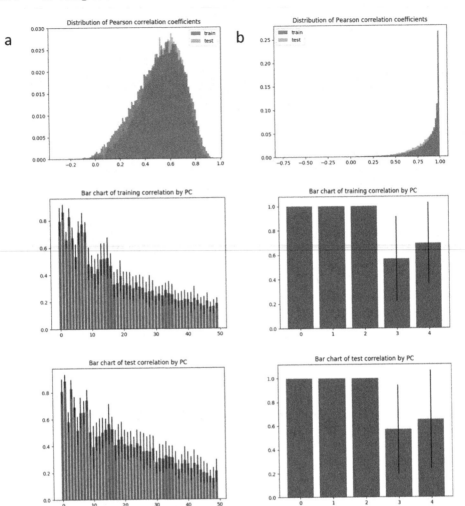

Fig. 2. Bar charts of reconstruction performance by PC (error bar is std across ensemble) and distributions of Pearson correlations between reconstructed PCs and actual PCs for the gene expression profiles in the training and test set for (a) ESC dataset. Training distribution mean = 0.53, std = 0.19, test mean = 0.54, std = 0.17; (b) Synthetic dataset. Training distribution mean = 0.83, std = 0.19, test mean = 0.81, std = 0.18.

We next evaluate how well the block structure in the latent code is enforced. We define the disentanglement ratio to be how much of the L1 norm of the latent code is attributable to units that are supposed to be nonzero. Mathematically, it is defined as:

$$disentanglement = \frac{\|z_i \cdot \hat{z}_i\|_1}{\|z_i\|_1} \tag{5}$$

Note this is different from the L1 ratio loss used to regularize the autoencoder, which could theoretically be infinite and inconvenient for analysis here. A ratio of 1 would indicate perfect disentanglement of the single treatment conditions in the latent code. From Fig. 3, we can see that for most cells in the training and test set, the disentanglement ratio is close to 1, suggesting good regularization. The disentanglement ratios are stable across training and test set, suggesting that overfitting does not occur. Note the training sets for both mESC and synthetic data have a spike at 1, which corresponds to the combination where all single conditions are active. This combination is not present in the test set, thus there is no spike in the corresponding plot.

We further evaluate the quality of our latent space structure by measuring how well we can predict s_i, the set of treatments given cell i, from the latent code z_i. Given our disentangled representation we select s_i based upon the $\hat{z}_i$ that has the least cosine distance from $|z_i|$. As we can see from the count plot of the rank of the correct $\hat{z}_i$ in Fig. 3, in most cases, the $\hat{z}_i$ given by the correct treatment combination has the least cosine distance with the actual latent code, or is ranked second or third. We see from the error bars that there is limited variation across different models. For our ESC dataset, our prediction of s_i from the PC coefficients using the logistic regression baseline method achieves 74% accuracy on the training set and 60% accuracy on the test set. For the synthetic dataset, our prediction achieves 56% accuracy on the training set and 64% on the test set. The accuracy of the latent code prediction is given in the caption of Fig. 3. We see that for ESC data, which is inherently more complex than the synthetic data, our model's latent code does not perform as well in prediction, especially in the test set. A potential explanation is that the cosine distance metric could be problematic in cases where one sampled block values are much larger than another sampled block. More importantly, our model is not designed for prediction and no supervised model is fit on the latent code. We are satisfied with the model's regularization performance in imposing block structure on the latent code in the test set and proceed to its main use case, generation.

3.2 Generation

We generated 100 gene expression profiles for each treatment condition in the test set. We first examine the correlation of the mean of the 100 profiles with the mean of the observed gene expression profiles. The results from our two baseline methods were also generated for comparison. Note that all correlations we computed are for the full generated gene expression profiles, not the PC coefficients.

We found that the autoencoder method outperforms the two baseline methods for all treatment conditions (Table 1). Model performance in the ensemble is heterogeneous. This may be caused by the random upsampling in the data preparation process and the stochasticity of model training via stochastic gradient descent. Our results show if we further average the mean profiles produced by all the models in the ensemble, the correlations with the actual mean gene expression profiles improve across all treatment combinations. Most importantly, the

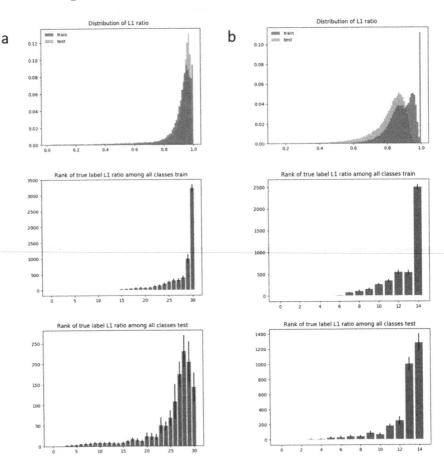

Fig. 3. Bar charts of the rank counts of the true labels of the training and test set cells (higher rank means the latent code is more aligned with the correct label) and distribution of the disentanglement ratio (blue for training, green for test) for (a) ESC dataset. Training distribution mean $= 0.92$, std $= 0.082$, test mean $= 0.89$, std $= 0.16$; training highest rank (correct prediction) mean $= 51\%$, std $= 0.02\%$, test mean $= 12\%$, std $= 0.03\%$. (b) Synthetic dataset. Training distribution mean $= 0.89$, std $= 0.083$, test mean $= 0.81$, std $= 0.11$; training highest rank mean $= 55\%$, std $= 1.3\%$, test mean $= 43\%$, std $= 3.8\%$. (Color figure online)

autoencoder's performance is relatively stable across the combinations tested, in contrast to the linear baseline methods, as seen in Fig. 4b.

We hypothesize that the autoencoder can successfully disentangle both linear and nonlinear interactions between treatment conditions within its latent space, causing it to perform well in both combinations with mostly linear interactions and combinations with nonlinear interactions. Results on this task for the synthetic dataset, as present in Table 2, confirm this hypothesis. We observe that the autoencoder method has superior performance when both combinatorial

Table 1. Table of mean reconstruction performance measured by correlation across all genes for different test combinations for ESC dataset

Treatment condition (active pathways)	Pearson r mean (std) of mean profiles (n = 21)	Pearson r of averaged mean profile	Pearson r of baseline 1 profile	Pearson r of baseline 2 profile
Bmp, Tgfb, Fgf	0.63 (0.045)	0.68	0.44	0.3
Wnt, Tgfb, Fgf	0.64 (0.041)	0.72	0.26	0.41
Wnt, Bmp, Fgf	0.61 (0.076)	0.67	0.22	0.37
Wnt, Bmp, Tgfb	0.63 (0.043)	0.72	0.15	0.49
RA, Tgfb, Fgf	0.73 (0.043)	0.79	0.10	0.55
RA, Bmp, Fgf	0.70 (0.036)	0.78	0.46	0.015
RA, Bmp, Tgfb	0.72 (0.034)	0.79	0.084	0.62
RA, Wnt, Fgf	0.57 (0.058)	0.66	0.10	0.43
RA, Wnt, Tgfb	0.67 (0.047)	0.75	0.030	0.6
RA, Wnt, Bmp	0.57 (0.1)	0.67	0.084	0.5

Table 2. Table of mean reconstruction performance measured by correlation across all genes for different test combinations for synthetic dataset.

Treatment condition	Pearson r mean (std) of mean profiles (n = 21)	Pearson r of averaged mean profile	Pearson r of baseline 1 profile	Pearson r of baseline 2 profile
1, 2	0.54 (0.065)	0.56	0.13	0.81
1, 3	0.90 (0.039)	0.92	1	0.36
2, 3	0.87 (0.051)	0.90	1	0.28
1, 4	0.85 (0.039)	0.88	1	0.36
2, 4	0.82 (0.090)	0.86	1	0.34
3, 4	0.90 (0.037)	0.96	1	1

interactions (combination $\{1, 2\}$) and linear interactions (all other combinations) exist between conditions. In comparison, the baseline methods either fail at one or the other.

We next evaluated for each gene the correlation of its predicted expression level and the actual expression level across different treatment conditions. An averaged mean expression profile from the models in the ensemble are used. This is perhaps the more relevant task in cellular engineering, where the desired information is the treatment combination that would upregulate or silence a gene. The results are shown in Fig. 4a. The autoencoder performs much better in this task than the control methods as well.

a
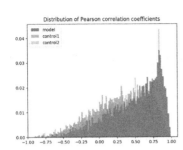

b
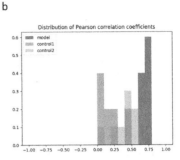

Fig. 4. ESC data. (a) Histogram of correlation across 10 different combinations in the test set between observed expression value and predicted expression value for each gene. Data is from averaged profile across ensemble. Autoencoder (blue) mean = 0.48 std = 0.36, control 1 (green) mean = 0.30 std = 0.38, control 2 (cyan) mean = 0.37 std = 0.39. (b) Histogram of correlation across all genes for different combinations in the test set. Data is from averaged profile across ensemble. Autoencoder (blue) mean = 0.73 std = 0.05, control 1 (green) mean = 0.19 std = 0.14, control 2 (cyan) mean = 0.43 std = 0.17. (Color figure online)

We have evaluated here the ability of the model to produce the mean gene expression values for an unseen treatment combination, which we judge to be the most relevant task for cell reprogramming and disease treatment. Future work would be aimed at reproducing other population-level statistics such as variance. This is challenging since the PCA step would need to be replaced by a denoising technique which preserves higher order statistics information.

3.3   Performance on Combinatorial Genes

To further test the model's performance at recovering combinatorial effects on gene expression, we examine the model's performance on a set of 67 genes in our ESC dataset where distinct treatment conditions interact nonlinearly [8]. We find that the autoencoder can learn disentangled latent representations of combinatorial effects and faithfully reproduce them in generating profiles for unseen combinations (Fig. 5). A heatmap of predicted gene expression for these 67 combinatorial genes further suggests that the autoencoder strongly outperforms linear baselines.

4   Discussion

Recent research has suggested that latent representations of data learned by deep generative models such as generative adversarial networks and variational autoencoders can be manipulated meaningfully. Algebraic operations in the latent space can correspond to semantic operations in the decoded data space [3,26]. Such latent space manipulations are commonly used now to generate

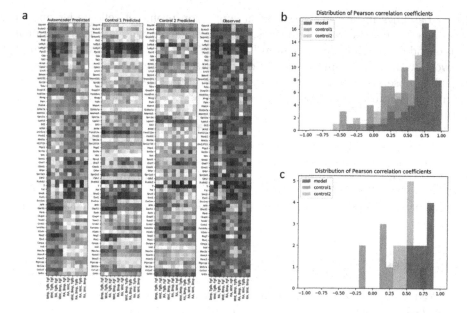

Fig. 5. Performance on combinatorial genes on the ESC dataset. (a) Heatmap of predicted gene expression for autoencoder method and two controls. The mean of the ensemble was used for the autoencoder method. From left to right: autoencoder, control 1, control 2, observed (b) Histogram of correlation across 10 different combinations in the test set between experimental expression value and generated expression value for the combinatorial genes. Autoencoder (blue) mean = 0.70 std = 0.21, control 1 (green) mean = 0.38 std = 0.33, control 2 (cyan) mean = 0.52 std = 0.32. (c) Histogram of correlation across the combinatorial genes for 10 different combinations in the test set. Data is from averaged profile across 21 models. Autoencoder (blue) mean = 0.73 std = 0.14, control 1 (green) mean = 0.30 std = 0.30, control 2 (cyan) mean = 0.49 std = 0.09. (Color figure online)

synthetic data examples with desired properties, such as chemical properties for molecules or content of images [7, 19].

Recently, generative approaches using deep learning models have also been applied to predict gene expression profiles from complementary information [27], generating full profiles from landmark genes [25], deriving latent representations to integrate datasets [2], and removing batch and technical variation effects [13]. This work, while informative, does not allow the generation of gene expression profiles "from scratch" for previously unseen treatment conditions, which holds the promise to accelerate cellular engineering and therapeutic development where a large number of treatment conditions need to be explored [5, 15].

Inspired by the advances in latent space manipulation and applications of generative models to gene expression, we introduce a new model of cellular identity based on a disentangled latent representation. This latent representation, together with a nonlinear decoder, enables our approach to vastly outperform linear baselines in predicting the mean expression profile of cells under unseen combinations of treatment conditions. We implement our model as a variational autoencoder. In principle, we can also implement it as a generative adversarial network with a disentangled (conditioned) noise representation. In contrast to sampling observed latent block values in generation, the distributions over the noise representation must be specified in advance. This exploration is left for future work.

We expect that the guiding principle of disentangled representations will find new applications in computational biology and other fields as they permit the separation of complex nonlinear responses to distinct perturbations. Our implementation can be found at http://giffordlab.mit.edu/disentangled.

Acknowledgements. We acknowledge the members of the Gifford and Sherwood labs for helpful discussion.

References

1. Al-Lazikani, B., Banerji, U., Workman, P.: Combinatorial drug therapy for cancer in the post-genomic era. Nat. Biotechnol. **30**(7), 679 (2012)
2. Ghahramani, A., Watt, F.M., Luscombe, N.M.: Generative adversarial networks simulate gene expression and predict perturbations in single cells. bioArXiv preprint (2018). https://doi.org/10.1101/262501
3. Bojanowski, P., Joulin, A., Lopez-Paz, D., Szlam, A.: Optimizing the latent space of generative networks. arXiv preprint arXiv:1707.05776 (2017)
4. Ding, J., Condon, A., Shah, S.P.: Interpretable dimensionality reduction of single cell transcriptome data with deep generative models. Nat. Commun. **9**(1), 2002 (2018)
5. Eguchi, A., et al.: Reprogramming cell fate with a genome-scale library of artificial transcription factors. Proc. National Acad. Sci. **113**(51), E8257–E8266 (2016)
6. Ferdous, M.M., Bao, Y., Vinciotti, V., Liu, X., Wilson, P.: Predicting gene expression from genome wide protein binding profiles. Neurocomputing **275**, 1490–1499 (2018)
7. Gómez-Bombarelli, R., et al.: Automatic chemical design using a data-driven continuous representation of molecules. ACS Cent. Sci. **4**(2), 268–276 (2018)
8. Yeo, G.H.T., Lin, L., Qi, Y.C., Gifford, D.K., Sherwood, R.I.: Elucidation of combinatorial signaling logic with multiplexed barcodelet single-cell RNA-seq (2018, in prep)
9. Jaitin, D.A., et al.: Massively parallel single-cell RNA-seq for marker-free decomposition of tissues into cell types. Science **343**(6172), 776–779 (2014)
10. Kendall, A., Gal, Y.: What uncertainties do we need in Bayesian deep learning for computer vision? In: Advances in Neural Information Processing Systems, pp. 5574–5584 (2017)
11. Kingma, D.P., Mohamed, S., Rezende, D.J., Welling, M.: Semi-supervised learning with deep generative models. In: Advances in Neural Information Processing Systems, pp. 3581–3589 (2014)
12. Li, H., Xu, Z., Taylor, G., Goldstein, T.: Visualizing the loss landscape of neural nets. arXiv preprint arXiv:1712.09913 (2017)

13. Lopez, R., Regier, J., Cole, M., Jordan, M., Yosef, N.: A deep generative model for gene expression profiles from single-cell RNA sequencing. arXiv preprint arXiv:1709.02082 (2017)
14. Lun, A.T., Bach, K., Marioni, J.C.: Pooling across cells to normalize single-cell RNA sequencing data with many zero counts. Genome Biol. **17**(1), 75 (2016)
15. Macarron, R., et al.: Impact of high-throughput screening in biomedical research. Nat. Rev. Drug Discov. **10**(3), 188 (2011)
16. Mohammadi, S., Ravindra, V., Gleich, D.F., Grama, A.: A geometric approach to characterize the functional identity of single cells. Nat. Commun. **9**(1), 1516 (2018)
17. Okawa, S., et al.: Transcriptional synergy as an emergent property defining cell subpopulation identity enables population shift. Nat. Commun. **9**(1), 2595 (2018)
18. Patel, A.P., et al.: Single-cell RNA-seq highlights intratumoral heterogeneity in primary glioblastoma. Science **344**(6190), 1396–1401 (2014)
19. Radford, A., Metz, L., Chintala, S.: Unsupervised representation learning with deep convolutional generative adversarial networks. arXiv preprint arXiv:1511.06434 (2015)
20. Salvatier, J., Wiecki, T.V., Fonnesbeck, C.: Probabilistic programming in python using PyMC3. PeerJ Comput. Sci. **2**, e55 (2016). https://doi.org/10.7717/peerj-cs.55
21. Satija, R., Farrell, J.A., Gennert, D., Schier, A.F., Regev, A.: Spatial reconstruction of single-cell gene expression data. Nat. Biotechnol. **33**(5), 495 (2015)
22. Singh, R., Lanchantin, J., Robins, G., Qi, Y.: DeepChrome: deep-learning for predicting gene expression from histone modifications. Bioinformatics **32**(17), i639–i648 (2016)
23. Takahashi, K., et al.: Induction of pluripotent stem cells from adult human fibroblasts by defined factors. Cell **131**(5), 861–872 (2007)
24. Wagner, A., Regev, A., Yosef, N.: Revealing the vectors of cellular identity with single-cell genomics. Nat. Biotechnol. **34**(11), 1145 (2016)
25. Wang, X., Ghasedi Dizaji, K., Huang, H.: Conditional generative adversarial network for gene expression inference. Bioinformatics **34**(17), i603–i611 (2018)
26. White, T.: Sampling generative networks. arXiv preprint arXiv:1609.04468 (2016)
27. Xie, R., Wen, J., Quitadamo, A., Cheng, J., Shi, X.: A deep auto-encoder model for gene expression prediction. BMC Genomics **18**(9), 845 (2017)
28. Zou, H., Hastie, T.: Regularization and variable selection via the elastic net. J. Roy. Stat. Soc. Ser. B (Stat. Methodol.) **67**(2), 301–320 (2005)

RENET: A Deep Learning Approach for Extracting Gene-Disease Associations from Literature

Ye Wu, Ruibang Luo, Henry C. M. Leung, Hing-Fung Ting,
and Tak-Wah Lam[✉]

Department of Computer Science, The University of Hong Kong,
Pokfulam, Hong Kong
twlam@cs.hku.hk

Abstract. Over one million new biomedical articles are published every year. Efficient and accurate text-mining tools are urgently needed to automatically extract knowledge from these articles to support research and genetic testing. In particular, the extraction of gene-disease associations is mostly studied. However, existing text-mining tools for extracting gene-disease associations have limited capacity, as each sentence is considered separately. Our experiments show that the best existing tools, such as BeFree and DTMiner, achieve a precision of 48% and recall rate of 78% at most. In this study, we designed and implemented a deep learning approach, named RENET, which considers the correlation between the sentences in an article to extract gene-disease associations. Our method has significantly improved the precision and recall rate to 85.2% and 81.8%, respectively. The source code of RENET is available at https://bitbucket.org/alexwuhkucs/gda-extraction/src/master/.

Keywords: Literature mining · Relation Extraction ·
Gene-disease association · Deep learning

1 Introduction

Knowledge of gene-disease associations is essential for clinical diagnosis, selecting preventive and therapeutic strategies against diseases, and developing new treatments for diseases [1, 2]. Many gene-disease associations have been discovered after decades of dedicated research, but the discoveries are spread throughout a vast amount of biomedical literature. Finding all the right associations is often tedious and difficult. Moreover, biomedical literature is growing at a rate of over one million articles per year; the public medical literature database MEDLINE has a growth rate of 0.195 and doubles every 5.1 years [3]. This makes manual extraction impossible. Thus, there is a pressing need for a tool that can extract gene-disease association information from biomedical literature accurately and efficiently.

There are two major tasks in the extraction process: (1) Name Entity Recognition (NER), for recognizing gene and disease entities in an article, and (2) Relation Extraction (RE), for determining whether there is an association between a recognized

© Springer Nature Switzerland AG 2019
L. J. Cowen (Ed.): RECOMB 2019, LNBI 11467, pp. 272–284, 2019.
https://doi.org/10.1007/978-3-030-17083-7_17

gene and a disease in the article. NER is relatively simple because there are a limited number of possible genes and diseases. The existing tool for NER is mature enough to recognize genes and diseases in an article. For example, the PubTator system [4] is a popular tool, which integrates several state-of-art NER algorithms to assist manual curation of biomedical associations. It pre-annotates biomedical entities in MEDLINE abstracts for human experts to read the text and determine the relationship between different entities. By automating NER, PubTator successfully improved the efficiency and accuracy of manual extraction of gene-disease associations. However, it is still labor-intensive and time-consuming to rely upon humans to perform the RE task.

RE is more difficult than NER since instead of working on single words, it needs to understand the meaning of a full sentence (sentence-level) or even a whole article (document-level). As it is difficult to understand an entire article, the existing tools can extract information only from each sentence separately without considering the context of the whole article. Using the following sentence as an example: "polymorphisms of the SYNGR1 and SYNII genes have been shown to be a risk factor for bipolar disorder" [5], the existing tools will derive an association between the genes "SYNGR1" and "SYNII" and the disease "bipolar disorders". A wide range of approaches have been applied to sentence-level RE, including co-occurrence based statistics [6], rule-based systems [7, 8], and machine-learning methods [9–11]. In particular, supervised learning with natural language processing (NLP) feature engineering techniques has achieved good performance. The BeFree system [12] achieved state-of-art results by using a support vector machine as a classifier and dependency parsing trees to extract deep semantic features. Although significant progress has been made in sentence-level RE, this approach overlooks crucial contextual information in the document. First, the associated gene and disease entities may not appear in the same sentence; they may instead be spread over multiple sentences. Sentence-level RE methods overlook these possibilities and result in a lower recall. Second, single sentences are not enough to represent the main idea of the whole article. Using the example above, the known association between SYNGR1 and SYNII genes and bipolar disorder might be mentioned only in the background, while the main article denies the association. Therefore, sentence-level RE is insufficient for extracting gene-disease associations from an article.

In this paper, we propose a document-level gene-disease association extraction approach. This approach works on understanding the context of the whole article and extracts gene-disease associations that are supported at the article level. We call these associations Gene-Disease-Article (G-D-A) associations. Our approach utilizes deep neural networks, which have been widely adopted in information extraction in recent years [13, 14]. Unlike traditional machine learning algorithms, which commonly require feature engineering, a deep neural network automatically learns from raw data the representations needed for feature detection or classification. This method is commonly used for modeling semantic composition in text [15].

Our method RENET not only captures the sentence-level relationships between gene and disease entities but also models the interaction across sentences to understand the context of the whole document. As shown in Fig. 1, RENET learns a representation of the document through two levels of abstraction. First, sentence representations are computed from word representations through a convolutional neural network (CNN).

Then sentence representations are transformed into a document representation through a recurrent neural network (RNN). Finally, the document representation is used as a collection of features for extracting G-D-A associations.

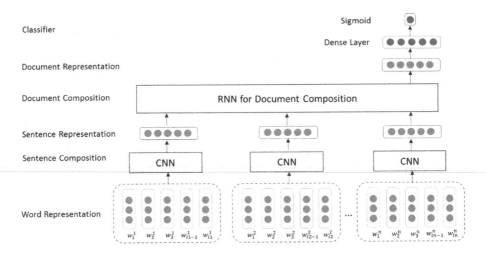

Fig. 1. The neural network for classification of G-D-A association. w_i^j stands for the i-th word in the j-th sentence, and l_j is the length of the j-th sentence.

Our experiments show that compared with the best existing tools, BeFree [12] and DTMiner [16] (which achieve at most 48% on precision and 78% on recall rate), our RENET method significantly improves the precision and recall rate to 85.2% and 81.8%, respectively.

Using RENET, we analyzed 1,032,790 abstracts stored in MEDLINE [17], a bibliographic database of life sciences and biomedical information, compiled by United States National Library of Medicine (NLM). We extracted about 869,000 G-D-As. The extracted G-D-As and their corresponding articles are available for download at https://bitbucket.org/alexwuhkucs/gda-extraction/src/master/.

2 Method

As shown in Fig. 2, the gene-disease extraction procedure has two major tasks: Name Entity Recognition (NER) and Relation Extraction (RE).

Fig. 2. Procedure for extracting true G-D-A associations

For NER, RENET uses the popular tool PubTator [4], which when given a text as input, returns the location of every gene and disease mentioned in the text and converts them to the corresponding IDs (e.g., see Fig. 3).

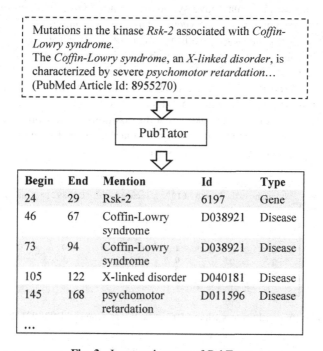

Fig. 3. Input and output of PubTator

The major contribution of this paper is an innovative and effective deep learning method for RENET to carry out the RE task, i.e., given the abstract of article A, and the target gene G and target disease D, decide whether G-D-A is a true association. We designed and implemented four deep neural networks for this task, and after comparing their performance, we selected the best one for RENET. All four neural networks adopted the same two-stage computation to classify the G-D-A association. For the first stage, we used a CNN to extract important signals from each sentence of the input abstract, and for the second stage, we used four variants of the standard RNN method, namely GRU, BiGRU, LSTM, and BiLSTM, to extract inter-sentence signals from the output of the CNN. These signals were fed to a simple feedforward neural network to determine their classification. We call the resulting networks CNN-GRU, CNN-BiGRU, CNN-LSTM, and CNN-BiLSTM. More details are given in Sects. 2.1, 2.2 and 2.3.

2.1 Word Representation

The input for the first stage of computation (i.e., for the CNN) is a sequence of word vectors constructed from the input abstract and the target gene and disease. We note that since a disease or gene may have synonyms or acronyms, we need the following preprocessing before constructing the vectors: based on the output of PubTator, replace every occurrence of a gene or disease in the abstract by its corresponding ID.

As shown in Fig. 4, every word in the abstract is represented by a vector of 204 numbers. The first 200 numbers capture important semantic features of the words [18], for which we used the tool Word2vec [19]. It is interesting to note that we tried to initialize these numbers randomly and found that the two approaches had similar results (see Sect. 3.4 for comparison). It is likely that the positions of the words are much more important than their semantic signals for G-D-A classification. In the end, we used the random number approach in RENET to construct this part of word vectors.

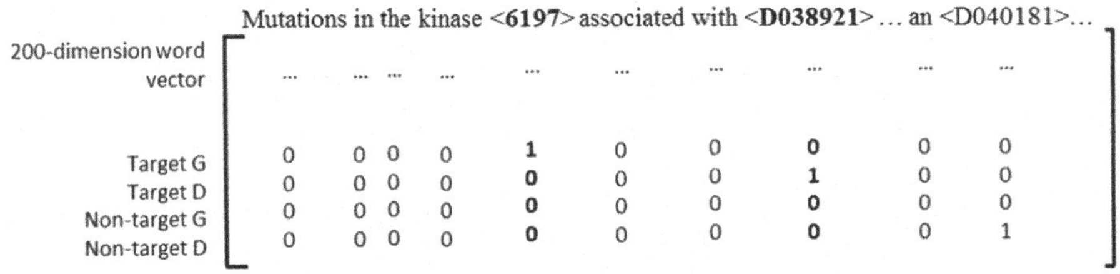

Mutations in the kinase <**6197**> associated with <**D038921**> ... an <D040181>...

200-dimension word vector	...	... ...	...	...	...	...	...	...	...
Target G	0	0 0	0	**1**	0	0	**0**	0	0
Target D	0	0 0	0	**0**	0	0	**1**	0	0
Non-target G	0	0 0	0	**0**	0	0	**0**	0	0
Non-target D	0	0 0	0	**0**	0	0	**0**	0	1

Fig. 4. A sequence of word vectors when 6197 is the target gene and D038921 the target disease

The remaining four numbers in a word vector are for marking whether the corresponding word is a target gene, a target disease, a non-target gene, or a non-target disease (see Fig. 4).

2.2 Sentence-Level Representation

After generating the word vectors, for each sentence in the abstract, we fed its sequence of word vectors to the CNN shown in Fig. 5 to generate its sentence-level representation, which is a vector of real numbers that captures important features of that sentence. Note that as in [20], we used filters of different widths to capture the features. Such a CNN can capture different local n-gram features in a sentence; this has been used successfully in various sentence-level text classification tasks [21, 22]. For our problem, we expected these filters to capture n-gram patterns around the target gene and target disease that may suggest a true association.

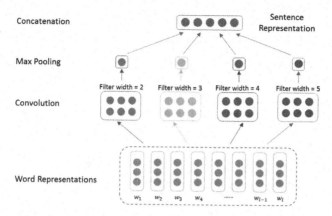

Fig. 5. Sentence-level representation generated by our CNN, which uses filters of widths 2, 3, 4, 5.

To provide more details about the CNN, let us denote a sentence consisting of l words as $w_1, w_2, ...w_l$. Each word w_i is represented by its vector e_i of 204 numbers. A convolutional operation acts on a window of k words. The input of the operator is the concatenation of the sequence of word vectors for the k words, which is denoted as $I = [e_i; e_{i+1}; ...; e_{i+k-1}]$. The output is calculated as

$$O = W \cdot I + b,$$

where W is the weight vector and b is the bias. The convolution operator is applied to every possible window of k words in the sentence, producing a list of features $O_1, O_2, ..., O_{l-k+1}$. A max pooling layer is applied to this sequence of features to keep only the output with the highest value. Finally, we concatenate the outputs of the different filters to get the representation of this sentence.

2.3 Document-Level Representation

After generating the sequence of sentence-level representations by the CNN, we constructed from this sequence a document-level representation that captures important signals from the whole document to help us make correct classifications.

Note that the output of the CNN captures signals only within a sentence, but this may not be enough, as it may overlook significant linguistic relationships between sentences. For example, relying only on these intra-sentence signals would not enable us to determine whether a sentence appears in the background section or in the conclusion section, but intuitively, one that appears in the conclusion section will give much stronger signal about the association.

To capture signals between sentences, we use RNN-based methods, which are designed for sequential data and work in time steps; the output for time step t is computed by combining and transforming (through a linear function) the input vector s_t for this time step and the output vector h_{t-1} of the previous time step. The output of the last time step is the document representation of the input abstract, as shown in Fig. 6.

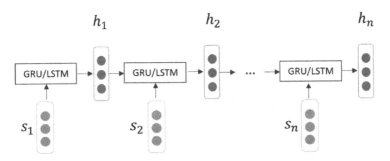

Document
Representation

h_1 h_2 h_n

Fig. 6. The process of generating document representation using GRU/LSTM.

To deal with the problem of gradient explosion and vanishing in training [23, 24], we did not use the original RNN; instead, we tried some of its variants. In particular, we tried the Long Short Term Memory model (LSTM) and the Gated Recurrent Unit (GRU) model [24, 25]. GRU and LSTM have similar architecture, in which special structures are introduced to decide to what extent information should be forgotten or updated in each time step. In this way, the contextual relationship between sentences can be modeled. For example, suppose the sentence "polymorphisms of the SYNGR1 and SYNII genes have been shown to be a risk factor for bipolar disorder" appears in the background section, while the main article denies the association between SYNGR1 and SYNII genes and bipolar disorder. With the special structures introduced by LSTM and GRU, the model can choose to forget about the previous memory from the background section and update the model with the denial of the association in the main article.

Although the special structures of GRU and LSTM enable them to model long-distance dependency, a single forward directional GRU/LSTM may be biased towards inputs in the later time steps. Therefore, we also compared GRU and LSTM with their bi-directional variants, namely BiGRU and BiLSTM [26]. In these bidirectional models, the outputs of forward and backward directions are concatenated as the representation for the document.

Finally, given the document-level representation, we applied a two-level feed-forward network for G-D-A classification. We used cross-entropy as the loss function and the Adam optimizer [27] to train our networks with back-propagation.

3 Experiments

3.1 Data Set

We conducted experiments on a large dataset generated from 30,192 abstracts in MEDLINE, maintained by DisGeNet [28], which is a platform that contains a large set of gene-disease associations collected from publicly available databases, such as CTD,

UniProt, and GAD. Most of the associations were manually curated. Our data included all the G-D-A associations where A is one of the 30,192 MEDLINE abstracts, and G and D appear in A. If such a G-D-A association appears in DisGeNet, we regard it as a true association; otherwise we regard it as false. For testing, we selected 1,000 MEDLINE abstracts randomly to form the test set, and the rest were used for the training and validation of RENET.

3.2 Comparison with Recent Tools

We compared RENET with the software tools BeFree [12] and DTMiner [16], two publicly available text-mining tools for extracting sentence-level gene-disease associations. Both BeFree and DTMiner use local lexical and global syntactic features with SVM as classifiers. To ensure a fair comparison, we used PubTator for the NER step for all three tools.

The experiment results are shown in Table 1. All our neural network methods outperformed BeFree and DTMiner by a significant margin in both precision (85.2% vs. 48.2%) and recall (81.8% vs. 78.4%). This indicates that our document-level analysis is more suitable than the sentence-level detection method used in BeFree and DTMiner.

Table 1. Comparison of different models

	Precision	Recall	F-Score
BeFree	48.2%	74.1%	58.4%
DTMiner	47.3%	78.4%	59.0%
CNN	84.2%	78.6%	81.3%
CNN-GRU (RENET)	85.2%	81.8%	**83.5%**
CNN-BiGRU	82.5%	83.6%	83.1%
CNN-LSTM	82.0%	83.2%	82.6%
CNN-BiLSTM	86.5%	79.5%	82.8%

The lower recall rate for DTMiner and BeFree can be explained by the missing of G-D pairs that do not appear in the same sentences in the abstract. The precision of BeFree and DTMiner is at least 30% lower than that of the CNN and CNN-RNN models, which indicates that information extracted from single sentences without contextual information is insufficient.

3.3 Comparison of Our Method with the Pure CNN Method

We compared the performance of our CNN-RNN method with the pure CNN method, which is our neural network without the second stage of computation, where we regard the whole abstract as one big sentence. Note from Table 1 that there is not much difference in the performance of the four different RNN methods, and they have only a small advantage over the pure CNN method. This indicates that CNN alone can capture the important features of the document, but the extra RNN layer boosts the performance of the CNN method.

We also note that of the CNN-RNN models, the CNN-GRU model (i.e., RENET) achieved the best performance. The reason may be that compared with GRU, LSTM has one more output structure, which enables it to forget more information in each time step. This feature enables LSTM to achieve better model sequences with long dependency along each time step, and even longer with the help of the bidirectional structure. But with sequences of shorter length, like those in our task, it may not perform better than a simpler structure like GRU.

3.4   Comparison Between Settings for Word Representations

We conducted experiments to find out whether pre-trained word-vectors can boost the performance of neural network models. In the experiments above, we used randomly initialized vectors drawn from a uniform distribution between -0.05 and 0.05. The randomly initialized vectors were fine-tuned during training via back-propagation [22]. We compared them with word2vec generated vectors that were pre-trained on PubMed and PMC texts [29]. We also experimented with the options to fine-tune vectors during the training. Since a CNN layer performs feature extraction from word representations in our neural network models, we chose the pure CNN model as the representative model. The results are shown in Table 2.

Table 2. Comparison of different settings of word representations

CNN		Precision	Recall	F-Score
Word2vec	Fine-tuned			
Yes	Yes	81.4%	81.1%	**81.3%**
Yes	No	75.9%	75.7%	75.8%
No	Yes	84.2%	78.6%	**81.3%**
No	No	75.2%	81.0%	78.0%

From these results, we can draw two conclusions. First, pre-trained word2vec vectors did not bring extra advantage to our task. Second, fine-tuning word vectors effectively improved the performance of the model. Therefore, there is no necessity to introduce pre-trained word2vec vectors in our task.

3.5   Fault Case Analysis

We performed a fault case analysis to determine why some of our predictions disagreed with the human-curated associations from DisGeNet. False negatives were a particular concern since they influence the comprehensiveness of the database RENET generates. Table 3 shows the confusion matrix for the predictions made by RENET.

Table 3. Confusion matrix

		Actual values (DisGeNet)		Total
		True G-D-A	False G-D-A	
Model predicts	True G-D-A	1229	**214**	1443
	False G-D-A	**273**	3506	3779
	Total	1502	3720	5222

For the 1000 MEDLINE abstracts we tested, we randomly sampled and analyzed 100 of the 273 false negative cases. We found that although the predictions did not agree exactly with the human-curated associations, some of them were not at all unreasonable. Following are four types of "false negative" faults we consider permissive.

(1) *RENET predicted a more specific disease.* In these cases, multiple diseases were mentioned in the abstract, with some being more general (e.g., cardiomyopathy) and some more specific (e.g., hypertrophic cardiomyopathy). DisGeNet reports associations of the general disease or both diseases, while RENET predicts the specific disease. In this case, RENET actually pinpoints a more accurate association between genes and diseases.

(2) *Errors of NER step propagated to RE.* In these cases, PubTator made errors in the NER step (e.g., it could not recognize a protein name for a gene), and this affected the results of the RE step.

(3) *Not enough information in the abstract.* In these cases, genes and diseases were mentioned in the abstract, but there was not enough evidence to suggest their association. We suppose human curators read both the abstract and the main body of the articles to curate the associations. The evidence is probably provided in the main body of the articles but not the abstract.

(4) *RENET predicted a more general disease.* This is similar to the first case above, but the model predicts a more general disease, while DisGeNet reports a specific disease or both. Although the RENET predictions are not as good as the sample answers, they are still relevant.

Detailed examples are provided in the Supplementary Materials (available at https://bitbucket.org/alexwuhkucs/gda-extraction/src/master/). Figure 7 shows a breakdown of the false negative cases. About 7.6% are other faults made by RENET without a specific reason.

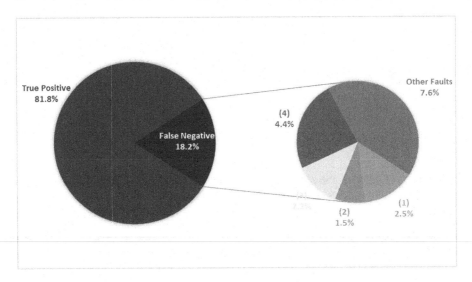

Fig. 7. Breakdown of false negatives

3.6   Large-Scale Extraction of Gene-Disease Article Association

We applied RENET on 1,032,790 MEDLINE abstracts pertaining to human diseases and genes, published from 1980 on. The abstracts were retrieved using the following query at https://www.ncbi.nlm.nih.gov/pubmed [12].

```
"Psychiatry and Psychology Category" [Mesh] AND "genetics" [Sub-
heading]) OR ("Diseases Category" [Mesh] AND "genetics" [Sub-
heading]) AND (hasabstract[text] AND ("1980" [PDAT]: "2018"
[PDAT]) AND "humans" [MeSH Terms] AND English[lang]).
```

We retrieved 869,152 G-D-A associations between 24,221 genes and 5,804 diseases, reported in 530,581 publications. The entire set of extracted G-D-A associations is available at https://bitbucket.org/alexwuhkucs/gda-extraction/src/master/.

4   Conclusion

In this paper, we introduce an innovative and efficient approach and implement a tool call RENET for extracting gene-disease association. RENET outperforms existing tools significantly because it uses deep learning to dig out important classification signals over the whole document while existing tools use SVM or rule-based methods to analyze sentences separately. Our tool learns a representation of a document through a two-stage computation: (1) from word representation to sentence representation using CNN; (2) from sentence representation to document representation using RNN-based method. We have done comprehensive experiments on a large-scale dataset generated from DisGeNet. RENET outperforms traditional sentence-level relation extraction methods by a large margin (more than 20% in F-Score).

In the analysis of the false negative cases, we found that RENET still makes sensible predictions despite not completely agreeing with the benchmark. We believe the methodology of this work is transferable to other association extraction tasks such as drug-disease associations, variant-disease associations and protein interactions as well.

Acknowledgments. This work was supported by Hong Kong ITF Grant ITS/331/17FP and General Research Fund No. 27204518.

References

1. Lu, Y.-F., Goldstein, D.B., Angrist, M., Cavalleri, G.: Personalized medicine and human genetic diversity. Cold Spring Harbor Perspect. Med. **4**, a008581 (2014)
2. Garraway, L.A., Verweij, J., Ballman, K.V.: Precision oncology: an overview. J. Clin. Oncol. **31**(15), 1803–1805 (2013)
3. Westergaard, D., Stærfeldt, H.-H., Tønsberg, C., Jensen, L.J., Brunak, S.: A comprehensive and quantitative comparison of text-mining in 15 million full-text articles versus their corresponding abstracts. PLoS Comput. Biol. **14**(2), e1005962 (2018)
4. Wei, C.-H., Kao, H.-Y., Lu, Z.: PubTator: a web-based text mining tool for assisting biocuration. Nucleic Acids Res. **41**(W1), W518–W522 (2013)
5. Wang, Y., et al.: No association between bipolar disorder and syngr1 or synapsin II polymorphisms in the Han Chinese population. Psychiatry Res. **169**(2), 167–168 (2009)
6. Hakenberg, J., et al.: A SNPshot of PubMed to associate genetic variants with drugs, diseases, and adverse reactions. J. Biomed. Inf. **45**(5), 842–850 (2012)
7. Song, M., Kim, W.C., Lee, D., Heo, G.E., Kang, K.Y.: PKDE4 J: entity and relation extraction for public knowledge discovery. J. Biomed. Inf. **57**, 320–332 (2015)
8. Thompson, P., Ananiadou, S.: Extracting gene-disease relations from text to support biomarker discovery. In: Proceedings of the 2017 International Conference on Digital Health, pp. 180–189. ACM (2017)
9. Bundschus, M., Dejori, M., Stetter, M., Tresp, V., Kriegel, H.-P.: Extraction of semantic biomedical relations from text using conditional random fields. BMC Bioinf. **9**(1), 207 (2008)
10. Chun, H.-W., et al.: Extraction of gene-disease relations from Medline using domain dictionaries and machine learning. In: Biocomputing, pp. 4–15. World Scientific (2006)
11. Peng, Y., Lu, Z.: Deep learning for extracting protein-protein interactions from biomedical literature. arXiv preprint arXiv:1706.01556 (2017)
12. Bravo, À., Piñero, J., Queralt-Rosinach, N., Rautschka, M., Furlong, L.I.: Extraction of relations between genes and diseases from text and large-scale data analysis: implications for translational research. BMC Bioinf. **16**(1), 55 (2015)
13. Miwa, M., Bansal, M.: End-to-end relation extraction using LSTMS on sequences and tree structures. arXiv preprint arXiv:1601.00770 (2016)
14. Nguyen, T.H., Grishman, R.: Relation extraction: perspective from convolutional neural networks. In: Proceedings of the 1st Workshop on Vector Space Modeling for Natural Language Processing, pp. 39–48 (2015)

15. Tang, D., Qin, B., Liu, T.: Document modeling with gated recurrent neural network for sentiment classification. In: Proceedings of the 2015 Conference on Empirical Methods in Natural Language Processing, pp. 1422–1432 (2015)

16. Xu, D., et al.: DTMiner: identification of potential disease targets through biomedical literature mining. Bioinformatics **32**(23), 3619–3626 (2016)

17. Roberts, R.J.: PubMed central: the GenBank of the published literature. Proc. Natl. Acad. Sci. U. S. A. **98**(2), 381–382 (2001)

18. Bengio, Y., Ducharme, R., Vincent, P., Jauvin, C.: A neural probabilistic language model. J. Mach. Learn. Res. **3**(Feb), 1137–1155 (2003)

19. Mikolov, T., Sutskever, I., Chen, K., Corrado, G.S., Dean, J.: Distributed representations of words and phrases and their compositionality. In: Advances in Neural Information Processing Systems, pp. 3111–3119 (2013)

20. Tang, D., Qin, B., Liu, T.: Learning semantic representations of users and products for document level sentiment classification. In: Proceedings of the 53rd Annual Meeting of the Association for Computational Linguistics and the 7th International Joint Conference on Natural Language Processing (Volume 1: Long Papers), pp. 1014–1023 (2015)

21. Denil, M., Demiraj, A., Kalchbrenner, N., Blunsom, P., de Freitas, N.: Modelling, visualising and summarising documents with a single convolutional neural network. arXiv preprint arXiv:1408.5882 (2014)

22. Kim, Y.: Convolutional neural networks for sentence classification. arXiv preprint arXiv: 1408.5882 (2014)

23. Bengio, Y., Simard, P., Frasconi, P.: Learning long-term dependencies with gradient descent is difficult. IEEE Trans. Neural Netw. **5**(2), 157–166 (1994)

24. Hochreiter, S., Schmidhuber, J.: Long short-term memory. Neural Comput. **9**(8), 1735–1780 (1997)

25. Cho, K., et al.: Learning phrase representations using RNN encoder-decoder for statistical machine translation. In: Proceeding of the Conference on Empirical Methods in Natural Language Processing, pp. 1724–1734 (2014)

26. Graves, A., Jaitly, N., Mohamed, A.-R.: Hybrid speech recognition with deep bidirectional LSTM. In: 2013 IEEE Workshop on Automatic Speech Recognition and Understanding (ASRU), pp. 273–278. IEEE (2013)

27. Kingma, D.P., Ba, J.: Adam: a method for stochastic optimization. arXiv preprint arXiv: 1412.6980 (2014)

28. Piñero, J., et al.: DisGeNET: a comprehensive platform integrating information on human disease-associated genes and variants. Nucleic Acids Res. **45**(D1), D833–D839 (2016)

29. Moen, S., Ananiadou, T.S.S.: Distributional semantics resources for biomedical text processing. In: Proceedings of the 5th International Symposium on Languages in Biology and Medicine, Tokyo, Japan, pp. 39–43 (2013)

Short Papers

APPLES: Fast Distance-Based Phylogenetic Placement

Metin Balaban[1], Shahab Sarmashghi[2], and Siavash Mirarab[2(✉)]

[1] Bioinformatics and Systems Biology, UC San Diego, La Jolla, CA 92093, USA
[2] Electrical and Computer Engineering, UC San Diego, La Jolla, CA 92093, USA
smirarab@ucsd.edu

Keywords: Phylogenetic placement · Distance-based methods · Genome-skimming

Extended Abstract

Methods for inferring phylogenetic trees from very large datasets exist, yet, large-scale tree reconstructions still require significant resources. New species are continually being sequenced, and as a result, even large trees can become outdated. Reconstructing the tree *de novo* each time new sequences become available is not practical. An alternative approach is phylogenetic placement where new sequence(s) are simply added to an existing *backbone* tree. Phylogenetic placement has applications other than updating trees, including sample identification, where the goal is to detect the identity of given *query* sequences of unknown origins. This problem arises [3] in the study of mixed environmental samples that make up much of the microbiome literature. Sample identification is also the essence of barcoding and meta-barcoding, methods used often in biodiversity studies.

Maximum Likelihood (ML) methods of phylogenetic placement are now available and in wide use (e.g., [4] and EPA(-ng) [2]). The ML approach is computationally demanding, and in particular requires large amounts of memory, and therefore, is limited in the size of the backbone tree it can use. More fundamentally, existing placement tools take as input alignments of assembled sequences for the backbone set, even when queries allowed to be unassembled reads. This reliance on assembled sequences makes them unsuitable for alignment and assembly-free scenarios. For example, sample identification using genome-skimming is fast becoming cost-effective. Methods like Skmer [5] (introduced in RECOMB 2018) can be used to infer k-mer-based estimates of phylogenetic distance from genome skims, and these distances can potentially be used for placement on phylogenetic trees. However, existing methods cannot be used for this purpose.

Distance-based phylogenetics has a rich methodological history, and yet, there are no existing tools for distance-based phylogenetic placement. Such methods, if developed, can be scalable to ultra-large backbone trees. Moreover,

L. J. Cowen (Ed.): RECOMB 2019, LNBI 11467, pp. 287–288, 2019.
https://doi.org/10.1007/978-3-030-17083-7

distance-based methods only need distances, not assembled sequences, and therefore, can be used for sample identification from reads in an assembly-free and alignment-free fashion.

We have developed a new method for distance-based phylogenetic placement called APPLES (Accurate Phylogenetic Placement using LEast Squares). APPLES finds the placement of a query sequence that minimizes the least square error of phylogenetic distances with respect to sequence distances. It can also operate on the minimum evolution principle, or a hybrid of minimum evolution and least square error. Using dynamic programming, APPLES is able to perform placement in time and memory that both scale linearly with the size of the backbone tree.

We have performed extensive studies on simulated and real datasets to evaluate APPLES. Our results show that in the alignment-based scenario, APPLES is much faster than ML tools, uses much less memory, and is very close to ML in the accuracy. Moreover, APPLES can handle much larger backbone trees (we have tested up to 200,000 leaves), and has *increased* accuracy when the backbone trees become larger and more densely sampled. In contrast, ML methods cannot handle backbones with several thousand species. For assembly-free scenarios, we study three genome skimming datasets of insects and show that APPLES applied to Skmer distances can accurately identify genome skim samples using coverage below 1X [1]. APPLES is open-source and freely available at https:// github.com/balabanmetin/apples.

References

1. Balaban, M., Sarmashghi, S., Mirarab, S.: Apples: Fast distance-based phylogenetic placement. bioRxiv (2018). https://doi.org/10.1101/475566. https://www.biorxiv. org/content/early/2018/11/23/475566
2. Barbera, P., et al.: EPA-ng: massively parallel evolutionary placement of genetic sequences. BioRxiv, 291658 (2018)
3. Janssen, S., et al.: Phylogenetic placement of exact amplicon sequences improves associations with clinical information. mSystems **3**(3), 00021–18 (2018). https://doi.org/10.1128/mSystems.00021-18. http://msystems.asm.org/ lookup/doi/10.1128/mSystems.00021-18
4. Matsen, F.A., Kodner, R.B., Armbrust, E.V.: pplacer: linear time maximum-likelihood and bayesian phylogenetic placement of sequences onto a fixed reference tree. BMC Bioinf. **11**(1), 538 (2010)
5. Sarmashghi, S., Bohmann, K., Gilbert, M.T.P., Bafna, V., Mirarab, S.: Assembly-free and alignment-free sample identification using genome skims. Genome Biology (abstract appeared at RECOMB 2018) (2018, in press). https://doi.org/10.1101/ 230409. https://www.biorxiv.org/content/early/2018/04/02/230409

De Novo Peptide Sequencing Reveals a Vast Cyclopeptidome in Human Gut and Other Environments

Bahar Behsaz[1], Hosein Mohimani[2], Alexey Gurevich[3],
Andrey Prjibelski[3], Mark F. Fisher[4], Larry Smarr[2,5,6],
Pieter C. Dorrestein[6,7], Joshua S. Mylne[4],
and Pavel A. Pevzner[2,3,6(✉)]

[1] Bioinformatics and Systems Biology Program,
University of California San Diego, La Jolla, USA
[2] Department of Computer Science and Engineering,
University of California San Diego, La Jolla, USA
ppevzner@ucsd.edu
[3] Center for Algorithmic Biotechnology, Institute of Translational Biomedicine,
St. Petersburg State University, St. Petersburg, Russia
[4] School of Molecular Sciences and The ARC Centre of Excellence in Plant
Energy Biology, The University of Western Australia, Crawley, Australia
[5] California Institute for Telecommunications and Information Technology,
University of California San Diego, La Jolla, USA
[6] Center for Microbiome Innovation,
University of California at San Diego, La Jolla, USA
[7] Department of Pharmacology,
University of California at San Diego, La Jolla, USA

Extended Abstract

Cyclic and branch cyclic peptides (cyclopeptides) represent an important class of bioactive natural products that include many antibiotics and anti-tumor compounds. However, little is known about cyclopeptides in the human gut, despite the fact that humans are constantly exposed to them. To address this bottleneck, we developed CycloNovo algorithm [1] for *de novo* cyclopeptide sequencing that employs de Bruijn graphs, the workhorse of DNA sequencing algorithms. Figure 1 illustrates the CycloNovo pipeline. CycloNovo reconstructed many new cyclopeptides that we validated with transcriptome, metagenome, and genome mining analyses.

We applied CycloNovo to high-resolution spectral dataset generated from daisy seeds (*Senecio vulgaris*), human microbiome (HUMANSTOOL), and a large dataset of 40 high-resolution spectra from GNPS (GNPS). CycloNovo reconstructed ten cyclopeptides in *S. vulgaris* including 4 known and 6 novel cyclopeptides that were further validated using assembled RNA-seq transcripts. Our analysis revealed 703 cyclospectra in HUMANSTOOL dataset corresponding to 79 unique putative cyclopeptides (identified by MS-Cluster) forming 69 spectral families (identified by molecular networking). Dereplicator search yielded only nine PSMs with 0% FDR and

L. J. Cowen (Ed.): RECOMB 2019, LNBI 11467, pp. 289–291, 2019.
https://doi.org/10.1007/978-3-030-17083-7

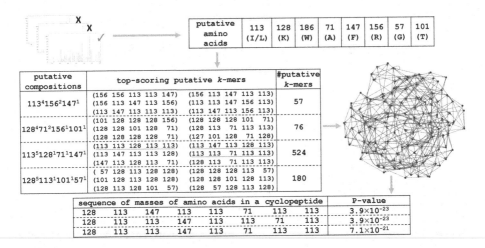

Fig. 1. CycloNovo outline illustrated using $Spectrum_{Surugamide}$. CycloNovo includes six steps: (*i*) recognizing cyclospectra using their spectral-convolution [2], (*ii*) predicting amino acids in a cyclopeptide, (*iii*) predicting amino acid composition of a cyclopeptide, (*iv*), predicting *k*-mers in a cyclopeptide, (*v*) constructing the de Bruijn graph of a spectrum, and (*vi*) generating cyclopeptide reconstructions and calculating P-values [3, 4]. Only six top-scoring putative *k*-mers for each putative amino acid composition are shown. Masses of amino acids occurring in surugamide are shown in red and *k*-mers occurring in surugamide are underlined. To simplify the de Bruijn graph (corresponding to the composition $71^1113^5128^1147^1$), all tips and isolated edges in the graph were removed. Red, blue and green feasible cycles in the graph spell out three cyclopeptides shown in the bottom table along with their P-values. The red cycle spells out surugamide.

P-value $< 10^{-15}$, seven that originated from Flax cyclolinopeptides A [5], B [6], C [7], D [7], H [7], E [7], and P [8] as well as Citrusin V and Massetolide F. Cyclolinopeptides belong to the family of flaxseed orbitides that are present in the seeds of *Linum usitatissimum*. We confirmed that the diet of the individual who provided the HUMANSTOOL sample (L.S., co-author) contained flaxseed eaten frequently as an ingredient in his cooking. Citrusin V belong to the citrusin family of antimicrobial orbitides found in the extracts of various species from the *Citrus* genus [9]. Massetolides are non-ribosomal lipopeptides produced by *Pseudomonas fluoresences*, an indigenous member of human and plant microbiota [10, 11]. Analysis of the metagenome assembly of reads paired with the HUMANSTOOL dataset confirmed that *P. fluoresences* is present in the stool samples where massetolide F was detected.

In addition to the nine identified cyclopeptides, CycloNovo reconstructed 32 cyclopeptides in the HUMANSTOOL dataset with P-values below 10^{-15} forming 26 cyclofamilies. Finding many bioactive cyclopeptides in our study that remain stable in the proteolytic environment of the human gut raises the question of how these bioactive antimicrobial cyclopeptides affect the bacterial composition of the human microbiota.

We analyzed cyclopeptide spectra identified in the GNPS dataset with the goal of estimating the number of still unknown cyclopeptides from spectra already deposited in GNPS. Dereplicator search of the entire GNPS dataset identified 80 unique known cyclopeptides containing 41 cyclofamilies. CycloNovo predicted a total of 12,004 cyclopeptide spectra representing 512 putative cyclopeptides forming 213 cyclofamilies. These putative cyclopeptides include 67 (37 cyclofamilies) of the 80 known cyclopeptides. We showed that even in the case of the phyla with extensively analyzed cyclopeptides (*Cyanobacteria*, *Pseudomonas*, and *Actinomyces*), only less than 30% of the predicted cyclopeptides are already known.

Link to preprint version: https://www.biorxiv.org/content/10.1101/521872v2

References

1. Behsaz, B., et al.: De novo peptide sequencing reveals a vast cyclopeptidome in human gut and other environments. bioRxiv (2019)
2. Ng, J., et al.: A: Dereplication and de novo sequencing of nonribosomal peptides. Nat Methods **6**, 596–599 (2009)
3. Mohimani, H., Kim, S., Pevzner, P.A.: A new approach to evaluating statistical significance of spectral identifications. J. Proteome Res. **12**, 1560–1568 (2013)
4. Mohimani, H., et al.: Dereplication of peptidic natural products through database search of mass spectra. Nat. Chem. Biol. **13**, 30–37 (2017)
5. Kaufmann, H.P., Tobschirbel, A.: Über ein oligopeptid aus leinsamen. Eur. J. Inorg. Chem. **92**, 2805–2809 (1959)
6. Morita, H., et al.: A new immunosuppressive cyclic nonapeptide, cycloinopeptide B from linum usitatissimum. Bioorg. Med. Chem. Letts. **7**, 1269–1272 (1997)
7. Morita, H., Shishido, A., Matsumoto, T., Itokawa, H., Takeya, K.: Cyclolinopeptides B-E, new cyclic peptides from Linum usitatissimum. Tetrahedron. **55**, 967–976 (1999)
8. Okinyo-Owiti, D.P., Young, L., Burnett, P.G.G., Reaney, M.J.T.: New flaxseed orbitides: Detection, sequencing, and15N incorporation. Biopolymers - Peptide Science Section. **102**, 168–175 (2014)
9. Noh, H.J., et al.: Anti-inflammatory activity of a new cyclic peptide, citrusin XI, isolated from the fruits of Citrus unshiu. J. Ethnopharmacol. **163**, 106–112 (2015)
10. Scales, B.S., Dickson, R.P., Lipuma, J.J., Huffnagle, G.B.: Microbiology, genomics, and clinical significance of the pseudomonas fluorescens species complex, an unappreciated colonizer of humans. Clin. Microbiol. Rev. **27**, 927–948 (2014)
11. O'Sullivan, D.J., O'Gara, F.: Traits of fluorescent Pseudomonas spp. involved in suppression of plant root pathogens. Microbiol. Rev. **56**, 662–676 (1992)

Biological Sequence Modeling with Convolutional Kernel Networks

Dexiong Chen[1(✉)], Laurent Jacob[2], and Julien Mairal[1]

[1] Univ. Grenoble Alpes, Inria, CNRS, Grenoble INP, LJK, 38000 Grenoble, France
{dexiong.chen,julien.mairal}@inria.fr
[2] Univ. Lyon, Université Lyon 1, CNRS, Laboratoire de Biométrie et Biologie Évolutive UMR 5558, Lyon, France
laurent.jacob@univ-lyon1.fr

Understanding the relationship between biological sequences and the associated phenotypes is a fundamental problem in molecular biology. Accordingly, machine learning techniques have been developed to exploit the growing number of phenotypic sequences in automatic annotation tools. Typical applications include classifying protein domains into superfamilies [6, 9], predicting whether a DNA or RNA sequence binds to a protein [1], its splicing outcome [3], or its chromatin accessibility [4], predicting the resistance of a bacterial strain to a drug [2], or denoising a ChIP-seq signal [5]. Choosing how to represent biological sequences is a critical part of methods that predict phenotypes from genotypes. Kernel-based methods [6, 9, 8] have often been used for this task. They have been proven efficient to represent biological sequences in various tasks but only construct fixed representations and lack scalability to large amount of data. By contrast, convolutional neural networks (CNN) [1] have recently shown scalable and able to optimize data representations for specific tasks. However, they typically lack interpretability and require large amounts of annotated data, which motivates us to introduce more data-efficient approaches.

In this work we introduce CKN-seq, a strategy combining kernel methods and deep neural networks for sequence modeling, by adapting the convolutional kernel network (CKN) model originally developed for image data [7]. CKN-seq relies on a convolutional kernel, a continuous relaxation of the mismatch kernel [6], and the Nyström approximation. The relaxation makes it possible to learn the kernel from data, and we provide an unsupervised and a supervised algorithm to do so—the latter leading to a special case of CNNs.

On a transcription factor binding prediction task and a protein remote homology detection task, both approaches show better performance than DeepBind, another existing CNN [1], especially when the amount of training data is small. On the other hand, the supervised algorithm produces task-specific and small-dimensional sequence representations while the unsupervised version dominates all other methods on small-scale problems but leads to higher dimensional representations. Consequently, we introduce a hybrid approach which enjoys the benefits of both supervised and unsupervised variants, namely the ability of learning low-dimensional models with good prediction performance in all data size regimes. Finally, the kernel point of view of our method provides us simple ways to visualize and interpret our models, and obtain sequence logos. On some

© Springer Nature Switzerland AG 2019
L. J. Cowen (Ed.): RECOMB 2019, LNBI 11467, pp. 292–293, 2019.
https://doi.org/10.1007/978-3-030-17083-7

simulated data, the logos given by CKN-seq are more informative and match better with the ground truth in terms of any probabilistic distance measures. We provide a free implementation of CKN-seq for learning from biological sequences, which can easily be adapted to other sequence prediction tasks and is available at https://gitlab.inria.fr/dchen/CKN-seq.

The fact that CKNs retain the ability of CNNs to learn feature spaces from large training sets of data while enjoying a reproducing kernel Hilbert space structure has other uncharted applications which we would like to explore in future work. First, it will allow us to leverage the existing literature on kernels for biological sequences to define the bottom kernel instead of the mismatch kernel, possibly capturing other aspects than sequence motifs. More generally, it provides a straightforward way to build models for non-vector objects such as graphs, taking as input molecules or protein structures. Finally, it paves the way for making deep networks amenable to statistical analysis, in particular to hypothesis testing. This important step would be complementary to the interpretability aspect, and necessary to make deep networks a powerful tool for molecular biology beyond prediction.

A full version of the paper is available at https://doi.org/10.1101/217257.

References

1. Alipanahi, B., Delong, A., Weirauch, M.T., Frey, B.J.: Predicting the sequence specificities of DNA-and RNA-binding proteins by deep learning. Nat. Biotechnol. **33**(8), 831–838 (2015)
2. Drouin, A., Giguère, S., Déraspe, M., Marchand, M., Tyers, M., Loo, V.G., Bourgault, A.M., Laviolette, F., Corbeil, J.: Predictive computational phenotyping and biomarker discovery using reference-free genome comparisons. BMC Genomics **17**(1), 754 (2016)
3. Jha, A., Gazzara, M.R., Barash, Y.: Integrative deep models for alternative splicing. Bioinformatics **33**(14), 274–282 (2017)
4. Kelley, D.R., Snoek, J., Rinn, J.L.: Basset: learning the regulatory code of the accessible genome with deep convolutional neural networks. Genome Res. **26**(7), 990–999 (2016)
5. Koh, P.W., Pierson, E., Kundaje, A.: Denoising genome-wide histone chip-seq with convolutional neural networks. Bioinformatics **33**(14), i225–i233 (2017)
6. Leslie, C.S., Eskin, E., Cohen, A., Weston, J., Noble, W.S.: Mismatch string kernels for discriminative protein classification. Bioinformatics **20**(4), 467–476 (2004)
7. Mairal, J.: End-to-end kernel learning with supervised convolutional kernel networks. In: Advances in Neural Information Processing Systems (NIPS), pp. 1399–1407 (2016)
8. Rangwala, H., Karypis, G.: Profile-based direct kernels for remote homology detection and fold recognition. Bioinformatics **21**(23), 4239–4247 (2005)
9. Saigo, H., Vert, J.P., Ueda, N., Akutsu, T.: Protein homology detection using string alignment kernels. Bioinformatics **20**(11), 1682–1689 (2004)

Dynamic Pseudo-time Warping of Complex Single-Cell Trajectories

Van Hoan Do[4], Mislav Blažević[1], Pablo Monteagudo[4], Luka Borozan[1],
Khaled Elbassioni[2], Sören Laue[3], Francisca Rojas Ringeling[4],
Domagoj Matijević[1], and Stefan Canzar[4(✉)]

[1] Department of Mathematics, University of Osijek, Osijek, Croatia
[2] Khalifa University of Science and Technology, Abu Dhabi, UAE
[3] Friedrich-Schiller-Universität Jena, Jena, Germany
[4] Gene Center, Ludwig-Maximilians-Universität München, Munich, Germany
`canzar@genzentrum.lmu.de`

1 Introduction

Single-cell RNA sequencing enables the construction of trajectories [1] describing the dynamic changes in gene expression underlying biological processes such as cell differentiation and development. The comparison of single-cell trajectories under two distinct conditions can illuminate the differences and similarities between the two and can thus be a powerful tool for analysis [2]. Recently developed methods for the comparison of trajectories [2, 3] rely on the concept of dynamic time warping (dtw), originally proposed for the comparison of two time series and consequently restricted to simple, linear trajectories. Here, we adopt and theoretically link arboreal matchings to dtw and implement a suite of exact and heuristic algorithms suitable for the comparison of complex trajectories of different characteristics in our tool Trajan (Fig. 1). Trajan's alignment enables the meaningful comparison of gene expression dynamics along a common pseudo-time scale. Trajan is available at https://github.com/canzarlab/Trajan.

2 Methods

Dynamic time warping (dtw) is the algorithmic workhorse underlying current methods that compare linear single-cell trajectories. We develop Trajan, the first method to compare and align complex trajectories (trees) with multiple branch points. Trajan aligns each path in one tree to at most one path in the second tree and vice versa and, similar to dtw, preserves the order of nodes along the paths. In [4] we have introduced *arboreal matchings* that formalize such a consistent path-by-path alignment of trees.

We devise scoring schemes for arboreal matchings that yield (guaranteed) similar distance measures between linear trajectories as dtw, but naturally

---

The full version of this paper is available as preprint at bioRxiv 522672.
Van Hoan Do and Mislav Blažević—equal contribution.

© Springer Nature Switzerland AG 2019
L. J. Cowen (Ed.): RECOMB 2019, LNBI 11467, pp. 294–296, 2019.
https://doi.org/10.1007/978-3-030-17083-7

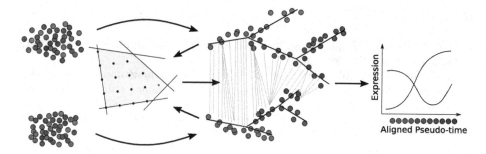

Fig. 1. Complex trajectories, reconstructed from single-cell RNA measurements using, e.g., Monocle 2, are aligned by Trajan based on arboreal matchings. The matching warps individual pseudo-time scales into a shared one along which expression kinetics can be compared.

extend to complex trajectories. Trajan implements a thoroughly engineered branch-and-cut algorithm that allows to practically compare complex single-cell trajectories. It repeatedly determines cutting planes that strengthen the LP relaxation in [4] in polynomial-time and uses an in-house developed, non-commercial, non-linear solver for all continuous optimization problems. For trajectories with a small number of cell fates k we employ a fixed-parameter tractable algorithm, parameterized by k, that applies a dynamic program similar to [5] to align them optimally.

3   Results

Adopting a strategy similar to [2], we re-analyzed two public single-cell datasets: human skeletal muscle myoblast (HSMM) differentiation and human fibroblasts undergoing MYOD-mediated myogenic reprogramming (hFib-MyoD). Trajan is able to align the core paths of each complex trajectory, without any previous knowledge of myoblast differentiation markers. From Trajan's alignment, we construct gene expression kinetics for a set of genes that were assessed in [2] and are able to reproduce their key findings, including the molecular barriers identified in [2] that hinder the efficient reprogramming of fibroblasts to myotubes.

In a perturbation experiment we demonstrate the benefits in terms of robustness and accuracy of our model which compares entire trajectories at once, as opposed to a pairwise application of dtw.

Acknowledgments. Sören Laue has been funded by Deutsche Forschungsgemeinschaft (DFG) under grant LA 2971/1-1. Mislav Blažević was supported in part by BAYHOST. Francisca Rojas Ringeling was supported by the Bavarian Gender Equality Grant (BGF).

References

1. Qiu, X.: Reversed graph embedding resolves complex single-cell trajectories. Nat. Methods **14**(10), 979–982 (2017)
2. Cacchiarelli, D.: Aligning single-cell developmental and reprogramming trajectories identifies molecular determinants of myogenic reprogramming outcome. Cell Syst. 1–18 (2016)
3. Alpert, A.: Alignment of single-cell trajectories to compare cellular expression dynamics. Nat. Methods **15**(4), 267–270 (2018)
4. Böcker, S., Canzar, S., Klau, G.W.: The generalized robinson-foulds metric. In: Darling, A., Stoye, J. (eds.) WABI 2013. LNCS, vol. 8126, pp. 156–169. Springer, Heidelberg (2013). https://doi.org/10.1007/978-3-642-40453-5_13
5. Zhang, K.: Simple fast algorithms for the editing distance between trees and related problems. SIAM J. Comput. **18**, 1245–1262 (1989)

netNMF-sc: A Network Regularization Algorithm for Dimensionality Reduction and Imputation of Single-Cell Expression Data

Rebecca Elyanow[1,2], Bianca Dumitrascu[3], Barbara E. Engelhardt[2,4], and Benjamin J. Raphael[2(✉)]

[1] Center for Computational Molecular Biology, Brown University,
Providence, RI 029012, USA
[2] Department of Computer Science, Princeton University,
Princeton, NJ 08540, USA
braphael@princeton.edu
[3] Lewis Sigler Institute for Integrative Genomics,
Princeton University, Princeton, NJ 029012, USA
[4] Center for Statistics and Machine Learning,
Princeton University, Princeton, NJ 08540, USA

Abstract

Motivation. Single-cell RNA-sequencing (scRNA-seq) enables high through-put measurement of RNA expression in individual cells. Due to technical and financial limitations, scRNA-seq datasets often contain zero counts for many transcripts in individual cells. These zero counts, or *dropout events*, complicate the analysis of scRNA-seq data using standard analysis methods developed for bulk RNA-seq data. Current methods for analysis of scRNA-seq data typically overcome dropout by combining information across cells, leveraging the observation that the cells measured in any scRNA-seq experiment generally occupy a small number of RNA expression states.

Results. We describe an algorithm to overcome dropout by combining information across *both* cells and genes. Our algorithm, netNMF-sc, combines network-regularized non-negative matrix factorization with a specialized procedure to handle the large fraction of zero entries in the transcript count matrix. The matrix factorization results in a low-dimensional representation of the transcript count matrix, while the network regularization encourages two genes connected in the network to be close in the low-dimensional representation. In addition, the two matrix factors can be used to cluster cells and to impute values for dropout events. While our netNMF-sc algorithm may use any type of network as prior information, a particularly promising approach is to leverage tissue-specific gene-coexpression networks derived from the vast repository of RNA-seq/microarray studies of bulk tissue.

We show that netNMF-sc outperforms existing methods in both clustering cells and imputing transcript counts on simulated data. netNMF-sc's advantages were especially pronounced at high dropout rates e.g. above 60%. Such high

© Springer Nature Switzerland AG 2019
L. J. Cowen (Ed.): RECOMB 2019, LNBI 11467, pp. 297–298, 2019.
https://doi.org/10.1007/978-3-030-17083-7

dropout rates are common in newer scRNA-seq technologies, such as from 10X Genomics, that measure large number of cells with low sequence coverage per cell. We also show that netNMF-sc outperforms existing methods on real scRNA-seq datasets, including the clustering of mouse embryonic stem cells into cell-cycle states and the clustering of mouse embryonic brain cells into known cell types. Finally, we show that gene-gene correlations computed from the netNMF-sc imputed data are more biologically meaningful than the gene-gene correlations obtained from existing algorithms.

Availability. netNMF-sc is available at https://github.com/raphael-group/netNMF-sc. The preprint is available at https://www.biorxiv.org/content/10.1101/544346v1.

Geometric Sketching of Single-Cell Data Preserves Transcriptional Structure

Brian Hie[1(✉)], Hyunghoon Cho[1], Benjamin DeMeo[2,4], Bryan Bryson[3], and Bonnie Berger[1,4(✉)]

[1] Computer Science and Artificial Intelligence Laboratory, MIT, Cambridge, MA 02139, USA
bab@mit.edu
[2] Department of Biomedical Informatics, Harvard University, Cambridge, MA 02138, USA
[3] Department of Biological Engineering, MIT, Cambridge, MA 02139, USA
[4] Department of Mathematics, MIT, Cambridge, MA 02139, USA

1 Introduction

Single-cell RNA-sequencing (scRNA-seq) experiments that profile hundreds of thousands of cells or more are becoming increasingly common. These large-scale data sets present a key computational bottleneck for conventional scRNA-seq analysis pipelines [1]. Standard methods of reducing the size of data sets, such as uniform downsampling, frequently remove rare transcriptional states, mitigating the advantage that large-scale experiments provide. Here we present *geometric sketching*, an efficient downsampling method that newly preserves the transcriptional heterogeneity of single-cell data sets by sampling evenly across transcriptomic space, thinning out dense clusters of common cells and preferentially selecting cells from sparser regions.

We empirically demonstrate that geometric sketches represent the geometry rather than the density of the original data set. We show that our sketches enhance and accelerate downstream analyses by: preserving rare cell types, producing visualizations that capture the full transcriptomic heterogeneity, and facilitating the identification of cell types via clustering. Geometric sketching downsamples from data sets with millions of cells in a matter of minutes, with an asymptotic runtime nearly linear in the size of the data set. As the size of single-cell data grows, geometric sketching will become increasingly crucial for broadening access to single-cell omics experiments even for researchers without expensive computational resources. The full version of this paper can be found at https://www.biorxiv.org/content/10.1101/536730v2.

2 Methods

Geometric sketching is based on the key insight that common cell types form dense clusters in transcriptomic space, while rare cell types may occupy larger

---

B. Hie and H. Cho—Contributed equally to this work.

L. J. Cowen (Ed.): RECOMB 2019, LNBI 11467, pp. 299–301, 2019.
https://doi.org/10.1007/978-3-030-17083-7

regions with much greater sparsity. To accurately summarize the transcriptomic landscape, geometric sketching first obtains a geometric approximation of the data set with equal-sized, non-overlapping, axis-aligned boxes (hypercubes), which we refer to as a plaid covering (Fig. 1). Once the geometry of the data is approximated with a set of covering boxes, we sample cells by uniformly sampling a covering box, then choosing a cell in the box also uniformly at random. The samples therefore more evenly cover the gene expression landscape, naturally diminishing the influence of densely populated regions and increasing the representation of rare transcriptional states.

The plaid covering generalizes grid-based approximation while maintaining computational efficiency in assigning points to their respective covering box. To obtain a plaid covering, we fix an interval length ℓ, and for each coordinate construct a minimal covering of the projected data with intervals of length ℓ. The Cartesian product of these coordinate-wise coverings yields a plaid covering of the original data set by axis-aligned boxes of side length ℓ. Note that after an $O(n \log(n))$ sorting operation, points can be assigned to boxes by rounding up or down, yielding an overall $O(n \log(n))$ runtime in each dimension. In practical scenarios where each coordinate requires only a small constant number of intervals to cover, we achieve $O(n)$ time complexity by using linear scans to find the next interval without sorting. We perform a binary search to find the value of ℓ that produces the number of covering boxes that match the number of samples to be taken. In addition, we use a fast random projection-based PCA to project the data to a relatively low-dimensional space (100 dimensions in the experiments below) before applying the sketching algorithm.

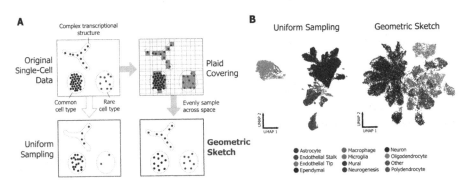

Fig. 1. Geometric sketches capture transcriptional heterogeneity. (**A**) An illustration of the geometric sketching algorithm. (**B**) Geometric sketches more evenly represent the transcriptomic landscape of a data set. Shown are the sketches of 20k cells sampled from a mouse brain data set with 666k cells.

3   Results

Visualizations of geometric sketches reflect the geometric "map" of the transcriptional variability within a data set, allowing researchers to more easily gain insight into rarer transcriptional states (Fig. 1). On data sets with three clusters of similar volumes but different densities, our algorithm samples each cluster with near equal probability (KL divergence = 0.063 versus $\geq$ 0.85 for other sampling methods). Our algorithm also detects rare cell types in a variety of settings: 293T cells mixed with Jurkat cells at a concentration of 0.66%, CD14+ monocytes at a concentration of 1.2%, and macrophages in a mouse brain data set at a concentration of 0.27%. In all cases, rare cell types are substantially better represented in geometric sketches than in subsamples made with other methods, which include spatial random sampling [2] and k-means++ [3], which have not been previously considered for the problem of subsampling scRNA-seq data. Finally, Louvain clustering on data subsampled via geometric sketching resulted in comparable or better agreement with known cell labels across a range of Louvain resolution parameters.

References

1. Angerer, P., et al.: Single cells make big data: new challenges and opportunities in transcriptomics. Curr. Opin. Syst. Biol. (2017)
2. Rahmani, M., Atia, G.K.: Spatial random sampling: a structure-preserving data sketching tool. IEEE Signal Process. Lett. (2017). 1705.03566
3. Arthur, D., Vassilvitskii, S.: K-Means++: the advantages of careful seeding. In: Proceedings of the ACM-SIAM Symposium on Discrete Algorithms (2007). 1212.1121

. Sketching Algorithms for Genomic Data Analysis and Querying in a Secure Enclave

Can Kockan[1], Kaiyuan Zhu[1], Natnatee Dokmai[1], Nikolai Karpov[1],
M. Oguzhan Kulekci[2], David P. Woodruff[3], and S. Cenk Sahinalp[1(✉)]

[1] Department of Computer Science, Indiana University, Bloomington, IN, USA
cenksahi@indiana.edu
[2] Informatics Institute, Istanbul Technical University, Istanbul, Turkey
[3] Department of Computer Science, Carnegie Mellon University, Pittsburgh, USA

Extended Abstract

Current practices in collaborative genomic data analysis (e.g. PCAWG [1]) necessitate all involved parties to exchange individual patient data and perform all analysis locally, or use a trusted server for maintaining all data to perform analysis in a single site (e.g. the Cancer Genome Collaboratory [2]). Since both approaches involve sharing genomic sequence data - which is typically not feasible due to privacy issues, collaborative data analysis remains to be a rarity in genomic medicine.

In order to facilitate efficient and effective collaborative or remote genomic computation we introduce SkSES (Sketching algorithms for Secure Enclave based genomic data analysiS), a computational framework for performing data analysis and querying on multiple, individually encrypted genomes from several institutions in an untrusted cloud environment. Unlike other techniques for secure/privacy preserving genomic data analysis, which typically rely on sophisticated cryptographic techniques with prohibitively large computational overheads, SkSES utilizes the secure enclaves supported by current generation microprocessor architectures such as Intel's SGX. The key conceptual contribution of SkSES is its use of *sketching* data structures that can fit in the limited memory available in a secure enclave.

While streaming/sketching algorithms have been developed for many applications, their feasibility in genomics has remained largely unexplored. On the other hand, even though privacy and security issues are becoming critical in genomic medicine, available cryptographic techniques based on, e.g. homomorphic encryption, secure multi-party computing or garbled circuits, can not always address the performance demands of this rapidly growing field [3–6]. The alternative offered by Intel's SGX, a combination of hardware and software solutions for secure data analysis, is severely limited by the relatively small size of a secure enclave, a private region of the memory protected from other processes [7]. SkSES addresses this limitation through the use of sketching data structures to support efficient secure and privacy preserving SNP analysis across individually

---

C. Kockan, K. Zhu and N. Dokmai—Joint first authors.

© Springer Nature Switzerland AG 2019
L. J. Cowen (Ed.): RECOMB 2019, LNBI 11467, pp. 302–304, 2019.
https://doi.org/10.1007/978-3-030-17083-7

encrypted VCF files from multiple institutions. In particular SkSES provides the users the ability to query for the "k most significant SNPs" among any set of user specified SNPs and any value of k - even when the total number of SNPs to be maintained is far beyond the memory capacity of the secure enclave.

SkSES processes individual genomic data presented as VCF files from participating parties who aim to perform collective statistical tests. For compacting the input VCF files, SkSES uses a simple scheme to filter out non-essential components of a VCF file and encode essential components efficiently - reducing the storage and communication needs and speeding up encryption/decryption within the framework. SkSES then builds a sketch of the compacted VCF files, based on either the count-min sketch [8] or the count sketch [9] structures in order to approximate the actual allele count distribution with respect to L_1 measure (the difference between case and control) - as a proxy to the χ^2 statistic.

Results: We tested SkSES on the extended iDASH-2017 competition data set comprised of 1000 case and 1000 control samples related to an unknown phenotype. SkSES was able to identify the top SNPs with respect to the χ^2 statistic, among any user specified subset of SNPs across this data set of 2000 individually encrypted complete human genomes quickly and accurately - significantly improving our iDASH-2017 (http://www.humangenomeprivacy.org/2017/) runner-up software for secure GWAS - demonstrating the feasibility of secure and privacy preserving computation at human genome scale via Intel's SGX.

Availability: https://github.com/ndokmai/sgx-genome-variants-search
Full Text: https://www.biorxiv.org/content/early/2018/11/12/468355

Acknowledgements. S.C.S. was supported in part by grant NSF CCF-1619081, NIH GM108348 and the Indiana University Grand Challenges Program, Precision Health Initiative; M.O.K. was partially supported by the TUBITAK grant number 114E293; Part of the work was done while D.P.W. was visiting the Simons Institute for the Theory of Computing.

References

1. Campbell, P.J., et al.: Pan-cancer analysis of whole genomes. bioRxiv (2017)
2. Yung, C.K., et al.: Abstract 378: the cancer genome collaboratory. Cancer Res. **77**(13 Supplement), 378–378 (2017)
3. Constable, S.D., et al.: Privacy-preserving gwas analysis on federated genomic datasets. BMC Med. Inform. Decis. Mak. **15**(5), S2 (2015)
4. Zhang, Y., et al.: Secure distributed genome analysis for gwas & sequence comparison computation. BMC Med. Inform. Decis. Mak. **15**(5), S4 (2015)
5. Xie, W., et al.: Securema: protecting participant privacy in genetic association meta-analysis. Bioinformatics **30**(23), 3334–3341 (2014)
6. Cho, H., et al.: Secure genome-wide association analysis using multiparty computation. Nat. Biotechnol. **36**(6), 547 (2018)
7. Chen, F., et al.: Princess: privacy-protecting rare disease international network collaboration via encryption through software guard extensions. Bioinformatics **33**(6), 871–878 (2017)

8. Cormode, G., et al.: An improved data stream summary: the count-min sketch and its applications. J. Algorithms **55**(1), 58–75 (2005)

9. Charikar, M., Chen, K., Farach-Colton, M.: Finding frequent items in data streams. In: Widmayer, P., Eidenbenz, S., Triguero, F., Morales, R., Conejo, R., Hennessy, M. (eds.) ICALP 2002. LNCS, vol. 2380, pp. 693–703. Springer, Heidelberg (2002). https://doi.org/10.1007/3-540-45465-9_59

Mitigating Data Scarcity in Protein Binding Prediction Using Meta-Learning

Yunan Luo[1], Jianzhu Ma[2], Xiaoming Zhao[1], Yufeng Su[3], Yang Liu[1], Trey Ideker[2], and Jian Peng[1(✉)]

[1] University of Illinois at Urbana-Champaign, Champaign, USA
jianpeng@illinois.edu
[2] University of California San Diego, San Diego, USA
[3] Shanghai Jiao Tong University, Shanghai, China

1 Introduction

A plethora of biological functions are performed through various types of protein-peptide binding, e.g., protein kinase phosphorylation on peptide substrates. Understanding the specificity of protein-peptide interactions is critical for unraveling the architectures of functional pathways and the mechanisms of cellular processes in human cells. A line of computational prediction methods has been recently proposed to predict protein-peptide bindings which efficiently provide rich functional annotations on a large scale. To achieve a high prediction accuracy, these computational methods require a sufficient amount of data to build the prediction model. However, the number of experimentally verified protein-peptide bindings is often limited in real cases. These methods are thus limited to building accurate prediction models for only well-characterized proteins with a large volume of known binding peptides and cannot be extended to predict new binding peptides for less-studied proteins.

2 Methods

We propose a new two phases *meta-learning* framework, named MetaKinase, for the prediction of kinase phosphorylation sites. In phase one, using multiple training kinase families, we train a model which can generate more adaptable representations which are broadly suitable for every kinase family (called meta-learning). In phase two, using a few (e.g., <10) known phosphorylation sites from a new target kinase family, we fine-tune the model on this target family to capture its specificity. With the general patterns captured in phase one, the adaption to the target family in phase two is very sample-efficient: we can tweak the model by only using a few data points to make it family-specific and accurately predict the specificity of the target family (called few-shot learning). With its transferability and fast adaptability, our framework can thus be applied to mitigate the data scarcity issue in characterizing specificities of less-studied kinases. Even with only a few known phosphorylation sites, the model is still able to accurately characterize the specificity of the target kinase family.

Y. Luo and J. Ma—Equal contribution.

© Springer Nature Switzerland AG 2019
L. J. Cowen (Ed.): RECOMB 2019, LNBI 11467, pp. 305–307, 2019.
https://doi.org/10.1007/978-3-030-17083-7

3   Results

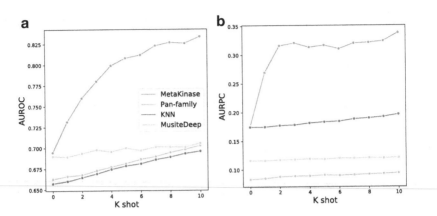

Fig. 1. Evaluation of few-shot learning. MetaKinase was trained with data of multiple kinase families in the meta-learning phase and fine-tuned in the few-shot learning phase using k samples of the test family for $k = 1, 2, \ldots, 10$.

We compared our framework with three baseline methods: pan-family approach (one prediction model for all kinase families), K-nearest neighbor, and MusiteDeep [1]. We varied the value of k-shot from 0 to 10 (0-shot means the model was trained on training family only), and for each value of k, we randomly sampled k samples from the target family and used the remaining samples as test data. The process was repeated for 50 times for each value of k. We used the AUROC and AUPRC scores as the evaluation metrics and showed the results in Fig. 1. We first observed that MetaKinase outperformed other methods for each value of k in terms of both AUROC and AUPRC scores. In addition, while other methods had relatively similar prediction performance as the number of k-shot increased, we observed that the improvement was clear for MetaKinase when more k-shot samples were provided. Our framework also achieved fast adaption to a target family. For example, the predictor had a 0.316 AUPRC score when using 2-shot samples in the few-shot learning phase, which was closed to AUPRC score achieved with 10-shot (0.338). These results demonstrated the transferability and fast-learning ability of MetaKinase. The full paper describing MetaKinase is available at [2].

Acknowledgements. This work was supported in part by the NSF CAREER Award, the Sloan Research Fellowship, the PhRMA Foundation Award in Informatics, and the CompGen Fellowship.

References

1. Wang, D., et al.: Musitedeep: a deep-learning framework for general and kinase-specific phosphorylation site prediction. Bioinformatics **33**, 3909–3916 (2017)
2. Luo, Y. et al.: Mitigating data scarcity in protein binding prediction using meta-learning. bioRxiv (2019). https://doi.org/10.1101/519413

Efficient Estimation and Applications of Cross-Validated Genetic Predictions

Joel Mefford[1]($\boxtimes$), Danny Park[1], Zhili Zheng[2], Arthur Ko[3],
Mika Ala-Korpela[4,5], Markku Laakso[6], Paivi Pajukanta[3], Jian Yang[2],
John Witte[7], and Noah Zaitlen[8]

[1] Program in Pharmaceutical Science and Pharmacogenomics, UCSF,
San Francisco, USA
joelmefford@gmail.com
[2] Institute for Molecular Bioscience, University of Queensland, Brisbane, Australia
[3] Department of Human Genetics, UCLA, Los Angeles, USA
[4] Systems Epidemiology, Baker Heart and Diabetes Institute,
Melbourne, Victoria, Australia
[5] Computational Medicine, Faculty of Medicine,
University of Oulu and Biocenter Oulu, Oulu, Finland
[6] School of Medicine, University of Eastern Finland, Kuopio, Finland
[7] Departments of Epidemiology and Biostatistics, and Urology, UCSF,
San Francisco, USA
[8] Department of Neurology, UCLA, Los Angeles, USA

While complex traits are highly heritable, individual genetic polymorphisms typically explain only a small proportion of the heritability [1]. Polygenic scores (PS), also known as polygenic risk scores for disease phenotypes, aggregate the contributions of multiple genetic variants to a phenotype [2]. These scores can be calculated using routinely recorded genotypes [1, 2], are strongly associated with heritable traits [1], and are independent of environmental exposures or other factors that are uncorrelated with germ line genetic variants. These properties have motivated a rapidly expanding list of applications from basic science (e.g. causal inference and Mendelian randomization [3], hierarchical disease models and identification of pleiotropy [4]) to translation (e.g. estimating disease risk [5], identifying patients who are likely to respond well to a particular therapy [6], or flagging subjects for modified screening [7]).

Polygenic scores are calculated as a weighted sum of genotypes. This may include all genotyped SNPs, but often only a small set is given nonzero weight – such as a genome spanning but uncorrelated (LD-pruned) set or SNPs with independent evidence of association with the phenotype of interest. Gene-specific polygenic score are also generated using selected sets of SNPs within a region of the genome, such as a window around the coding region of a particular gene [8]. The weights on the SNPs included in a polygenic score are often derived from the regression coefficients of an external GWAS [9, 10], but they may instead be based on predictive models using all SNPs. Joint predictive models include LMMs and their sparse extensions and other regularized regression models such as the lasso or elastic net [8, 11]. The predictions from these joint analyses using genome wide variation are also approximated by post-processing of GWAS summary statistics [8, 11].

© Springer Nature Switzerland AG 2019
L. J. Cowen (Ed.): RECOMB 2019, LNBI 11467, pp. 308–310, 2019.
https://doi.org/10.1007/978-3-030-17083-7

For these SNP-weights to accurately reflect the SNPs' joint association with the phenotype and to generate informative and interpretable polygenic scores, the reference data set must match the target data set in many ways: the populations must have similar ancestry; the trait of interest must be measured; and identical genotypes must be assayed or imputed. Further, the reference data must be large enough to accurately learn the PS weights. An alternative approach is to use the studied data set to build a reference-free PS. This eliminates the need for an external reference data set with matched genotypes, phenotypes, and populations. However, as we show below, naive approaches can easily overfit genetic effects. This overfitting results in PS correlated with non-genetic components of phenotype, that will induce bias or other errors in downstream applications.

Here we report an efficient method to generate PS by using the out-of-sample predictions from a cross-validated linear mixed model (LMM). Our approach generates leave-one-out (LOO) polygenic scores, which we call *cvBLUPs* after a single LMM fit, with computational complexity linear in sample size. In addition to eliminating the reliance on external data and guaranteeing the PS are generated from a relevant population and phenotype, we describe several applications that are only feasible with cvBLUPs. We first demonstrate several desirable statistical properties of cvBLUPs and then consider applications including evidence of polygenicity across metabolic phenotypes, estimation of the shrink term in linear mixed models, a novel formulation of mixed model association studies, and selection of relevant principal components for downstream analyses. To make the results of this work accessible to the community, we have implemented them in the GCTA software package [12].

Full paper on bioRxiv at: https://doi.org/10.1101/517821

References

1. Nolte, I.M., et al.: Missing heritability: is the gap closing? an analysis of 32 complex traits in the lifelines cohort study. Eur. J. Hum. Genet. **25**, 877 (2017)
2. Torkamani, A., Wineinger, N.E., Topol, E.J.: The personal and clinical utility of polygenic risk scores. Nature Rev. Genet. 1 (2018)
3. Burgess, S., Thompson, S.G.: Use of allele scores as instrumental variables for mendelian randomization. Int. J. Epidemiol. **42**, 1134–1144 (2013)
4. Cortes, A., et al.: Bayesian analysis of genetic association across tree-structured routine healthcare data in the UK Biobank. Nature Genet. **49**, 1311 (2017)
5. Maas, P., et al.: Breast cancer risk from modifiable and nonmodifiable risk factors among white women in the United States. JAMA Oncol. **2**, 1295–1302 (2016)
6. Natarajan, P., et al.: Polygenic risk score identifies subgroup with higher burden of atherosclerosis and greater relative benefit from statin therapy in the primary prevention setting. Circulation **135**, 2091–2101 (2017)
7. Seibert, T.M., et al.: Polygenic hazard score to guide screening for aggressive prostate cancer: development and validation in large scale cohorts. bmj **360**, j5757 (2018)
8. Gusev, A., et al.: Integrative approaches for large-scale transcriptome-wide association studies. Nature Genet. **48**, 245 (2016)
9. Dudbridge, F.: Polygenic epidemiology. Genet. Epidemiol. **40**, 268–272 (2016)

10. Wray, N.R., Goddard, M.E., Visscher, P.M.: Prediction of individual genetic risk to disease from genome-wide association studies. Genome Res. **17** (2007)
11. Vilhjálmsson, B.J., et al.: Modeling linkage disequilibrium increases accuracy of polygenic risk scores. Am. J. Hum. Genet. **97**, 576–592 (2015)
12. Yang, J., Lee, S.H., Goddard, M.E., Visscher, P.M.: GCTA: a tool for genome-wide complex trait analysis. Am. J. Hum. Genet. **88**, 76–82 (2011)

Inferring Tumor Evolution from Longitudinal Samples

Matthew A. Myers[1], Gryte Satas[1,2], and Benjamin J. Raphael[1(✉)]

[1] Department of Computer Science, Princeton University, Princeton, NJ 08540, USA
braphael@princeton.edu
[2] Department of Computer Science, Brown University, Providence, RI 02912, USA

Abstract

Background: Determining the clonal composition and somatic evolution of a tumor greatly aids in accurate prognosis and effective treatment for cancer. In order to understand how a tumor evolves over time and/or in response to treatment, multiple recent studies have performed longitudinal DNA sequencing of tumor samples from the same patient at several different time points. However, none of the existing algorithms that infer clonal composition and phylogeny using several bulk tumor samples from the same patient integrate the information that these samples were obtained from longitudinal observations.

Results: We introduce a model for a longitudinally-observed phylogeny and derive constraints that longitudinal samples impose on the reconstruction of a phylogeny from bulk samples. These constraints form the basis for a new algorithm, Cancer Analysis of Longitudinal Data through Evolutionary Reconstruction (CALDER), which infers phylogenetic trees from longitudinal bulk DNA sequencing data. We show on simulated data that constraints from longitudinal sampling can substantially reduce ambiguity when deriving a phylogeny from multiple bulk tumor samples, each a mixture of tumor clones. On real data, where there is often considerable uncertainty in the clonal composition of a sample, longitudinal constraints yield more parsimonious phylogenies with fewer tumor clones per sample. We demonstrate that CALDER reconstructs more plausible phylogenies than existing methods on two longitudinal DNA sequencing datasets from chronic lymphocytic leukemia patients. These findings show the advantages of directly incorporating temporal information from longitudinal sampling into tumor evolution studies.

Availability: CALDER is available at https://github.com/raphael-group.

Preprint: Preprint version of the full manuscript is available at https://www.biorxiv.org/content/10.1101/526814v1.

© Springer Nature Switzerland AG 2019
L. J. Cowen (Ed.): RECOMB 2019, LNBI 11467, p. 311, 2019.
https://doi.org/10.1007/978-3-030-17083-7

Scalable Multi-component Linear Mixed Models with Application to SNP Heritability Estimation

Ali Pazokitoroudi[1], Yue Wu[1], Kathryn S. Burch[2], Kangcheng Hou[3,4], Bogdan Pasaniuc[3,5,6], and Sriram Sankararaman[1,5,6(✉)]

[1] Department of Computer Science, UCLA, Los Angeles, CA, USA
[2] Bioinformatics Interdepartmental Program, UCLA, Los Angeles, CA, USA
[3] Department of Pathology and Laboratory Medicine,
David Geffen School of Medicine, UCLA, Los Angeles, CA, USA
[4] College of Computer Science and Technology, Zhejiang University,
Hangzhou, Zhejiang, China
[5] Department of Human Genetics, David Geffen School of Medicine,
UCLA, Los Angeles, CA, USA
[6] Department of Computational Medicine, David Geffen School of Medicine,
UCLA, Los Angeles, CA, USA
sriram@cs.ucla.edu

A central question in human genetics is to find the proportion of variation in a trait that can be explained by genetic variation [1]. A number of methods have been developed to estimate this quantity, termed narrow-sense heritability, from genome-wide SNP data [2–6]. Recently, it has become clear that estimates of narrow-sense heritability are sensitive to modeling assumptions that relate the effect sizes of a SNP to its minor allele frequency (MAF) and linkage disequilibrium (LD) patterns [6, 7]. A principled approach to estimate heritability while accounting for variation in SNP effect sizes involves the application of linear Mixed Models (LMMs) [8] with multiple variance components where each variance component represents the fraction of genetic variance explained by SNPs that belong to a given range of MAF and LD values. Beyond their importance in accurately estimating genome-wide SNP heritability, multiple variance component LMMs are useful in partitioning the contribution of genomic annotations to trait heritability which, in turn, can provide insights into biological processes that are associated with the trait.

Existing methods for fitting multi-component LMMs rely on maximizing the likelihood of the variance components. These methods pose major computational bottlenecks that makes it challenging to apply them to large-scale genomic datasets such as the UK Biobank which contains half a million individuals genotyped at tens of millions of SNPs.

We propose a scalable algorithm, RHE-reg-mc, to jointly estimate multiple variance components in LMMs. RHE-reg-mc is a randomized method-of-moments estimator with a runtime that is observed to scale as $\mathcal{O}(\frac{NMB}{\max(\log_3(N), \log_3(M))} + k^3)$ for N individuals, M SNPs, k variance components, and $B \approx 10$, a parameter that controls the number of random matrix-vector multiplication. RHE-reg-mc also efficiently computes asymptotic and jackknife standard errors. We evaluate the accuracy and scalability of RHE-reg-mc for estimating the total heritability as well as in partitioning heritability.

© Springer Nature Switzerland AG 2019
L. J. Cowen (Ed.): RECOMB 2019, LNBI 11467, pp. 312–313, 2019.
https://doi.org/10.1007/978-3-030-17083-7

The ability to fit multiple variance components to SNPs partitioned according to their MAF and local LD allows RHE-reg-mc to obtain relatively unbiased estimates of SNP heritability. On the UK Biobank dataset consisting of $\approx 300,000$ individuals and $\approx 500,000$ SNPs, RHE-reg-mc can fit 250 variance components, corresponding to genetic variance explained by 10 MB blocks, in ≈ 40 minutes on standard hardware. The full version of the paper is available at: http://biorxiv.org/cgi/content/short/522003v2.

References

1. Visscher, P.M., Hill, W.G., Wray, N.R.: Heritability in the genomics era: concepts and misconceptions. Nat. Rev. Genet. **9**(4), 255 (2008)
2. Yang, J., et al.: Common snps explain a large proportion of the heritability for human height. Nat. Genet. **42**(7), 565 (2010)
3. Zhou, X.: A unified framework for variance component estimation with summary statistics in genome-wide association studies. Ann. Appl. Stat. **11**(4), 2027 (2017)
4. Lee, S.H., Wray, N.R., Goddard, M.E., Visscher, P.M.: Estimating missing heritability for disease from genome-wide association studies. Am. J. Hum. Genet. **88**(3), 294–305 (2011)
5. Golan, D., Lander, E.S., Rosset, S.: Measuring missing heritability: inferring the contribution of common variants. Proc. Nat. Acad. Sci. **111**(49), E5272–E5281 (2014)
6. Speed, D., Hemani, G., Johnson, M.R., Balding, D.J.: Improved heritability estimation from genome-wide snps. Am. J. Hum. Genet. **91**(6), 1011–1021 (2012)
7. Evans, L.M., et al.: Comparison of methods that use whole genome data to estimate the heritability and genetic architecture of complex traits. Nat. Genet. **50**(5), 737 (2018)
8. McCulloch, C.E., Searle, S.R.: Generalized, linear, and mixed models. John Wiley & Sons, (2004)

A Note on Computing Interval Overlap Statistics

Shahab Sarmashghi[1] and Vineet Bafna[2]

[1] Department of Electrical and Computer Engineering, University of California, San Diego, La Jolla, CA 92093, USA
ssarmash@ucsd.edu

[2] Department of Computer Science and Engineering, University of California, San Diego, La Jolla, CA 92093, USA
vbafna@cs.ucsd.edu

Extended Abstract

We consider the following problem: Let I and I_f each describe a collection of n and m non-overlapping intervals on a line segment of finite length. Suppose that k of the m intervals of I_f are intersected by some interval(s) in I. Under the null hypothesis that intervals in I are randomly arranged w.r.t I_f, what is the significance of this overlap? This is a natural abstraction of statistical questions that are ubiquitous in the post-genomic era. The interval collections represent annotations that reveal structural or functional regions of the genome, and overlap statistics can provide insight into the correlation between different structural and functional regions. However, the statistics of interval overlaps have not been systematically explored. We propose a combinatorial algorithm for a constrained interval overlap problem that can accurately compute very small p-values. Specifically, we define $N(i, h, k, a)$ as the number of randomized arrangements of the first i intervals in I such that the i-th interval ends at genomic location h, and k intervals in I_f are hit by the first i intervals in I (a is an auxiliary binary variable). Assuming that the order of intervals in I is retained, $N(i, h, k, a)$ is computed using a dynamic programming algorithm in pseudo-polynomial time $\mathcal{O}(ngm)$ [1], where n and m are the number of intervals in I and I_f, and g is the genome length. The p-value of the overlap is then given by

$$P\text{-value}(k) = \frac{\sum_{\kappa=k}^{m} N_1(n, g, \kappa, 0)}{\sum_{\kappa=0}^{m} N_1(n, g, \kappa, 0)}.$$

We have also provided a fast approximate method based on Poisson binomial distribution to facilitate problems consisted of very large number of intervals, and have introduced parameter η as a measure of the spread of intervals to estimate the closeness of approximated p-values.

We tested our tool, ISTAT, on simulated interval data to obtain precise estimates of low p-values, and characterize the performance of our methods. We also applied ISTAT to four cases of interval overlap problem from previous studies, and showed that ISTAT can estimate very small p-values, considering the length

© Springer Nature Switzerland AG 2019
L. J. Cowen (Ed.): RECOMB 2019, LNBI 11467, pp. 314–315, 2019.
https://doi.org/10.1007/978-3-030-17083-7

and structure of intervals, while avoiding inflated p-values reported from basic permutation or parametric tests. The ISTAT software is made publicly available on Github (https://github.com/shahab-sarmashghi/ISTAT.git).

Reference

1. Sarmashghi, S., Bafna, V.: A Note on Computing Interval Overlap Statistics. bioRxiv, p. 517987, January 2019. https://doi.org/10.1101/517987, https://www. biorxiv.org/content/early/2019/01/11/517987.full.pdf+html

Distinguishing Biological from Technical Sources of Variation Using a Combination of Methylation Datasets

Mike Thompson[1(✉)], Zeyuan Johnson Chen[1], Elior Rahmani[1], and Eran Halperin[1,2,3,4(✉)]

[1] Department of Computer Science, University of California Los Angeles, Los Angeles, CA, USA
mjthompson@ucla.edu, ehalperin@cs.ucla.edu
[2] Department of Human Genetics, University of California Los Angeles, Los Angeles, CA, USA
[3] Department of Anesthesiology and Perioperative Medicine, University of California Los Angeles, Los Angeles, CA, USA
[4] Department of Biomathematics, University of California Los Angeles, Los Angeles, CA, USA

DNA methylation remains one of the most widely studied epigenetic markers. One of the major challenges in population studies of methylation is the presence of global methylation effects that may mask local signals [1, 2]. Such global effects may be due to either technical effects (e.g., batch effects) or biological effects (e.g., cell type composition, genetics). Many methods have been developed for the detection of such global effects, typically in the context of Epigenome-wide association studies [3–9]. However, current unsupervised methods do not distinguish between biological and technical effects, resulting in a loss of highly relevant information. Though supervised methods can be used to estimate known biological effects, it remains difficult to identify and estimate unknown biological effects that globally affect the methylome.

Here, we propose CONFINED (CCA ON Features for INter- dataset Effect Detection), a reference-free method based on sparse canonical correlation analysis (CCA) that captures replicable sources of variation across multiple methylation datasets such as age, sex, and cell-type composition and distinguishes them from dataset-specific sources of variability (e.g., technical effects). Our method is based on the observation that the same biological sources of variation typically affect different studies that are performed under the same conditions (e.g., on the same tissue type), while technical variability is study-specific. Thus, unlike previous unsupervised methods that utilize single-matrix decomposition techniques to account for covariates in methylation data, we propose the use of canonical correlation analysis, which captures shared signal across multiple datasets. Nonetheless, there are two substantial differences between CONFINED and traditional uses of CCA in genomic studies. First, CONFINED looks for shared structure of one methylation profile across two sets of individuals rather than looking for shared structure in one set of individuals across two sets of genomic measurements. Second, CONFINED performs a feature selection procedure that is critical to detect the shared sources of variability across the different datasets.

© Springer Nature Switzerland AG 2019
L. J. Cowen (Ed.): RECOMB 2019, LNBI 11467, pp. 316–317, 2019.
https://doi.org/10.1007/978-3-030-17083-7

Across several datasets we demonstrate that CONFINED accurately captures global biological sources of variability. Specifically, we shrow through simulated and real data that our approach captures replicable sources of biological variation such as age, sex, and cell-type composition better than the state-of-the-art methods and is considerably more robust to technical noise than previous reference-free methods. Additionally, we demonstrate that the features selected by CONFINED recapitulate biological functionality inherent to both datasets. For example, when pairing two whole-blood datasets together, the sites best ranked by CONFINED were significantly enriched for immune cell function.

CONFINED is available at https://github.com/cozygene/CONFINED as an R package. The calculations in the R package were optimized with C++ code using Rcpp and RcppArmadillo. Also included in the package is an ultra-fast function for performing CCA. The preprint of the manuscript can be found at https://www.biorxiv.org/content/early/2019/01/16/521146.

References

1. Schmidt, F., List, M., Cukuroglu, E., Köhler, S., Göke, J., Schulz, M.H.: An ontology-based method for assessing batch effect adjustment approaches in heterogeneous datasets. Bioinformatics 34(17), i908–i916 (2018)
2. Maksimovic, J., Gagnon-Bartsch, J.A., Speed, T.P., Oshlack, A.: Removing unwanted variation in a differential methylation analysis of illumina humanmethylation450 array data. Nucleic Acids Res. 43, e106–e106 (2015)
3. Rahmani, E., et al.: Sparse PCA corrects for cell type heterogeneity in epigenome-wide association studies. Nat. Methods 13, 443 (2016)
4. Zou, J., Lippert, C., Heckerman, D., Aryee, M., Listgarten, J.: Epigenome-wide association studies without the need for cell-type composition. Nat. Methods 11, 309 (2014)
5. Houseman, E.A., Kile, M.L., Christiani, D.C., Ince, T.A., Kelsey, K.T., Marsit, C.J.: Reference-free deconvolution of DNA methylation data and mediation by cell composition effects. BMC Bioinf. 17, 259 (2016)
6. Lutsik, P., Slawski, M., Gasparoni, G., Vedeneev, N., Hein, M., Walter, J.: Medecom: discovery and quantification of latent components of heterogeneous methylomes. Genome Biol. 18, 55 (2017)
7. Rahmani, E., et al.: Bayescce: a bayesian framework for estimating cell-type composition from dna methylation without the need for methylation reference. Genome Biol. 19, 141 (2018)
8. Houseman, E.A., Molitor, J., Marsit, C.J.: Reference-free cell mixture adjustments in analysis of dna methylation data. Bioinformatics 30(10), 1431–1439 (2014)
9. Rahmani, E., et al.: Correcting for cell-type heterogeneity in DNA methylation: a comprehensive evaluation. Nat. Methods 14, 218 (2017)

GRep: Gene Set Representation via Gaussian Embedding

Sheng Wang[2,3], Emily Flynn[1], and Russ B. Altman[1,2,3]($\boxtimes$)

[1] Biomedical Informatics Training Program, Stanford University,
Stanford, CA 94035, USA
[2] Department of Bioengineering, Stanford University, Stanford, CA 94035, USA
[3] Department of Genetics, Stanford University, Stanford, CA 94035, USA
`russ.altman@stanford.edu`

1 Introduction

Molecular interaction networks are our basis for understanding functional inter-dependencies among genes. Network embedding approaches analyze these complicated networks by representing genes as low-dimensional vectors based on the network topology. These low-dimensional vectors have recently become the building blocks for a larger number of systems biology applications. Despite the success of embedding genes in this way, it remains unclear how to effectively represent gene sets, such as protein complexes and signaling pathways. The direct adaptation of existing gene embedding approaches to gene sets cannot model the diverse functions of genes in a set. Here, we propose GRep, a novel gene set embedding approach, which represents each gene set as a multivariate Gaussian distribution rather than a single point in the low-dimensional space. The diversity of genes in a set, or the uncertainty of their contribution to a particular function, is modeled by the covariance matrix of the multivariate Gaussian distribution. By doing so, GRep produces a highly informative and compact gene set representation. Using our representation, we analyze two major pharmacogenomics studies and observe substantial improvement in drug target identification from expression-derived gene sets. Overall, the GRep framework provides a novel representation of gene sets that can be used as input features to off-the-shelf machine learning classifiers for gene set analysis. A full version of the paper can be found on bioRxiv https://www.biorxiv.org/content/early/2019/01/13/519033.

2 Methods

Biologically meaningful gene sets, such as signaling pathways and protein complexes, aggregate gene level information into higher level patterns. A key observation behind our approach is that gene sets can have diverse molecular functions and/or biological processes. GRep explicitly models this diversity as a low-dimensional Gaussian distribution which summarizes both location and uncertainty of each dimension. To summarize, GRep takes a network and a collection

L. J. Cowen (Ed.): RECOMB 2019, LNBI 11467, pp. 318–319, 2019.
https://doi.org/10.1007/978-3-030-17083-7

of gene sets as input. It first calculates the diffusion states of each gene and gene set to characterize their topological information in the network. GRep then finds the low-dimensional representations for genes and gene sets according to these diffusion states. Each gene is represented as a single point in the low-dimensional space. Each gene set is represented as a multivariate Gaussian distribution which is parameterized by a mean vector and a covariance matrix. In this paper, we present GRep (Gene set Representation), a novel computational method that represents each gene set as a highly informative and compact multivariate Gaussian distribution. GRep takes a biological network and a collection of gene sets as input. It represents each gene as a single point and each gene set as a multivariate Gaussian distribution parameterized by a low-dimensional mean vector and a low-dimensional covariance matrix. The mean vector of each gene set describes the joint contribution of genes in this gene set, and the covariance matrix characterizes the agreement among individual genes in each dimension. By using this representation, GRep is able to differentiate between gene sets that would be considered equivalent by average embedding. The key idea of GRep is to use the prior knowledge in gene sets and group genes in the same set closely as a multivariate Gaussian distribution in the low-dimensional space. To achieve this, GRep solves an optimization problem to preserve the network topology according to diffusion states. We evaluate GRep on a collection of drug response correlated gene sets derived from Genomics of Drug Sensitivity in Cancer (GDSC) and The Cancer Therapeutic Portal (CTRP). We demonstrate that representing those gene sets using GRep substantially outperforms comparison approaches on drug-target identification in both datasets.

3   Results

To evaluate GRep, we performed large-scale drug target identification on two pharmacogenomics studies, GDSC and CTRP. Our approach significantly outperforms comparison approaches on both datasets. In CTRP, our method achieved 0.8667 AUROC, which is much higher than 0.7102 AUROC of plain average embedding, 0.7104 of weighted gene set average embedding and 0.7319 AUROC of weighted average embedding. The same improvement was observed on GDSC where our method achieved 0.8890 AUROC, which is again substantially higher than 0.6870 AUROC of plain average embedding, 0.7325 of weighted average embedding and 0.6870 AUROC of weighted gene set average embedding. All improvements were statistically significant ($P < 0.05$; paired Wilcoxon signed-rank test). The above results suggest that representing a gene set through simple averaging is not able to modeling uncertainty, leading to worse performance. By incorporating prior knowledge about gene sets and jointly optimizing the gene and gene set representations, our method substantially improved drug target identification.

Accurate Sub-population Detection and Mapping Across Single Cell Experiments with PopCorn

Yijie Wang, Jan Hoinka, and Teresa M. Przytycka[✉]

National Center of Biotechnology Information, National Library of Medicine, NIH,
Bethesda, MD 20894, USA
przytyck@ncbi.nlm.nih.gov

Extended Abstract

Recent technological advances have facilitated unprecedented opportunities for studying biological systems at single-cell level resolution. One notable example is single-cell RNA sequencing (scRNA-seq), which enables the measurement of transcriptomic information of thousands of individual cells in one experiment. Single cell measurements open the ability of capturing the heterogeneity of a population of cells and thus provide information that is not accessible using bulk sequencing. Among its many applications, scRNA-seq is more prominently employed in the identification of sub-populations of cells present in a sample, and for comparative analysis of such sub-populations across samples [3–6, 8–11].

We report PopCorn (single cell Populations Comparison)– a new method allowing for the identification of sub-populations of cells present within individual experiments and their mapping across experiments. PopCorn uses several innovative ideas to perform this task accurately. First, in contrast to previous approaches, PopCorn performs the two tasks (sub-population identification and mapping) simultaneously by optimizing a function that combines both objectives. This allows for integrating information across experiments and reducing noise. The second key innovation consists of a new approach to identify sub-populations of cells within a given experiment. Specifically, PopCorn utilizes Personalized PageRank vectors [1] and a quality measure of cohesiveness of a cell population to perform this task. Finally, the simultaneous identification of sub-populations within each experiment and their mapping across experiments uses a graph theoretical approach.

We tested the performance of PopCorn in two distinct settings. We demonstrated its potential in identifying and aligning sub-populations informed by single cell data from human and mouse pancreatic singe cell data [2]. In addition, we applied PopCorn to the task of aligning biological replicates of mouse kidney single cell data [7]. In both scenarios PopCorn achieved a striking improvement over alternative tools.

Taken together, our results demonstrate that PopCorn's novel approach provides a powerful tool for comparative analysis of single-cells sub-populations.

L. J. Cowen (Ed.): RECOMB 2019, LNBI 11467, pp. 320–321, 2019.
https://doi.org/10.1007/978-3-030-17083-7

The preprint of the manuscript is available at https://www.biorxiv.org/content/early/2018/12/28/485979.article-metrics.

References

1. Andersen, R., Chung, F., Lang, K.: Local graph partitioning using pagerank vectors. In: FOCS, pp. 475–486 (2006)
2. Butler, A., Hoffman, P., Smibert, P., Papalexi, E., Satija, R.: Integrating single-cell transcriptomic data across different conditions, technologies, and species. Nat. Biotechnol. **36**(5), 411–420 (2018)
3. Byrnes, L.E., et al.: Lineage dynamics of murine pancreatic development at single-cell resolution. Nat. Commun. **9**(1), 3922 (2018)
4. Duan, L., et al.: PDGFRβ Cells Rapidly Relay Inflammatory Signal from the Circulatory System to Neurons via Chemokine CCL2. Neuron **100**(1), 183–200 (2018)
5. Mayer, C., et al.: Developmental diversification of cortical inhibitory interneurons. Nature **555**(7697), 457–462 (2018)
6. Ordovas-Montanes, J., et al.: Allergic inflammatory memory in human respiratory epithelial progenitor cells. Nature **560**(7720), 649–654 (2018)
7. Park, J., et al.: Single-cell transcriptomics of the mouse kidney reveals potential cellular targets of kidney disease. Science **360**(6390), 758–763 (2018)
8. Paulson, K.G., et al.: Acquired cancer resistance to combination immunotherapy from transcriptional loss of class I HLA. Nat. Commun. **9**(1), 3868 (2018)
9. Shrestha, B.R., Chia, C., Wu, L., Kujawa, S.G., Liberman, M.C., Goodrich, L.V.: Sensory neuron diversity in the inner ear is shaped by activity. Cell **174**(5), 1229–1246 (2018)
10. Sun, S., et al.: Hair cell mechanotransduction regulates spontaneous activity and spiral ganglion subtype specification in the auditory system. Cell **174**(5), 1247–1263 (2018)
11. Verma, M., et al.: Muscle satellite cell cross-talk with a vascular niche maintains quiescence via VEGF and notch signaling. Cell Stem Cell **23**(4), 530–543 (2018)

Fast Estimation of Genetic Correlation
for Biobank-Scale Data

Yue Wu[1], Anna Yaschenko[3], Mohammadreza Hajy Heydary[4],
and Sriram Sankararaman[1,2(✉)]

[1] Department of Computer Science, UCLA, Los Angeles, USA
[2] Department of Human Genetics, UCLA, Los Angeles, USA
sriram@cs.ucla.edu
[3] Department of Computer Science and Electrical Engineering,
University of Maryland, Baltimore County, Baltimore, USA
[4] Department of Computer Science, California State University, Fullerton,
Fullerton, USA

Genetic correlation, *i.e.*, the proportion of phenotypic correlation across a pair of traits that can be explained by genetic variation, is an important parameter in efforts to understand the relationships among complex traits [1]. The observation of substantial genetic correlation across a pair of traits, can provide insights into shared genetic pathways as well as providing a starting point to investigate causal relationships. Attempts to estimate genetic correlations among complex phenotypes attributable to genome-wide SNP variation data have motivated the analysis of large datasets as well as the development of sophisticated methods.

Bi-variate Linear Mixed Models (LMMs) have emerged as a key tool to estimate genetic correlation from datasets where individual genotypes and traits are measured [2]. The bi-variate LMM jointly models the effect sizes of a given SNP on each of the pair of traits being analyzed. The parameters of the bi-variate LMM, *i.e.*, the variance components, are related to the heritability of each trait as well as correlation across traits attributable to genotyped SNPs. The most commonly used method for estimating genetic correlation as well as trait heritabilities in a bi-variate LMM relies on the restricted maximize likelihood method, termed genomic restricted maximum likelihood (GREML) [3–6] However, GREML poses serious computational burdens. GREML is a non-convex optimization problem that relies on an iterative optimization algorithm.

Another state-of-the-art method, LD-score regression (LDSC), requires only summary statistics from genome-wide association studies (GWAS) to estimate genetic correlations [1]. As LD-score preserves privacy and has substantially reduced computational requirements (assuming that the summary statistics have been computed), LDSC has some drawbacks: its estimates tend to have large standard errors and is prone to bias in some settings [7].

We propose, RG-Cor, a scalable randomized Method-of-Moments (MoM) estimator of genetic correlations in bi-variate LMMs. RG-Cor leverages the structure of genotype data to obtain runtimes that scale sub-linearly with the number of individuals in the input dataset (assuming the number of SNPs is held constant). We perform extensive simulations to validate the accuracy and scalability of RG-Cor. Compared to GREML estimators, we show that the loss in

© Springer Nature Switzerland AG 2019
L. J. Cowen (Ed.): RECOMB 2019, LNBI 11467, pp. 322–323, 2019.
https://doi.org/10.1007/978-3-030-17083-7

statistical inefficiency of RG-Cor is fairly modest. On the other hand, RG-Cor is several orders of magnitude faster than other methods. RG-Cor can compute the genetic correlations on the UK biobank dataset consisting of 430,000 individuals and 460,000 SNPs in 3 hours on a stand-alone compute machine.

Link to the full paper: https://www.biorxiv.org/content/early/2019/01/20/525055

References

1. Bulik-Sullivan, B., Finucane, H.K., Anttila, V., et al.: An atlas of genetic correlations across human diseases and traits. Nat. Genet. **47**(11), 1236 (2015)
2. Yang, J., Lee, S.H., Goddard, M.E., Visscher, P.M.: GCTA: a tool for genome-wide complex trait analysis. Am. J. Hum. Genet. (2010)
3. Lee, S.H., Yang, J., Goddard, M.E., Visscher, P.M., Wray, N.R.: Estimation of pleiotropy between complex diseases using single-nucleotide polymorphism-derived genomic relationships and restricted maximum likelihood. Bioinformatics **28**(19), 2540–2542 (2012)
4. Chen, G.-B.: Estimating heritability of complex traits from genome-wide association studies using ibs-based haseman-elston regression. Front. Genet. **5**, 107 (2014)
5. Loh, P.-R., Tucker, G., Bulik-Sullivan, B.K., et al.: Efficient bayesian mixed-model analysis increases association power in large cohorts. Nat. Genet. **47**(3), 284 (2015)
6. Loh, P.-R., Bhatia, G., Gusev, A., et al.: Contrasting genetic architectures of schizophrenia and other complex diseases using fast variance-components analysis. Nat. Genet. **47**(12), 1385 (2015)
7. Ni, Guiyan, Moser, Gerhard, Ripke, Stephan, et al.: Estimation of genetic correlation via linkage disequilibrium score regression and genomic restricted maximum likelihood. Am. J. Hum. Genet. (2018)

Distance-Based Protein Folding Powered by Deep Learning

Jinbo Xu[✉]

Toyota Technological Institute at Chicago, Chicago, USA
jinboxu@gmail.com

Accurate description of protein structure and function is a fundamental step towards understanding biological life and highly relevant in the development of therapeutics. Although greatly improved, experimental protein structure determination is still low-throughput and costly, especially for membrane proteins. Predicting the structure of a protein with a new fold is very challenging and usually needs a large amount of computing power. We show that we can accurately predict the distance matrix of a protein by deep learning (DL), even for proteins with few sequence homologs. Using only the geometric constraints given by the resulting distance matrix we may construct 3D models without involving any folding simulation.

This work is an extension of our previous CASP-winning deep learning method RaptorX-Contact [1] that uses deep and global (or fully) convolutional residual neural network (ResNet) to predict protein contacts. ResNet is one type of DCNN (deep convolutional neural network), but much more powerful than the traditional DCNN. RaptorX-Contact is the first DL method that greatly outperforms DCA (direct coupling analysis) and shallow learning methods such as the CASP11 winner MetaPSICOV. The accuracy of RaptorX-Contact decreases much more slowly than DCA when more predicted contacts are evaluated even when the protein under study has thousands of sequence homologs (see Table 1 in the paper [1]). As reported in [1, 2], without folding simulation, RaptorX-Contact may produce much better 3D models than DCA methods such as CCMpred and shallow methods such as MetaPSICOV. RaptorX-Contact also works well for membrane proteins even trained by soluble proteins [2] and for complex contact prediction even trained by single-chain proteins [3]. Inspired by the success of RaptorX-Contact, many CASP13 participants have adopted global ResNet or DCNN into their prediction pipeline, as shown in the CASP13 abstract book, and made very good progress. As a result, CASP13 has achieved the largest progress in the history of CASP.

Instead of contact prediction, here we study distance prediction. The distance matrix contains finer-grained information than contact matrix and provides more physical constraints of a protein structure, e.g., distance is metric while contact is not. A distance matrix can determine a protein structure (except mirror image) much more accurately than a contact matrix. Different from DCA that aims to predict only a small number of contacts and then use them to assist folding simulation, we predict the whole distance matrix and then directly construct protein 3D models without invoking any folding simulation at all. This significantly reduces running time needed for protein folding, especially for a large protein. Distance prediction is not totally new. In addition

© Springer Nature Switzerland AG 2019
L. J. Cowen (Ed.): RECOMB 2019, LNBI 11467, pp. 324–325, 2019.
https://doi.org/10.1007/978-3-030-17083-7

to few previous studies, my group employed a probabilistic neural network to predict inter-residue distance and then derived protein- and position-specific statistical potential from predicted distance distribution [4]. We have also studied folding simulation using this distance-based statistical potential [5]. Recently, we showed that protein-specific distance potential derived from deep ResNet may improve by a large margin protein threading with weakly similar templates [6].

We feed our predicted distance into CNS to generate 3D models for a protein under prediction. Our method successfully folded 21 of the 37 CASP12 hard targets with a median family size of 58 effective sequence homologs within 4 h on a Linux computer of 20 CPUs. In contrast, DCA cannot fold any of these hard targets in the absence of folding simulation, and the best CASP12 group folded only 11 of them by integrating DCA-predicted contacts into complex, fragment-based folding simulation. Rigorous experimental validation in CASP13 shows that our distance-based folding server successfully folded 17 of 32 hard targets (with a median family size of 36 sequence homologs) and obtained 70% precision on top L/5 long-range predicted contacts. In CASP13, our method was officially ranked first in terms of contact prediction accuracy among all CASP13 groups and our server was ranked second among all CASP13-participating servers in terms of tertiary structure prediction.

An extended version of this abstract is available at https://www.biorxiv.org/content/early/2018/12/20/465955 and https://arxiv.org/abs/1811.03481.

References

1. Wang, S., Sun, S., Li, Z., Zhang, R., Xu, J.: Accurate de novo prediction of protein contact map by ultra-deep learning model. PLoS Comput. Biol. **13**, e1005324 (2017)
2. Wang, S., Li, Z., Yu, Y., Xu, J.: Folding membrane proteins by deep transfer learning. Cell Syst. **5**, 202–211 (2017). e203
3. Zeng, H., et al.: ComplexContact: a web server for inter-protein contact prediction using deep learning. Nucleic Acids Res. **46**, W432–W437 (2018)
4. Zhao, F., Xu, J.: A position-specific distance-dependent statistical potential for protein structure and functional study. Structure **20**, 1118–1126 (2012)
5. Wang, Z.: Knowledge-based machine learning methods for macromolecular 3D structure prediction. Ph.D. thesis (2016)
6. Zhu, J.W., Wang, S., Bu, D.B., Xu, J.B.: Protein threading using residue co-variation and deep learning. Bioinformatics **34**, 263–273 (2018)

Comparing 3D Genome Organization in Multiple Species Using Phylo-HMRF

Yang Yang[1], Yang Zhang[1], Bing Ren[2], Jesse Dixon[3], and Jian Ma[1(✉)]

[1] Computational Biology Department, School of Computer Science,
Carnegie Mellon University, Pittsburgh, USA
jianma@cs.cmu.edu

[2] Ludwig Institute for Cancer Research, Department of Cellular and Molecular
Medicine, Moores Cancer Center and Institute of Genomic Medicine,
UCSD School of Medicine, San Diego, USA

[3] Salk Institute for Biological Studies, San Diego, USA

Recent developments in whole-genome mapping approaches for the chromatin interactome (such as Hi-C) have facilitated the identification of genome-wide three-dimensional (3D) chromatin organizations comprehensively, and offered new insights into 3D genome architecture. However, our knowledge of the evolutionary patterns of 3D genome structures in mammalian species remains surprisingly limited. In particular, there are no existing phylogenetic-model based methods to analyze chromatin interactions as continuous features across different species to uncover evolutionary patterns of 3D genome organization.

Here we develop a new probabilistic model, named phylogenetic hidden Markov random field (Phylo-HMRF), to identify evolutionary patterns of 3D genome structures based on multi-species Hi-C data by jointly utilizing spatial constraints among genomic loci and continuous-trait evolutionary models. Specifically, Phylo-HMRF integrates the continuous-trait evolutionary constraints (based on Ornstein-Uhlenbeck process in this work) with the hidden Markov random field (HMRF) model, enabling the joint modeling of general types of spatial dependencies among genomic loci and evolutionary temporal dependencies among species. The overview of Phylo-HMRF is shown in Fig. 1. The effectiveness of Phylo-HMRF is demonstrated in both simulation evaluation and application to real Hi-C data. We used Phylo-HMRF to uncover cross-species 3D genome patterns based on Hi-C data from the same cell type in four primate species (human, chimpanzee, bonobo, and gorilla). Phylo-HMRF identified genome-wide evolutionary patterns of Hi-C contact frequency across the four species, including conserved patterns and lineage-specific patterns. The identified evolutionary patterns of 3D genome organization correlate with other features of genome structure and function, including long-range interactions, topologically-associating domains (TADs), and replication timing patterns.

This work provides a new framework that utilizes general types of spatial constraints to identify evolutionary patterns of continuous genomic features and has the potential to reveal the evolutionary principles of 3D genome organization.

Link to the bioRxiv preprint: doi: http://doi.org/10.1101/552505.

© Springer Nature Switzerland AG 2019
L. J. Cowen (Ed.): RECOMB 2019, LNBI 11467, pp. 326–327, 2019.
https://doi.org/10.1007/978-3-030-17083-7

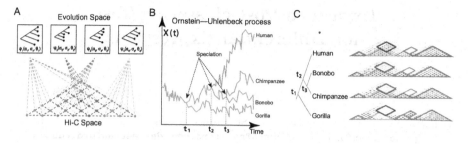

Fig. 1. Overview of Phylo-HMRF. (A) Illustration of the possible evolutionary patterns of chromatin interaction. The Hi-C space is a combined multi-species Hi-C contact map, which integrates aligned Hi-C contact maps of each species. Each node represents the multi-species observations of Hi-C contact frequency between a pair of genomic loci, with a hidden state assigned. Nodes with the same color have the same hidden state and are associated with the same type of evolutionary pattern represented by a parameterized phylogenetic tree ψ_i. The parameters of ψ_i include the selection strengths α_i, Brownian motion intensities σ_i, and the optimal values θ_i based on the Ornstein-Uhlenback (OU) process assumption. (B) Illustration of the OU process over a phylogenetic tree with four observed species. Time axis represents the evolution history. $X(t)$ represents the trait at time t. The trajectories reflect the evolution of the continuous-trait features in different lineages, where the time points t_1, t_2, t_3 represent the speciation events. (C) A cartoon example of the possible evolutionary patterns (partitioned with different colors). Phylo-HMRF aims to identify evolutionary Hi-C contact patterns among four primate species in this work. The four Hi-C contact maps represent the observations from the four species, which are combined into one multi-species Hi-C map as the input to Phylo-HMRF, as shown in (A). The phylogenetic tree of the four species in this study is on the left. The partitions with green borders are conserved Hi-C contact patterns. The partitions with red or blue borders represent lineage-specific Hi-C contact patterns.

Towards a Post-clustering Test
for Differential Expression

Jesse M. Zhang(iD), Govinda M. Kamath, and David N. Tse(✉)(iD)

Department of Electrical Engineering, Stanford University, Palo Alto 94304, USA
jessez@stanford.edu, gkamath@stanford.edu, dntse@stanford.edu

Extended Abstract

Single-cell technologies have seen widespread adoption in recent years. The datasets generated by these technologies provide information on up to millions or more individual cells; however, the identities of the cells are often only determined computationally. Single-cell computational pipelines involve two critical steps: organizing the cells in a biologically meaningful way (clustering) and identifying the markers driving this organization (differential expression analysis). Because clustering algorithms *force* separation, performing differential expression analysis after clustering on the same dataset will generate artificially low p-values, potentially resulting in false discoveries.

While several differential expression methods exist, as a motivating example we consider the classic Student's t-test introduced in 1908 [2]. The t-test was devised for controlled experiments where the hypothesis to be tested was defined before the experiments were carried out. For example, to test the efficacy of a drug, the researcher would randomly assign individuals to case and control groups, administer the placebo or the drug, and take a set of measurements. Because the populations were clearly defined a priori, so was the null hypothesis. Therefore, under the null hypothesis where no effect exists, the mean measurement should be the same across the two populations, and the p-value should be uniformly distributed between 0 and 1.

For single-cell analysis, however, the populations are often obtained *after* the measurements are taken, via clustering, and therefore we can expect the t-test to return significant p-values even if the null hypothesis was true. Figure 1 shows how a measurement, such as expression of a gene, is deemed significantly different between two clusters even though all samples came from the same normal distribution. The clustering introduces a **selection bias** [1, 3] that would result in several false discoveries if uncorrected.

In this work, we introduce the truncated normal (TN) test, an approximate test based on the truncated normal distribution that corrects for a significant portion of the selection bias generated by clustering. We condition on the clustering event using the hyperplane that separates the clusters. By incorporating this

Full paper available at https://www.biorxiv.org/content/early/2018/11/05/463265.
Code provided at https://github.com/jessemzhang/tn_test.

L. J. Cowen (Ed.): RECOMB 2019, LNBI 11467, pp. 328–329, 2019.
https://doi.org/10.1007/978-3-030-17083-7

hyperplane into our null model, we can obtain a uniformly distributed p-value even in the presence of clustering (Fig. 1). To our knowledge, the TN test is the first test to correct for clustering bias while addressing the differential expression question: *is this feature significantly different between the two clusters?* Based on the TN test, we provide a data-splitting based framework that allows us to generate valid p-values for differential expression of genes for clusters obtained from any clustering algorithm. We validate the method using both synthetic and real data, such as the peripheral blood mononuclear cell (PBMC) dataset generated using recent techniques developed by 10x Genomics [4], and we compare the method to several existing differential expression methods.

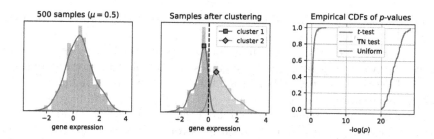

Fig. 1. Artificially low p-values due to clustering. Although the 500 samples are drawn from the same $\mathcal{N}(\mu, 1)$ distribution, our simple clustering approach will always generate two clusters that seem significantly different under the t-test. In this work, we explore an approach for correcting the selection bias due to clustering. In other words, we attempt to close the gap between the blue and green curves in the rightmost plot. We introduce the TN test, which generates significantly more reasonable p-values.

Acknowledgements. We thank Jonathan Taylor, Martin Zhang, and Vasilis Ntranos of Stanford University and Aaron Lun of the CRUK Cambridge Institute for helpful discussions about selective inference and applications of the method. GMK and JMZ are supported by the Center for Science of Information, an NSF Science and Technology Center, under grant agreement CCF-0939370. JMZ and DNT are supported in part by the National Human Genome Research Institute of the National Institutes of Health under award number R01HG008164.

References

1. Fithian, W., Sun, D., Taylor, J.: Optimal inference after model selection (2014). arXiv preprint, http://arxiv.org/abs/1410.2597
2. Student: The probable error of a mean. Biometrika pp. 1–25 (1908)
3. Taylor, J., Tibshirani, R.J.: Statistical learning and selective inference. Proc. Nat. Acad. Sci. **112**(25), 7629–7634 (2015)
4. Zheng, G.X., et al.: Massively parallel digital transcriptional profiling of single cells. Nat. Commun. **8**, 14049 (2017)

AdaFDR: A Fast, Powerful and Covariate-Adaptive Approach to Multiple Hypothesis Testing

Martin J. Zhang[1], Fei Xia[1], and James Zou[1,2,3]([✉])

[1] Department of Electrical Engineering, Stanford University, Palo Alto 94304, USA
{jinye,feixia,jamesz}@stanford.edu
[2] Department of Biomedical Data Science, Stanford University,
Palo Alto 94304, USA
[3] Chan-Zuckerberg Biohub, San Francisco 94158, USA

Introduction. Multiple hypothesis testing is an essential component in many modern data analysis workflows. A very common objective is to maximize the number of discoveries while controlling the fraction of false discoveries. For example, we may want to identify as many genes as possible that are differentially expressed between two populations such that less than, say, 10% of these identified genes are false positives.

In the standard setting, the data for each hypothesis is summarized by a p-value, with a smaller value presenting stronger evidence against the null hypothesis that there is no association. Commonly-used procedures such as the Benjamini-Hochberg procedure (BH) [1] works solely with this list of p-values [3, 7]. Despite being widely used, these multiple testing procedures fail to utilize additional information that is often available in modern applications that are not directly captured by the p-value.

For example, in expression quantitative trait loci (eQTL) mapping or genome-wide association studies (GWAS), single nucleotide polymorphism (SNP) in active chromatin state are more likely to be significantly associated with the phenotype [2]. Such chromatin information is readily available in public databases, but is not used by standard multiple hypothesis testing procedures—it is sometimes used for post-hoc biological interpretation. Similarly, the location of the SNP, its conservation score, etc., can alter the likelihood for the SNP to be an eQTL. Together such additional information, called covariates, forms a feature representation of the hypothesis; this feature vector is ignored by the standard multiple hypothesis testing procedures.

In this paper, we present `AdaFDR`, a fast and flexible method that adaptively learns the decision threshold from covariates to significantly improve the detection power while having the false discovery proportion (FDP) controlled at a user-specified level. A schematic diagram for `AdaFDR` is shown in Fig. 1.

Full paper available at https://www.biorxiv.org/content/early/2018/12/13/496372.

© Springer Nature Switzerland AG 2019
L. J. Cowen (Ed.): RECOMB 2019, LNBI 11467, pp. 330–333, 2019.
https://doi.org/10.1007/978-3-030-17083-7

AdaFDR takes as input a list of hypotheses, each with a p-value and a covariate vector. Conventional methods like BH use only p-values and have the same p-value threshold for all hypotheses (Fig. 1 top right). However, as illustrated in the bottom-left panel, the data may have an enrichment of small p-values for certain values of the covariate, which suggests an enrichment of alternative hypotheses around these covariate values. Intuitively, allocating more FDR budget to hypothesis with such covariates could increase the detection power. AdaFDR adaptively learns such pattern using both p-values and covariates, resulting in a covariate-dependent threshold that makes more discoveries under the same FDP constraint (Fig. 1 bottom right).

Methods. AdaFDR extends conventional procedures like BH and Storey-BH [1, 7] by considering multiple hypothesis testing with side information on the hypotheses. The input of AdaFDR is a set of hypotheses each with a p-value and a vector of covariates, whereas the output is a set of selected (also called rejected) hypotheses. For eQTL analysis, each hypothesis is one pair of SNP and gene, and the p-value tests for association between their values across samples. The covariate can be the location, conservation, and chromatin status at the SNP and the gene. The standard assumption of AdaFDR and all the related methods is that the covariates should not affect the p-values under the null hypothesis. AdaFDR learns the covariate-dependent p-value selection threshold by first fit-

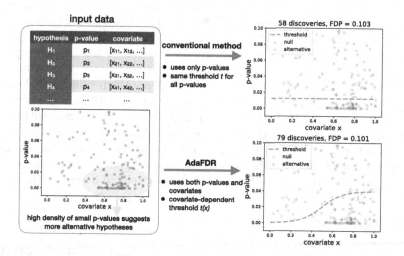

Fig. 1. Intuition of AdaFDR. Top-left: As input, AdaFDR takes a list of hypotheses, each with a p-value and a covariate that may be multi-dimensional. Bottom-left: A toy example with a univariate covariate. The enrichment of small p-values in the bottom right corner suggests more alternative hypotheses there. Leveraging this structure can lead to more discoveries. Top-right: Conventional method uses only p-values and has the same threshold for all hypotheses. Bottom-right: AdaFDR adaptively learns the uneven distribution of the alternative hypotheses, and makes more discoveries while controlling the false discovery proportion (FDP) at the desired level (0.1 in this case).

ting a mixture model using expectation maximization (EM) algorithm, where the mixture model is a combination of a generalized linear model (GLM) and Gaussian mixtures. Then it makes local adjustments to the p-value threshold by optimizing for more discoveries. We prove that AdaFDR controls FDP under standard statistical assumptions. AdaFDR is designed to be fast and flexible — it can simultaneously process more than 100 million hypotheses within an hour and allows multi-dimensional covariates with both numeric and categorical values. In addition, AdaFDR provides exploratory plots visualizing how each covariate is related to the significance of the hypotheses, allowing users to interpret their findings.

Results. We systematically evaluate the performance of AdaFDR across multiple datasets. We first consider the problem of eQTL discovery using the data from the Genotype-Tissue Expression (GTEx) project [2]. As covariates, we consider the distance between the SNP and the gene, the gene expression level, the alternative allele frequency as well as the chromatin states of the SNP. Across all 17 tissues considered in the study, AdaFDR has an improvement of 32% over BH and 27% over the state-of-art covariate-adaptive method independent hypothesis weighting (IHW) [4]. We next consider other applications, including three RNA-Seq datasets with the gene expression level as the covariate, two microbiome datasets with ubiquity (proportion of samples where the feature is detected) and the mean nonzero abundance as covariates, a proteomics dataset with the peptides level as the covariate, and two fMRI datasets with the Brodmann area label as the covariate that represents different functional regions of human brain. In all experiments, AdaFDR shows a similar improvement. Finally, we perform extensive simulations, including ones from a very recent benchmark paper [5], to demonstrate that AdaFDR has the highest detection power while controlling the false discovery proportion in various cases where the p-values may be either independent or dependent. The default parameters of AdaFDR are used for every experiment in this paper, both real data analysis and simulations, without any tuning. In addition to the experiments, we theoretically prove that AdaFDR controls FDP with high probability when the null p-values, conditional on the covariates, are independently distributed and stochastically greater than the uniform distribution, a standard assumption also made by related literature [1, 6].

References

1. Benjamini, Y., Hochberg, Y.: Controlling the false discovery rate: a practical and powerful approach to multiple testing. J. Royal Stat. Soc. B (Methodol.) 289–300 (1995)
2. GTEx Consortium: Genetic effects on gene expression across human tissues. Nature **550**(7675), 204 (2017)
3. Dunn, O.J.: Multiple comparisons among means. J. Am. Stat. Assoc. **56**(293), 52–64 (1961)
4. Ignatiadis, N., Klaus, B., Zaugg, J.B., Huber, W.: Data-driven hypothesis weighting increases detection power in genome-scale multiple testing. Nat. Methods **13**(7), 577–580 (2016)

5. Korthauer, K., et. al.: A practical guide to methods controlling false discoveries in computational biology. bioRxiv (2018). https://doi.org/10.1101/458786, https://www.biorxiv.org/content/early/2018/10/31/458786
6. Lei, L., Fithian, W.: Adapt: an interactive procedure for multiple testing with side information. J. Roy. Stat. Soc. B (Stat. Methodol.) **80**(4), 649–679 (2018)
7. Storey, J.D.: A direct approach to false discovery rates. J. Roy. Stat. Soc. B (Stat. Methodol.) **64**(3), 479–498 (2002)

Author Index

Printed in the United States
By Bookmasters